DSP System Design:
Using the TMS320C6000

DSP System Design: Using the TMS320C6000

Nasser Kehtarnavaz
Texas A&M University

Mansour Keramat
University of Connecticut

Prentice Hall, Upper Saddle River, New Jersey 07458

Library of Congress Cataloging-in-Publication Data

Kehtarnavaz, Nasser and Keramat, Mansour
  DSP System Design: Using the TMS320C6000
  p. cm.
  ISBN 0-13-091031-7
  1. C (Computer program language) I. Title.
  CIP DATA AVAILABLE

Vice President and Editorial Director, ECS: *Marcia J. Horton*
Publisher: *Tom Robbins*
Associate Editor: *Alice Dworkin*
Vice President and Director of Production and Manufacturing, ESM: *David W. Riccardi*
Executive Managing Editor: *Vince O'Brien*
Managing Editor: *David A. George*
Production Editor: *Audri Anna Bazlen*
Director of Creative Services: *Paul Belfanti*
Creative Director: *Carole Anson*
Art Director: *Jayne Conte*
Cover Designer: *Bruce Kenselaar*
Cover Art: *Robert Tinney*
Art Editor: *Adam Velthaus*
Manufacturing Manager: *Trudy Pisciotti*
Manufacturing Buyer: *Pat Brown*
Senior Marketing Manager: *Holly Stark*

© 2001 by Prentice Hall
Prentice-Hall, Inc.
Upper Saddle River, New Jersey 07458

All rights reserved. No part of this book may be reproduced, in any form or by any means, without permission in writing from the publisher.

The author and publisher of this book have used their best efforts in preparing this book. These efforts include the development, research, and testing of the theories and programs to determine their effective-ness. The author and publisher make no warranty of any kind, expressed or implied, with regard to these programs or the documentation contained in this book. The author and publisher shall not be liable in any event for incidental or consequential damages in connection with, or arising out of, the furnishing, performance, or use of these programs.

Printed in the United States of America
10 9 8 7 6 5 4 3 2 1

ISBN 0-13-091031-7

Prentice-Hall International (UK) Limited, *London*
Prentice-Hall of Australia Pty. Limited, *Sydney*
Prentice-Hall Canada Inc., *Toronto*
Prentice-Hall Hispanoamericana, S.A., *Mexico*
Prentice-Hall of India Private Limited, *New Delhi*
Prentice-Hall of Japan, Inc., *Tokyo*
Pearson Education Asia Pte. Ltd., *Singapore*
Editora Prentice-Hall do Brasil, Ltda., *Rio de Janeiro*

To my family and parents
N. K.

Contents

PREFACE AND ACKNOWLEDGMENTS

1 INTRODUCTION   1
    1.1 Examples of DSP systems   3
    1.2 Organization of Chapters   6
    1.3 Required Software and Hardware   7

2 ANALOG-TO-DIGITAL SIGNAL CONVERSION   8
    2.1 Sampling   8
        2.1.1 Fast Fourier Transform   12
        2.1.2 Amplitude Statistics   13
        2.1.3 Harmonics of a Distorted Sinewave   16
    2.2 Quantization   18
        2.2.1 Signal-to-Noise Ratio   22
        2.2.2 Sampling Time Jitter   25
    2.3 Signal Reconstruction   29
    2.4 Matlab Toolbox for Data Conversion   33

3 DATA CONVERTER SPECIFICATIONS   34
    3.1 Signal Conditioning   34
        3.1.1 Linearization and Amplification   35
        3.1.2 Limiting Bandwidth   36
        3.1.3 Output Buffer   37
    3.2 Sample and Hold   37
        3.2.1 Track-and-Hold Signal Distortion   38
        3.2.2 Input Impedance of A/D Converters   41
    3.3 Performance Metrics of A/D Converters   42
        3.3.1 A/D Static Metrics   43
        3.3.2 A/D Dynamic Metrics   49
    3.4 Performance Metrics of D/A Converters   53
        3.4.1 D/A Static Metrics   54
        3.4.2 D/A Dynamic Metrics   57

4 ARCHITECTURES OF DATA CONVERTERS 60

    4.1 A/D Architectures 60
- 4.1.1 Flash 61
- 4.1.2 Subranging 61
- 4.1.3 Pipelined 63
- 4.1.4 Folding 64
- 4.1.5 Successive Approximation 65
- 4.1.6 Interleaved 66
- 4.1.7 Sigma-Delta 67

    4.2 D/A Architectures 70
- 4.2.1 Resistor Ladder 71
- 4.2.2 Current Steering 72
- 4.2.3 Charge Redistribution 72
- 4.2.4 Sigma-Delta 73

    4.3 Selection of Data Converters for DSP Systems 76

5 TMS320C6X ARCHITECTURE 80

- 5.1 CPU Operation (Dot-product Example) 85
- 5.2 Pipelined CPU 87
- 5.3 VelociTI 90
- 5.4 C64x DSP 90

6 SOFTWARE TOOLS 94

- 6.1 EVM–DSK Target C6x Board 96
- 6.2 Assembly File 96
  - 6.2.1 Directives 98
- 6.3 Memory Management 99
  - 6.3.1 Linking 101
- 6.4 Compiler Utility 102
- 6.5 Code Initialization 104
  - 6.5.1 Data Alignment 105

Lab 1: GETTING FAMILIAR WITH CODE COMPOSER STUDIO 110

- L1.1 Creating Projects 110
- L1.2 Debugging Tools 115

7 INTERRUPT DATA PROCESSING — 125

Lab 2: AUDIO SIGNAL SAMPLING — 127
L2.1 Initialization of EVM and Codec   128
L2.2 Interrupt Service Routine   133
L2.3 DSK   137

8 FIXED POINT VS. FLOATING POINT — 138
8.1 Q-Format Number Representation on Fixed-point DSPs   138
8.2 Finite Word Length Effects on Fixed-point DSPs   141
8.3 Floating-point Number Representation   142
8.4 Overflow and Scaling   143
8.5 Some Useful Arithmetic Operations   146
    8.5.1 Division   146
    8.5.2 Sine and Cosine   149
    8.5.3 Square Root   149
    8.5.4 Lookup Table   150

Lab 3: INTEGER ARITHMETIC — 151
L3.1 Overflow Handling   151
L3.2 Scaling Approach   152
L3.3 Simulator   157

9 CODE OPTIMIZATION — 158
9.1 Word-wide Optimization   160
9.2 Mixing C and Assembly   161
9.3 Software Pipelining   161
    9.3.1 Linear Assembly   162
    9.3.2 Hand-coded Software Pipelining   164
9.4 C64x Improvements   170

Lab 4: REAL-TIME FILTERING — 175
L4.1 Design of FIR Filter   175
L4.2 FIR Filter Implementation   178
    L4.2.1 Handwritten Software-pipelined Assembly   187
    L4.2.2 Assembler Optimizer Software-pipelined Assembly   190
L4.3 Floating-point Implementation   193

| 10 | **CIRCULAR BUFFERING** | **194** |

Lab 5:   ADAPTIVE FILTERING — 196

L5.1 Design of IIR Filter   196
L5.2 IIR Filter Implementation   198
L5.3 Adaptive FIR Filter   200

| 11 | **FRAME PROCESSING** | **206** |

11.1 Direct Memory Access   206
11.2 DSP–Host Communication   208

Lab 6:   FAST FOURIER TRANSFORM — 209

L6.1 DFT Implementation   212
L6.2 FFT Implementation   215
L6.3 Real-time FFT   216

| 12 | **REAL-TIME ANALYSIS AND SCHEDULING** | **219** |

12.1 Real-time Analysis   222
12.2 Real-time Scheduling   223
12.3 Real-time Data Exchange   229

Lab 7:   DSP/BIOS — 230

L7.1 A DSP/BIOS-based Program Example   231
L7.2 DSP/BIOS Analysis and Instrumentation   232
L7.3 Multithread Scheduling   235

Lab 8:   DATA SYNCHRONIZATION AND COMMUNICATION — 243

L8.1 Prioritization of Threads   250
L8.2 RTDX   253

Appendix A: QUICK REFERENCE GUIDE   255

BIBLIOGRAPHY   265

INDEX   266

Preface

Digital signal processing (DSP) has experienced an enormous growth in the last 20 years. Nowadays, DSP systems such as cell phones and high-speed modems have become an integral part of our lives. The three major components of DSP systems are DSP processors, and analog-to-digital converters, and digital-to-analog data converters. DSP processors are expected to play a major role in the next generation of high-speed communication systems and networks. The TMS320C6000 processor family has been introduced by Texas Instruments to meet such high-performance demands.

This book has evolved from a DSP laboratory course I have taught at Texas A&M University. The objective of the book is twofold: (a) to provide DSP system designers with the knowledge needed to select an appropriate data converter for a specific DSP system of interest and (b) to provide the know-how for the implementation and optimization of computationally intensive signal-processing algorithms on the family of TMS320C6x DSP processors. The book is also written for the purpose of providing a textbook for a real-time DSP laboratory course using the TMS320C6x DSP. Such a course is meant to be a follow-up to a first course in DSP. The material presented here is written primarily for those who are already familiar with DSP concepts and are interested in designing DSP systems based on TI data converters and TI C6x DSP products. Note that a great deal of the information in the book appears in the TI reference manuals on the C6000 DSP family [9–19]. However, this information has been restructured, modified, and condensed to be used for teaching a DSP laboratory course in a semester period. It is recommended that these manuals be used in conjunction with the book in order to make full use of the information presented.

A data converter Matlab toolbox and eight lab exercises are discussed and included on an attached CD to take the reader through the entire process of analog-to-digital signal conversion and C6x code writing. As a result, the book can be used as a self-study guide for designing C6x-based DSP systems. The chapters are organized to create a close correlation between the topics and lab exercises if they are used as lecture materials for a DSP lab course. Knowledge of the C programming language and Matlab is required for understanding and performing the lab exercises.

ACKNOWLEDGMENTS

We would like to express our gratitude to Texas Instruments for allowing us to use the materials in their manuals. All the figures marked by † in their captions are redrawn or modified courtesy of Texas Instruments. We wish to extend our appreciation to Gene Frantz at Texas Instruments, who brought to our attention the need for writing this book. The book would not have materialized without his encouragement. We are grateful to Leon Adams and Maria Ho at Texas Instruments, who provided the necessary administrative support to write the book in a short period of time. We would like to thank Tom Robbins and Alice Dworkin at Prentice Hall for expediting the publication process. Among many students who provided feedback on the lab exercises, we are deeply indebted to Sooncheol Baeg, who, as the laboratory teaching assistant, made many useful suggestions.

Nasser Kehtarnavaz

DSP System Design:
Using the TMS320C6000

CHAPTER 1

Introduction

In general, sensors generate analog signals in response to various physical phenomena that occur in an analog manner (i.e., in continuous time and amplitude). Processing of signals can be done either in analog or digital domain. In order to process an analog signal in digital domain, it is required that a digital signal be formed by sampling and quantizing (digitizing) the analog signal. Hence, in contrast to an analog signal, a digital signal is discrete in both time and amplitude. The digitization process is achieved through an analog-to-digital (A/D) converter.

Digital Signal Processing (DSP) involves the manipulation of digital signals in order to extract useful information from them. Although an increasing amount of signal processing is being done in digital domain, there remains the need for interfacing to the analog world in which we live. A/D and D/A (digital-to-analog) data converters are the devices that make this interfacing possible. Figure 1-1 illustrates the main components of a DSP system, consisting of A/D, DSP, and D/A devices.

There are many reasons why one would want to process an analog signal in a digital fashion by converting it into a digital signal. The main reason is that digital processing allows programmability. The same DSP hardware can be used for many different applications simply by changing the code residing in memory. Another reason is that digital circuits provide a more stable and tolerant output than analog circuits—for instance, when subjected to temperature changes. In addition, the advantage of operating in digital domain may be intrinsic. For example, a linear phase filter or a steep-cutoff notch filter can only be realized by using digital signal processing techniques, and many adaptive systems are achievable in a practical product only through digital manipulation of signals. In essence, digital representation (0s and 1s) allows voice, audio, and video data

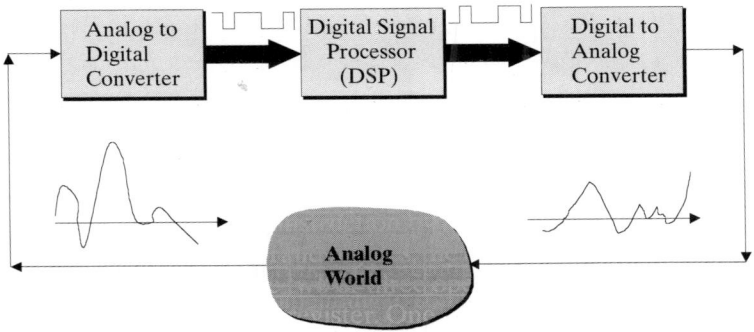

FIGURE 1-1   Main components of a DSP system.

to be treated the same for error-tolerant digital transmission and for storage purposes. As a result, digital processing, and hence digital signal processors (also called DSPs), are expected to play a major role in the next generation of telecommunication infrastructure, including 3G (third generation) wireless, cable (cable modems) and telephone lines (DSL—Digital Subscriber Line modems).

The processing of a digital signal can be implemented on various platforms, such as a DSP processor, a customized VLSI (Very Large Scale Integrated) circuit, or a general-purpose microprocessor. Some of the differences between a DSP and a single function VLSI implementation are as follows:

1. There is a fair amount of application flexibility associated with DSP implementation, since the same DSP hardware can be utilized for different applications. In other words, DSP processors are programmable. This is not the case for a hard-wired digital circuit.
2. DSP processors are cost-effective because they are mass-produced and can be used for many applications. A customized VLSI chip is normally built for a single application and a specific customer.
3. In many situations, new features constitute a software upgrade on a DSP processor not requiring new hardware. In addition, bug fixes are generally easier to perform.
4. Often, very high sampling rates can be achieved by a customized chip, whereas there are sampling rate limitations associated with DSP chips due to peripheral constraints and architectural design.

DSP processors share some common characteristics that also separate them from general-purpose microprocessors. Some of these characteristics include the following:

1. They are optimized to cope with the repetition or looping of operations common in signal processing algorithms. Relatively speaking, instruction sets of DSPs are smaller and are optimized for signal processing operations, such as single-cycle multiplication and accumulation.
2. DSPs allow specialized addressing modes, like indirect and circular addressing. These are efficient addressing mechanisms for implementing many signal processing algorithms.
3. DSPs possess appropriate peripherals that allow efficient input/output (I/O) interfacing to other devices.
4. In DSP processors, it is possible to execute several accesses to memory in a single instruction cycle. In other words, DSPs have a relatively high bandwidth between their CPUs (central processing units) and memory.

It should be kept in mind that, due to the growing features being placed on processors, one should be cautious of features that divide DSPs and general-purpose microprocessors.

Most of the market share of DSPs belong to real-time, cost-effective, embedded systems, like cellular phones, modems, and disk drives. *Real-time* means completing the processing within the allowable or available time between samples. This available time, of course, depends on the application. As illustrated in Figure 1-2, the number of in-

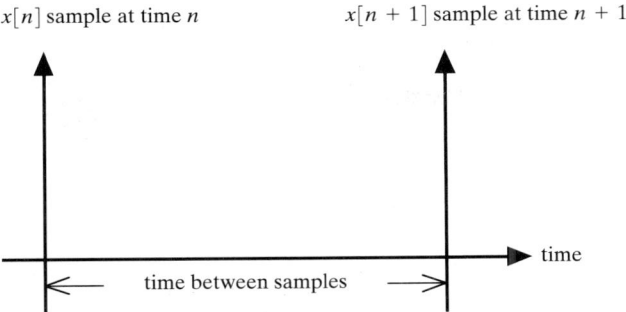

FIGURE 1-2 Maximum number of instructions to meet real-time = time between samples/instruction cycle time.

structions used for an algorithm running in real-time must be less than the number of instructions that can be executed between two consecutive samples. For example, for audio processing operating at 44.1 kHz sampling frequency, or approximately 22.6 μs sampling time interval, the number of instructions must be fewer than nearly 4500 instructions, assuming an instruction cycle time of 5 ns. There are two aspects of real-time processing: (a) sampling rate, and (b) system latencies (delays). Typical sampling rates and latencies for several different applications are shown in Table 1–1.

TABLE 1–1 Typical sampling rates and latencies for select applications.

Application	I/O Sampling Rate	Latency
Instrumentation	1 Hz	*system dependent
Control	>0.1 kHz	*system dependent
Voice	8 kHz	<50 ms
Audio	44.1 kHz	*<50 ms
Video	1–14 MHz	*<50 ms

*Many times, a signal may not need to be concerned with latency—for example, a TV signal is more dependent on synchronization with audio than the latency. In each of these cases, the latency is dependent on the application.

1.1 EXAMPLES OF DSP SYSTEMS

For the reader to appreciate the usefulness of DSPs and data converters, several examples of DSP systems currently in use are presented here.

During the past few years, there has been tremendous growth in the wireless market. Figure 1-3 illustrates a cellular phone wireless communication DSP system. As can be seen from this figure, there are two sets of data converters. On the voice band side, a low sampling rate [(e.g., 8 kSPS (kilo samples per second)] and a high-resolution (e.g., 13 bits) converter is required, whereas on the RF modulation side, a relatively high-speed (e.g., 20 MSPS), low-resolution (e.g., 8 bits) converter is used. System designers prefer to integrate more functionalities in DSP components rather than in analog

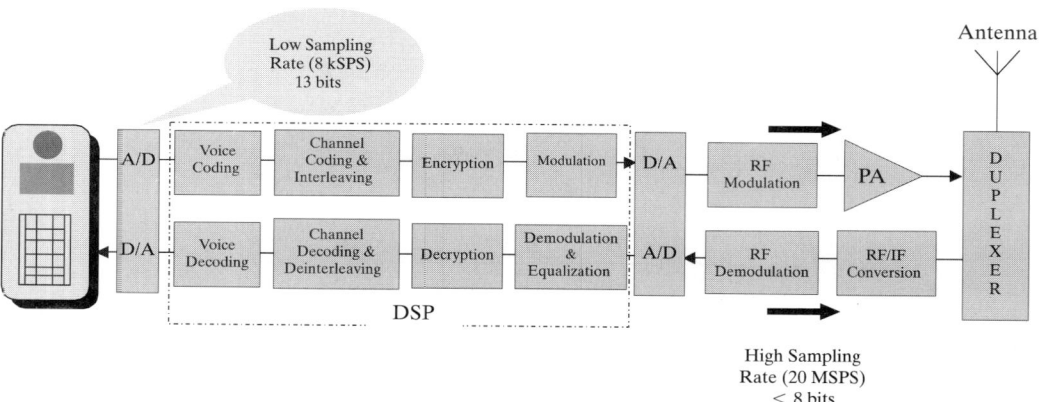

FIGURE 1-3   Cellular phone wireless communication DSP system.

components in order to lower the number of components and hence the overall cost. This strategy of more integration in DSP depends on specifications achievable for low power consumption in portable devices.

In wired communications, various types of modems are used to convert analog/digital signals to digital signals appropriate for error-tolerant transmission over wires or cables. Currently available modem types include high-speed voiceband [56 kbps (kilo bits per second)], ISDN (Integrated Services Digital Network), DSL, and cable modems. For example, DSL-type modems have data rates in the range of 1–52 Mbps. DSL makes use of the existing twisted-pair wires between residential homes and the phone company's central office. For example, the asymmetric version of DSL (ADSL) uses the frequency ranges of 25–138 kHz for upstream and 200 kHz–1.1 MHz for downstream data transmission, without interfering with the existing 0–4 kHz voiceband range. Figure 1-4 shows an ADSL system based on TI data converters and DSP products. The indicated A/D and D/A converters have a high-speed, high-resolution specification to cope with the multilevel nature of the transmitted signal. The transceiver is a dedicated DSP performing the ADSL modulation/demodulation.

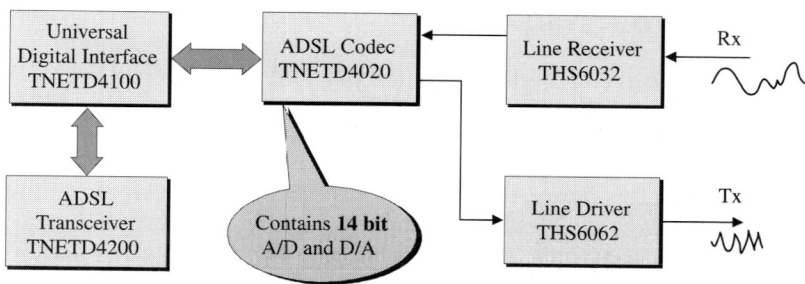

FIGURE 1-4   TI chipset for ADSL wired communication DSP system.

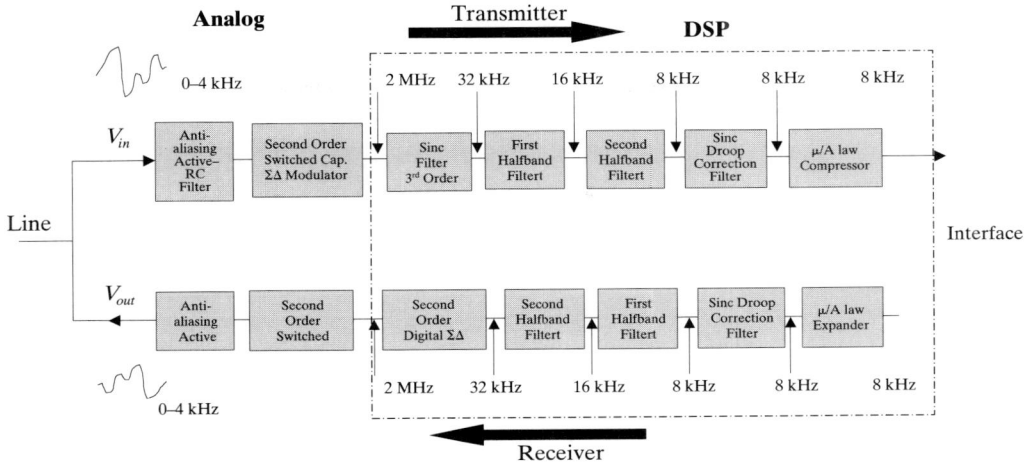

FIGURE 1-5 PCM voiceband DSP system.

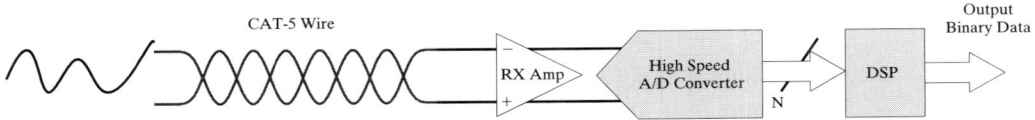

FIGURE 1-6 Gigabit Ethernet DSP system.

Considering that communication networks in use today are digital, an analog signal reaching the phone company's central office must be conditioned and converted to a digital signal for transmission through the network. Figure 1-5 shows the PCM (pulse code modulation) voiceband codec used in communications networks. As can be seen, a fair amount of the signal processing is done in the digital domain by the DSP component.

Figure 1-6 shows a gigabit Ethernet DSP system. The analog signal is sent through category-5 twisted-pair wires. Four 8-bit, high-speed A/D converters are used for data conversion. The dynamic range of the converters must be high enough to overcome noise, interference, and attenuation through an Ethernet link. A DSP is then used to execute echo cancellation, equalization, and demodulation signal processing.

Data stored on a compact disk (CD) or a computer hard drive is in binary format. However, the signal generated by a read head is analog and is corrupted by noise and distortion. This demands a fair amount of signal conditioning and filtering after reading data. As shown in Figure 1-7, this is achieved by using a DSP-based hard disk drive system.

Motor control is another area in which DSPs are making an impact. For example, as illustrated in Figure 1-8, DSPs are used to control induction motors by monitoring feedback signals, including current, voltage, and position. Such motors are widely used because of their low cost, high reliability, and high efficiency.

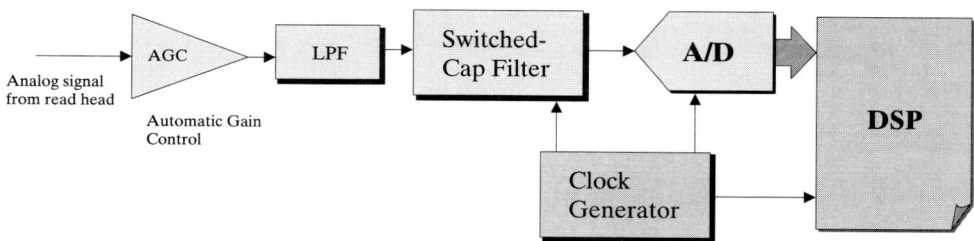

FIGURE 1-7 Hard disk drive DSP system.

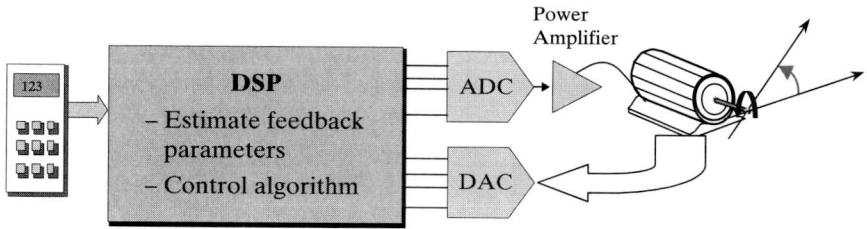

FIGURE 1-8 Motor control DSP system.

Smart sensors or devices are another example of DSP systems. These sensors are capable of both data acquisition and data processing. For example, such sensors are used in the airbag activation system in automobiles. Vehicle acceleration is measured by a suspension-mass sensor and converted into a digital signal by an A/D converter. This signal is then processed by a DSP to detect an accident by comparing features of the signal with those of the accident.

1.2 ORGANIZATION OF CHAPTERS

The three chapters that follow are dedicated to data converters in DSP systems to provide the reader with a system view of DSP-based products. Chapter 2 gives a discussion of the differences and relationships between analog and digital signals. It also includes a Matlab toolbox that can be used to acquire hands-on exposure to the issues related to A/D and D/A data conversion. Chapter 3 discusses the specifications associated with data converters. It provides the information needed for selecting data converters in various DSP applications of interest. The architectures of A/D and D/A converters are then presented in Chapter 4.

The DSP part of the book starts in Chapter 5, in which an overview of the TMS320C6000 architecture is presented. The focus is placed on the architectural features one needs to be aware of to implement algorithms on the C6x processor. In Chapter 6, the software tools are presented, and the steps used in taking a source file to an executable file are discussed. Lab 1 Chapter 6 provides a hands-on approach to familiarize oneself with the Code Composer Studio integrated development environment. Chapter 7 presents the concept of interrupt data processing. Lab 2 Chapter 7 shows how to sample an analog signal in real-time on a C6x target board. In Chapter 8, fixed-point and floating-point number representations are discussed and their differences are pointed out. Lab 3 Chapter 8 gives suggestions on how one may cope with the overflow or scaling problem. Code efficiency issues appear in Chapter 9, in which optimization techniques, as well as linear assembly and hand-coded pipelined assembly, are discussed. Lab 4 Chapter 9 covers FIR (finite impulse response) filtering while deploying different optimization techniques. Chapter 10 covers circular buffering. Lab 5 Chapter 10 shows how circular buffering is used to perform adaptive filtering. Frame processing is covered in Chapter 11. Lab 6 Chapter 11 provides an example of frame processing involving FFT (fast Fourier transform) implementation and the use of DMA (Direct Memory Access). Finally, Chapter 12, Labs 7 and 8 address the DSP–BIOS real-time analysis and scheduling features of Code Composer Studio.

1.3 REQUIRED SOFTWARE AND HARDWARE

The software tool needed to generate TMS320C6000 executable files is called Code Composer Studio (CCS). CCS incorporates the assembler, linker, compiler, simulator, and debugger. In the absence of a target board, which allows one to run an executable file on an actual C6x processor, the simulator can be used to verify code functionality by using data already stored in a datafile. However, when using the simulator, an interrupt service routine (ISR) cannot be used to read in signal samples from a signal source. To be able to process digital signals in real-time on an actual C6x processor, an EVM (evaluation module) or a DSK (DSP Starter Kit) board is needed for code development. The interfacing equipment may consist of a function generator, oscilloscope, microphone, boombox, and cables with audio jacks.

An EVM board can be easily installed in a full-length PCI slot inside a PC host. Refer to the *TI TMS320C6x Evaluation Module Reference Guide* [16] for the installation details. The PC host ought to have a minimum of 16M RAM. The interfacing with the EVM board is done through three standard audio jacks appearing at the back of the EVM board. A DSK board can also be easily connected to a PC host through its parallel port. The interfacing with the board is done through two audio jacks located on the side of the DSK board.

For studying the conversion of analog to digital signals, and for designing the digital filters in the lab, access and familiarity with Matlab are assumed. It is also assumed that the reader is familiar with C programming.

CHAPTER 2

Analog-to-Digital Signal Conversion

The process of analog-to-digital signal conversion consists of converting a continuous time and amplitude signal into discrete time and amplitude values. Sampling and quantization constitute the steps needed to achieve analog-to-digital signal conversion. To minimize any loss of information that may occur as a result of this conversion, it is important to understand the underlying principles behind sampling and quantization.

2.1 SAMPLING

Sampling is the process of generating discrete time samples from an analog signal. First, it is helpful to see the relationship between analog and digital frequencies. Let us consider an analog sinusoidal signal $x(t) = A\cos(\omega t + \phi)$. Sampling this signal at $t = nT_s$, with the sampling time interval of T_s, generates the discrete time signal

$$x[n] = A\cos(\omega n T_s + \phi) = A\cos(\theta n + \phi), \qquad n = 0, 1, 2, \ldots, \tag{2.1}$$

where $\theta = \omega T_s = \dfrac{2\pi f}{f_s}$ denotes digital frequency with units radians (as compared to analog frequency ω with units radians/sec).

The difference between analog and digital frequencies is more evident by observing that the same discrete time signal is obtained for different continuous time signals if the product ωT_s remains the same. (An example is shown in Figure 2-1.) Likewise, different discrete time signals are obtained for the same analog or continuous time signal when the sampling frequency is changed. (An example is shown in Figure 2-2.) In other words, both the frequency of an analog signal and the sampling frequency define the frequency of the corresponding digital signal.

It helps to understand the constraints associated with the above sampling process by examining signals in frequency domain. The Fourier transform pairs in analog and digital domains are given by

Fourier transform pair for analog signals
$$\begin{cases} X(j\omega) = \displaystyle\int_{-\infty}^{\infty} x(t) e^{-j\omega t}\, dt \\ x(t) = \dfrac{1}{2\pi} \displaystyle\int_{-\infty}^{\infty} X(j\omega) e^{j\omega t}\, d\omega \end{cases} \tag{2.2}$$

2.1 Sampling

and

Fourier transform pair for discrete signals

$$\begin{cases} X(e^{j\theta}) = \sum_{n=-\infty}^{\infty} x[t] e^{-jn\theta}, \; \theta = \omega T_s \\ x[t] = \dfrac{1}{2\pi} \displaystyle\int_{-\pi}^{\pi} X(e^{j\theta}) e^{jn\theta} \, d\theta. \end{cases}$$ (2.3)

As illustrated in Figure 2-3, when an analog signal with a maximum frequency of f_{max} (or bandwidth of W) is sampled at a rate of $T_s = \dfrac{1}{f_s}$, its corresponding frequency response is repeated every 2π radians, or f_s. In other words, Fourier transform in digital domain becomes a periodic version of Fourier transform in analog domain. That is why, for discrete signals, we are only interested in the frequency range 0–$f_s/2$.

Therefore, in order to avoid any aliasing or distortion of the frequency content of the discrete signal, and hence to be able to recover or reconstruct the frequency content of the original analog signal, we must have $f_s \geq 2f_{max}$. This is known as the *Nyquist rate*; that is, the sampling frequency should be at least twice the highest frequency in the signal. Normally, before any digital manipulation, an antialiasing analog lowpass filter is used to limit the highest frequency of the analog signal.

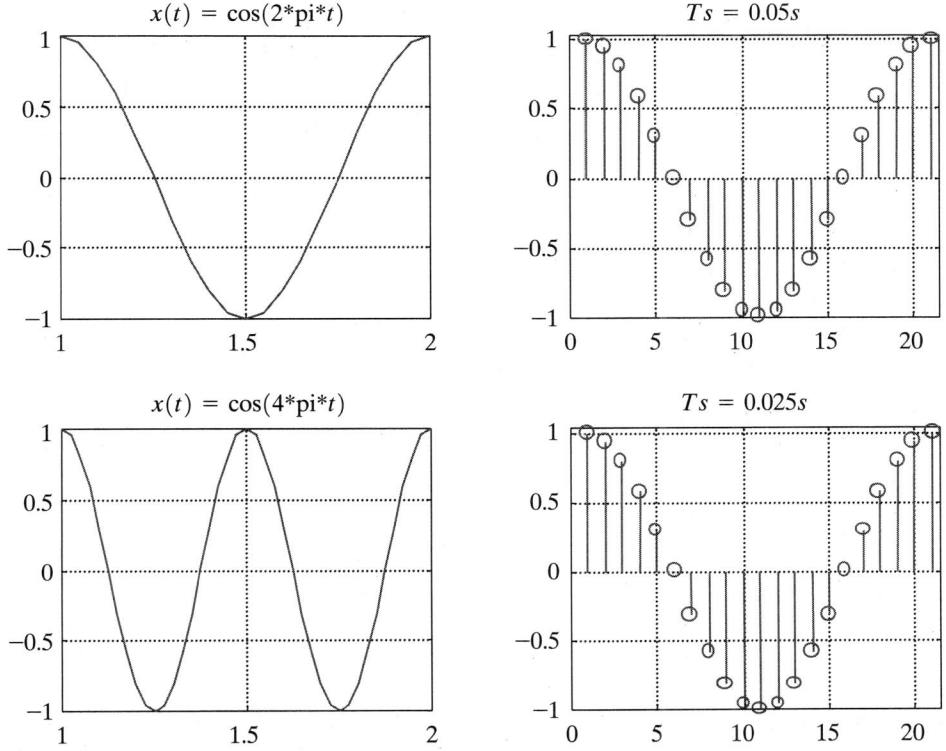

FIGURE 2-1 Different sampling of two different analog signals leading to the same digital signal.

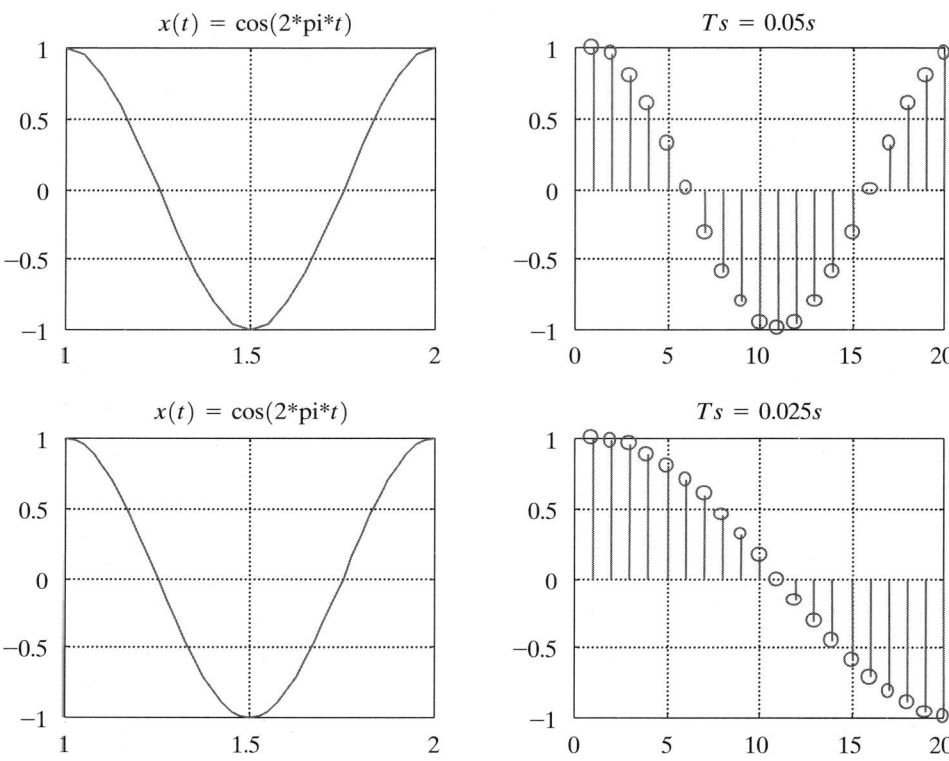

FIGURE 2-2 Different sampling of the same analog signal leading to two different digital signals.

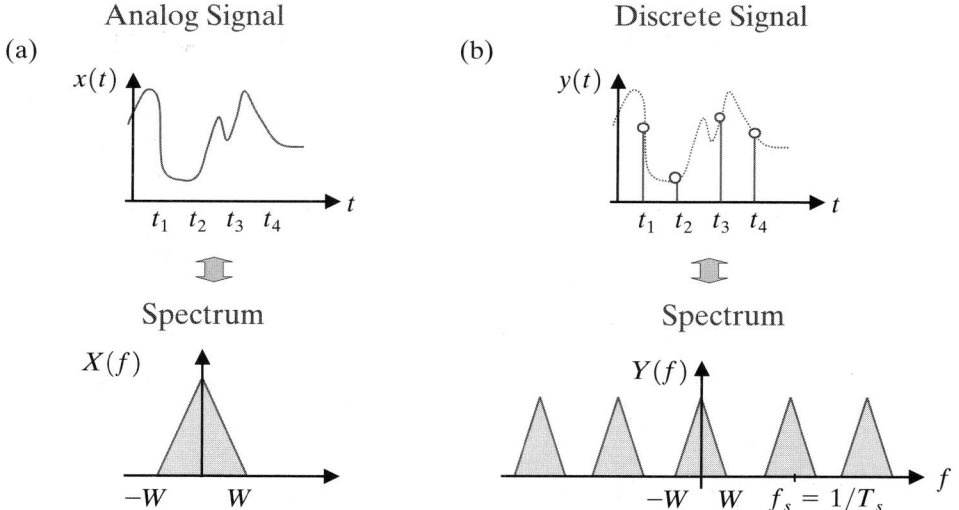

FIGURE 2-3 (a) Fourier transform of a continuous-time signal and (b) its discrete time version.

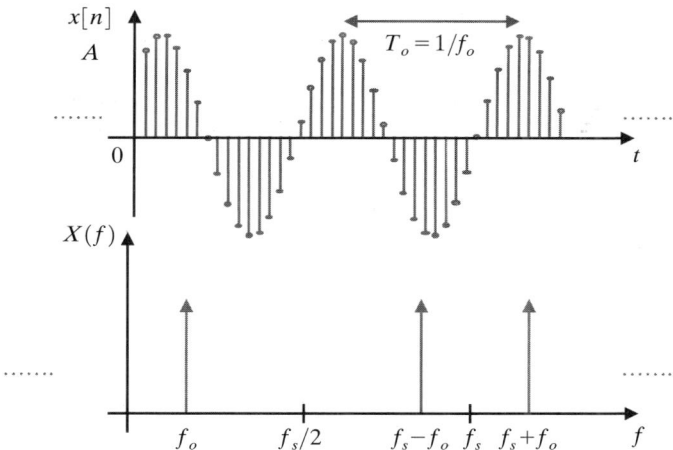

FIGURE 2-4 Fourier transform of a sampled sinusoidal signal.

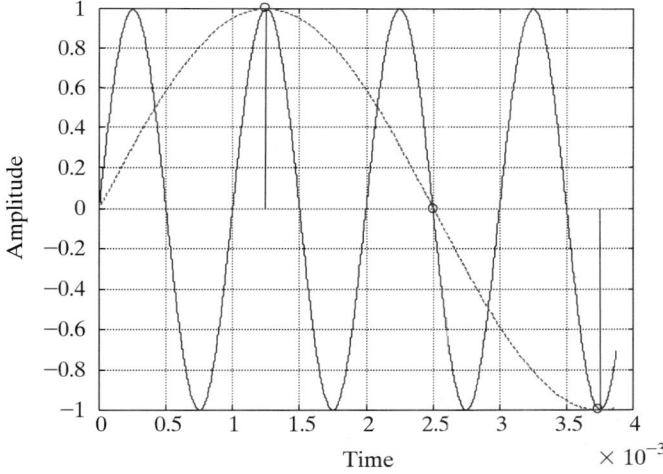

FIGURE 2-5 Ambiguity caused by aliasing.

Figure 2-4 shows the Fourier tranform of a sampled sinusoid with a frequency of f_o. As can be seen, there is only one frequency component at f_o. The aliasing problem can be further illustrated by considering an undersampled sinusoid as depicted in Figure 2-5. In this figure, a 1-kHz sinusoid is sampled at $f_s = 0.8$ kHz, which is less than the Nyquist rate. The dashed-line signal is a 200-Hz sinusoid passing through the same sample points. Thus, at this sampling frequency, the output of an A/D converter would be the same if either of the sinusoids were the input signal. On the other hand, oversampling a signal provides a richer description than that of the same signal sampled at the Nyquist rate.

12 Chapter 2 Analog-to-Digital Signal Conversion

To experiment with undersampling and oversampling, a Matlab function sinesample.m is provided on the attached CD. The synopsis of this function is as follows:

```
[Vs,t] = sinesample(fo,fs,Ns,phi,shg,option);
%    INPUTS:
%       -fo  :  input signal frequency    (kHz)
%       -fs  :  sampling frequency        (kHz)
%       -Ns  :  number of samples
%       -phi :  initial phase of sinewave (degree)
%       -shg :  show graphics
%                        0-> No
%                        1-> Yes
%       -option:  0-> zero DC sinewave
%                 1-> sinewave fits in zero to full scale
%
%    OUTPUTS:
%       -t   :  sampling time
%       -Vs  :  sampled signal
```

2.1.1 Fast Fourier Transform

Fourier transform of discrete signals is continuous over the frequency range 0–$f_s/2$. Thus, from a computational standpoint, this transform is not suitable to use. In practice, discrete Fourier transform (DFT) is used in place of Fourier transform. DFT is the equivalent of Fourier series in analog domain. However, it should be remembered that DFT and Fourier series pairs are defined for periodic signals. These transform pairs are expressed as

Fourier series for periodic analog signals

$$\begin{cases} X_k = \dfrac{1}{T} \int_{-\frac{T}{2}}^{\frac{T}{2}} x(t) e^{-j\omega_0 k t}\, dt \\ x(t) = \sum_{k=-\infty}^{\infty} X_k e^{j\omega_0 k t} \\ \text{where } T \text{ denotes period and} \\ \omega_0 \text{ fundamental frequency} \end{cases} \quad (2.4)$$

and

Discrete Fourier transform (DFT) for periodic discrete signals

$$\begin{cases} X[k] = \sum_{n=0}^{N_s-1} x[n] e^{-j\frac{2\pi}{N_s} n k}, \, k = 0, 1, \ldots, N_s - 1 \\ x[n] = \dfrac{1}{N_s} \sum_{k=0}^{N_s-1} X[k] e^{j\frac{2\pi}{N_s} n k}, \, n = 0, 1, \ldots, N_s - 1. \end{cases} \quad (2.5)$$

Hence, when computing DFT, it is required to assume periodicity with a period of N_s samples. Figure 2-6 illustrates a sampled sinusoid which is no longer periodic. In order to make sure that the sampled version remains periodic, the analog frequency should satisfy this condition [5]

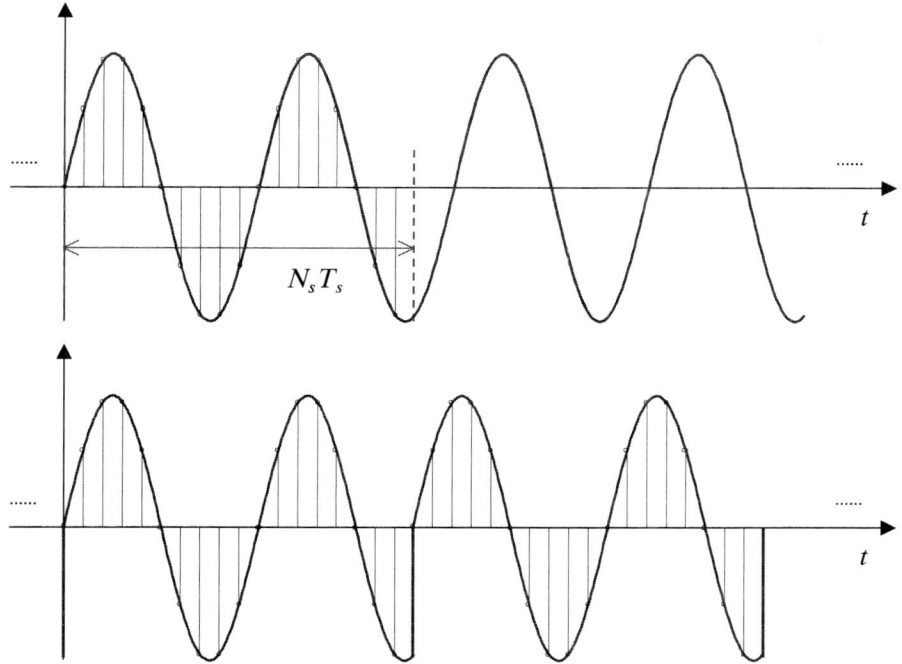

FIGURE 2-6 Periodicity condition of sampling.

$$f_o = \frac{m}{N_s} f_s, \qquad (2.6)$$

where m denotes the number of cycles over which DFT is computed.

The computational complexity (number of additions and multiplications) of DFT is reduced from N_s^2 to $N_s \log N_s$ by using fast Fourier transform (FFT) algorithms. In these algorithms, N_s is considered to be a power of two. Figure 2-7 shows the effect of the periodicity constraint on the FFT computation. In this figure, the FFTs of two sinusoids with respective frequencies of 250 Hz and 251 Hz are shown. The amplitudes of the sinusoids are unity. Although there is only a 1-Hz difference between the sinusoids, the FFT outcomes are significantly different due to improper sampling.

2.1.2 Amplitude Statistics

An important property used in signal analysis is amplitude statistic. This statistic reflects the probability density function (pdf) associated with amplitudes of a randomly sampled signal. In other words, the pdf shows the histogram of sample points if the signal is sampled with an infinitesimal sampling period. For example, for a sinewave $x(t) = a \sin(2\pi f_o t)$, its amplitude pdf is given by

$$f(x) = \frac{1}{\pi \sqrt{a^2 - x^2}}, \qquad |x| < a. \qquad (2.7)$$

This pdf is illustrated in Figure 2-8.

FIGURE 2-7 FFTs of (a) a 250- and (b) a 251-Hz sinusoid.

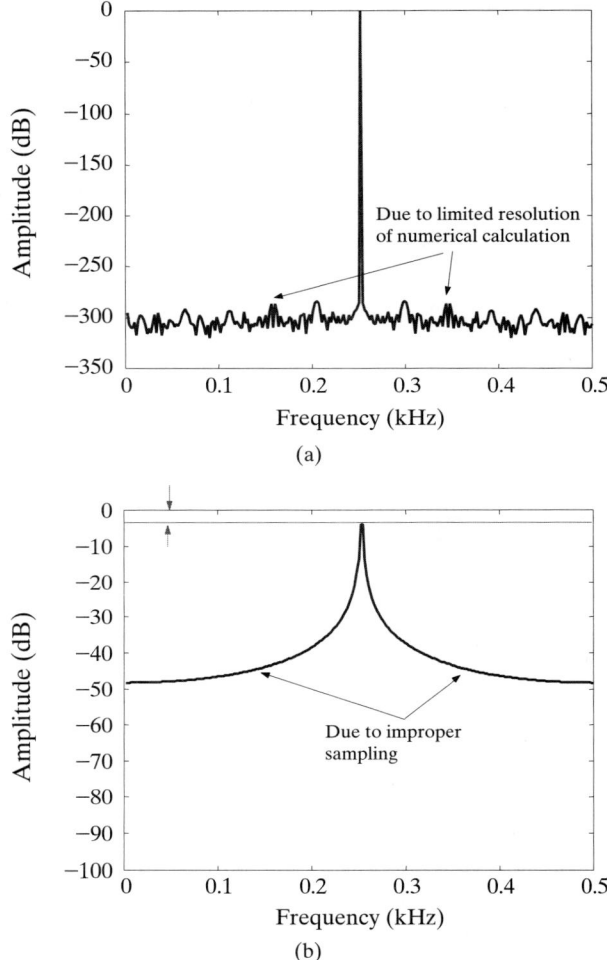

FIGURE 2-8 Amplitude pdf of sinewave.

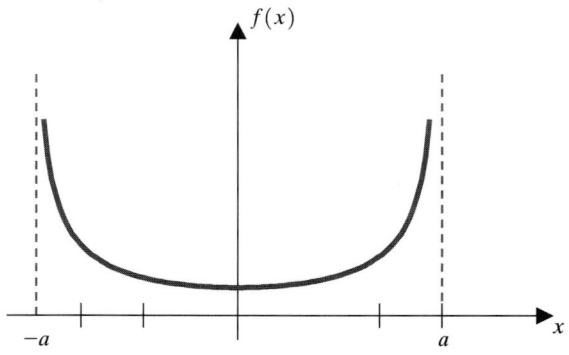

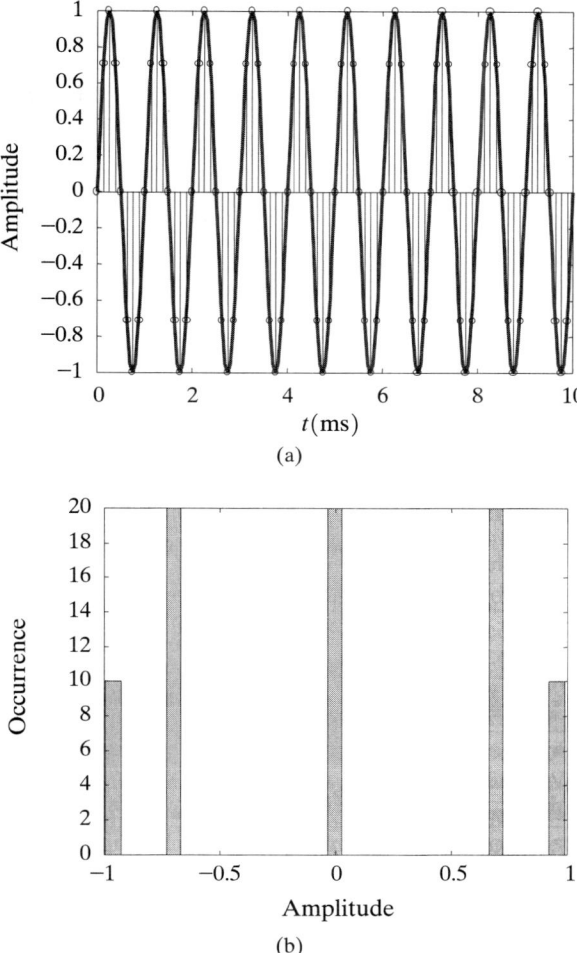

FIGURE 2-9 Histogram of amplitude levels when $m = 10$ and $N_s = 80$ are **not** mutually prime: (a) sampled signal and (b) histogram of sample points.

Consider a sinewave with $f_o = 1$ kHz, $N_s = 80$, and $m = 10$. As shown in Figure 2-9, the amplitude histogram is not correct because of improper sampling, caused by repeatedly sampling the same level. In order to avoid this sampling outcome and obtain a proper amplitude statistics, the number of cycles m and the number of samples N_s must be mutually prime. Figure 2-10 shows the amplitude histogram for $m = 13$.

FIGURE 2-10 Histogram of amplitude levels when $m = 13$ and $N_s = 80$ are mutually prime: (a) sampled signal and (b) histogram of sample points.

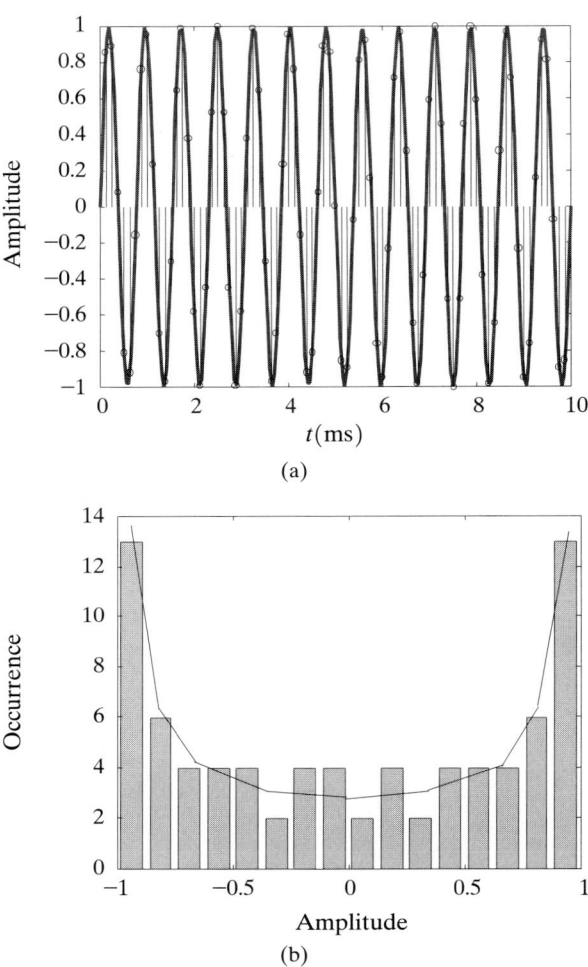

2.1.3 Harmonics of a Distorted Sinewave

Linear circuits, in general, are not perfectly linear, leading to distortion of output signals. Circuit designers measure this distortion by using a sinewave input. A distorted sinewave, as indicated by Fourier series, contains harmonics. In analog domain, the locations of harmonics on the frequency axis are easy to predict. These locations are at kf_o, where k denotes harmonic index. However, as a result of sampling, the locations of harmonics in digital domain are not so easy to predict because of aliasing. The Nyquist rate condition is usually held for the fundamental frequency. Consequently, this sampling frequency may not be sufficient for higher harmonics. Knowledge of the locations of harmonics is of great importance in the interpretation of FFT results, especially for diagnostic purposes. Figure 2-11 shows the effect of sampling on the harmonics index. It is seen that the sampling of a distorted sinewave results in the consecutive folding of harmonics between 0 and $f_s/2$. Figure 2-12 shows the FFT result of a distorted sinewave with $f_o = 1.5$ kHz and $f_s = 10$ kHz.

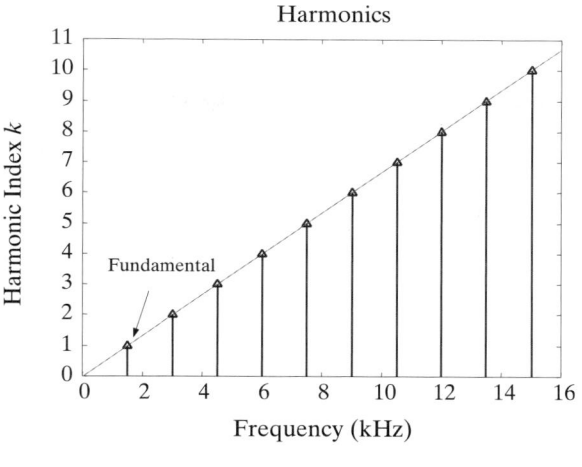

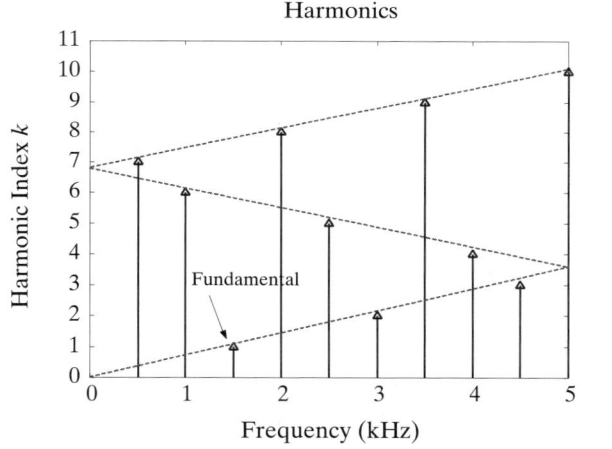

FIGURE 2-11 Effect of sampling on harmonic index; $f_o = 1.5$ kHz with 10 harmonics and $f_s = 10$ kHz: (a) before sampling and (b) after sampling.

The Matlab function `harmonicd.m` on the attached CD can be used to examine the locations of harmonics. The synopsis of this function appears below.

```
[fh,fhs,Nhs] = harmonicd(fo,fs,Nh,shg) ;
%
%   INPUTS:
%      - fo : input fundamental frequency (kHz)
%      - fs : sampling frequency (kHz)
%      - Nh : vector of desired harmonics, e.g., [ 1 3 5]
%      -shg : show results on the screen;
%             0-> don't show; 1-> show
%
%   OUTPUTS:
%      - fh  : location of harmonics
%      - fhs : sorted location of harmonics
%      - Nhs : sorted harmonic index
```

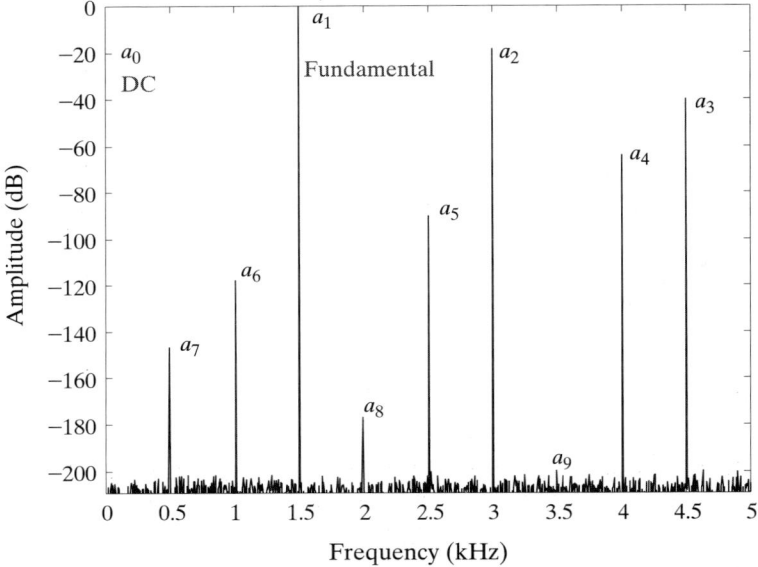

FIGURE 2-12 Sampling a distorted sinewave with $f_o = 1.5$ kHz and $f_s = 10$ kHz.

2.2 QUANTIZATION

An A/D converter has a finite number of bits (or resolution). As a result, continuous amplitude values get represented or approximated by discrete amplitude levels. The process of converting continuous into discrete amplitude levels is called *quantization*. This approximation leads to an error called *quantization noise*. The input/output characteristic of a 3-bit A/D converter is shown in Figure 2-13 to illustrate how analog voltage values are approximated by discrete voltage levels.

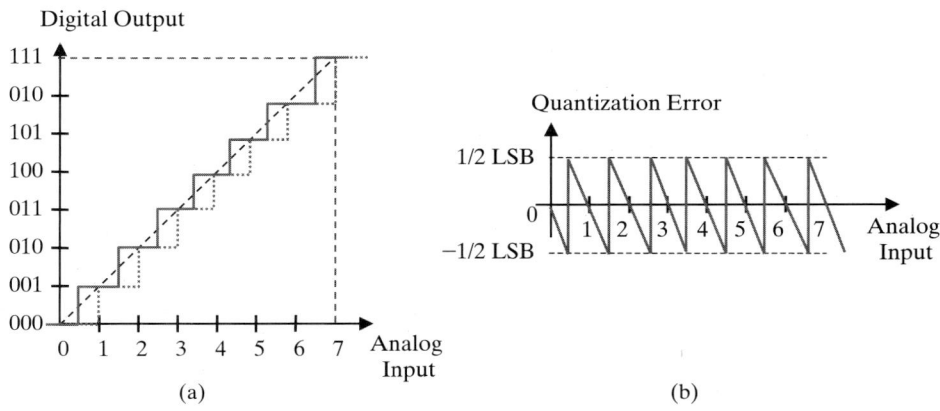

FIGURE 2-13 Characteristic of a 3-bit A/D converter: (a) input/output static transfer function and (b) additive quantization noise.

2.2 Quantization

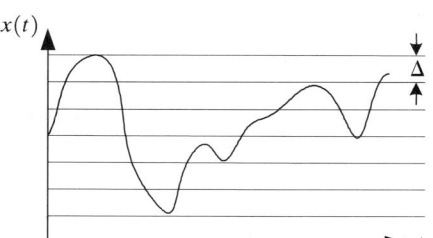

FIGURE 2-14 Quantization levels.

The quantization interval depends on the number of quantization or resolution levels, as illustrated in Figure 2-14. Clearly, the amount of quantization noise generated by an A/D converter depends on the size of quantization interval. More quantization bits translate into a narrower quantization interval and hence into a lower amount of quantization noise.

To avoid saturation or out of range distortion, the input voltage must be between V_{ref-} and V_{ref+}. The full-scale (FS) voltage or V_{ref} is defined as

$$V_{FS} = V_{ref} = V_{ref+} - V_{ref-}, \tag{2.8}$$

and one least significant bit (LSB) is given by

$$1 \text{ LSB} = \Delta = \frac{V_{ref}}{2^N}, \tag{2.9}$$

where N is the number of bits of the A/D converter. Table 2-1 lists 1 LSB in volts for different numbers of bits and reference voltages. It is interesting to note that a couple of microvolts, which is 1 LSB in a high-resolution A/D converter, can be generated by a dozen electrons in a 1-pF capacitor!

TABLE 2-1 LSB of A/D converter.

N	8	10	12	14	16	20
$V_{ref} = 5 V$	19.5 mV	4.9 mV	1.2 mV	305 μV	76 μV	4.8 μ
$V_{ref} = 3 V$	11.7 mV	2.9 mV	732 μV	183 μV	45.8 μV	2.8 μV
$V_{ref} = 1.8 V$	7.0 mV	1.7 mV	439 μV	110 μV	27.5 μV	1.7 μV

Usually, it is assumed that quantization noise is signal independent and is uniformly distributed over −0.5 LSB and 0.5 LSB. Figure 2-15 shows the quantization noise of an analog signal quantized by a 3-bit A/D converter. It is seen that, although the histogram of the quantization noise is not exactly uniform, it is reasonable to accept the uniformity assumption.

The Matlab function a2d.m on the attached CD simulates the behavior of an A/D converter. The synopsis of this function is as follows:

```
[Vout,q,t] = a2d(Vin,nBit,t,Vref,noise,INL,shg);
%   INPUTS:
%       -Vin   : input signal ( [] -> ramp input )
%       -nBit  : number of bits
%       -t     : time vector or [tmin tmax]
%       -Vref  : reference voltage (Full Scale = 2^Bit - 1)
```

FIGURE 2-15 Quantization of an analog signal by a 3-bit A/D converter: (a) output signal and quantization error, (b) histogram of quantization error, and (c) bit stream.

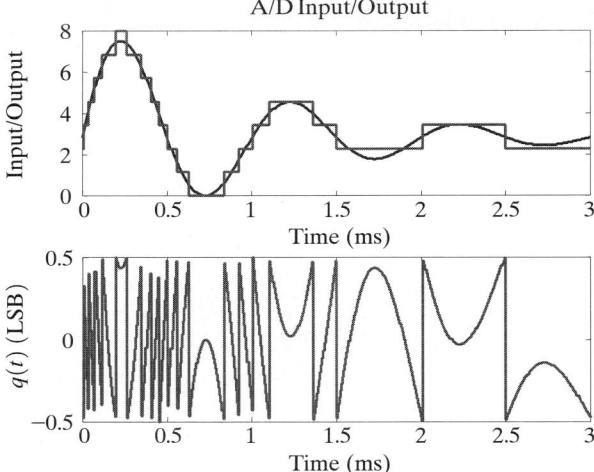

(a)

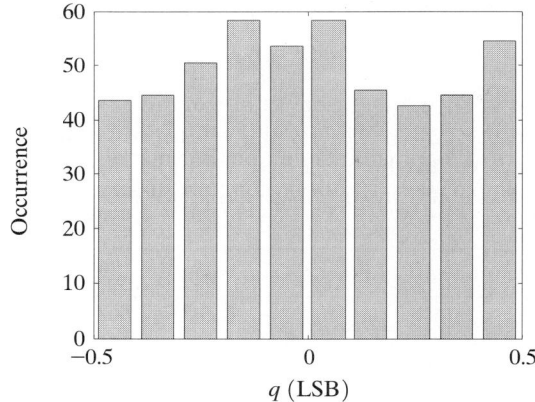

(b)

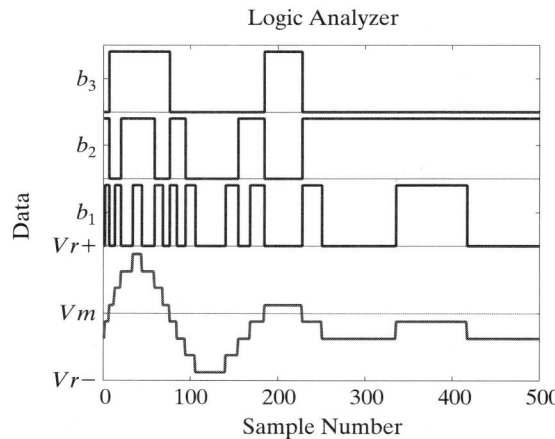

(c)

```
%           -noise: electronic noise at different level (Gaussian noise)
%                   noise (in dB) = 10*log ( Pn / Pgaus(+-1/2LSB) )
%                   (1/2 LSB = 3*sigma)
%                   Note: -inf means no noise
%           -INL  : Integral Nonlinearity (in +-LSB); vector or a max
%                   INL. In case of max INL, vector is generated by using
%                   uniformly distributed random variable.
%           -shg  : show results on the screen;
%                       0-> don't show; 1->  show
% OUTPUTS:
%           -Vout : output signal of the A/D converter
%           -q    : quantization error in LSB
```

Figure 2-16 shows the FFT of a sinewave before and after the digitization process. The input sinewave is at 250 Hz, with unity amplitude, $f_s = 1$ kHz, and $N_s = 512$. It is seen that the quantization error raises the noise level. The following

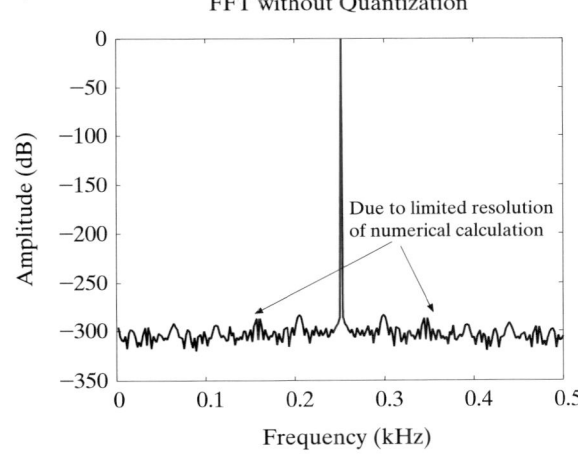

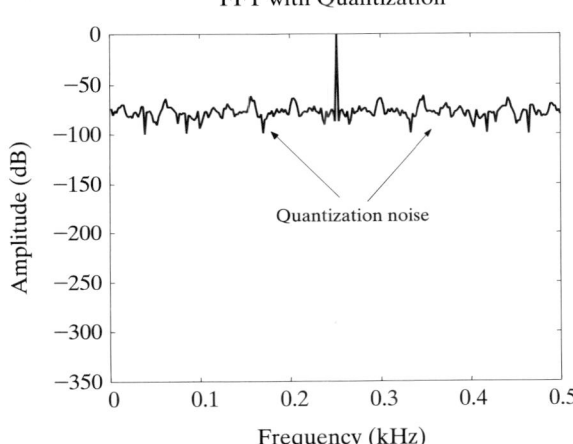

FIGURE 2-16 Sinewave before and after digitization, $f_o = 250$ Hz, $f_s = 1$ kHz, $N_s = 512$, $N = 8$-bit: (a) FFT before digitization and (b) FFT after digitization.

Chapter 2 Analog-to-Digital Signal Conversion

Matlab function sinfft.m is written to help one examine the spectrum of quantization noise. Its synopsis is as follows:

```
[x,xq,Xs,Xsq,f] = sinfft(fo,fs,N,nBit,noise,INL,shg);
% INPUTS:
%      -fo   : input sinewave frequency (kHz)
%      -fs   : sampling frequency (kHz)
%      -N    : number of samples
%      -nBit : number of bits
%      -noise: electronic noise at different level (Gaussian noise)
%              noise (in dB) = 10*log ( Pn / Pgaus(+-1/2LSB) )
%              (1/2 LSB = 3*sigma)
%              Note: -inf means no noise
%      -INL  : Integral Nonlinearity (in +-LSB); vector or a max INL.
%              In case of max INL, vector is generated by using
%              uniformly distributed random variable.
%      -shg  : show results on the screen;
%              0-> don't show; 1-> show
% OUTPUTS:
%      - x   : samples of sinewave
%      - xq  : quantized samples of the sinewave
%      - Xs  : FFT of samples
%      - Xsq : FFT of the quantized samples
%      - f   : frequency vector
```

2.2.1 Signal-to-Noise Ratio

Resolution is the term used to describe the minimum resolvable signal level by an A/D converter. The fundamental limit of an A/D converter is governed by quantization noise, which is caused by the A/D converter's finite resolution. If the output digital word consists of N bits, the minimum step that the converter can resolve is 1 LSB. If we assume that quantization error n_q is a random variable uniformly distributed and independent of the input signal, then we have

$$\delta_q^2 = E[n_q^2] = \frac{1}{\Delta} \int_{-\frac{\Delta}{2}}^{\frac{\Delta}{2}} n_q^2 dn_q = \frac{\Delta^2}{12}, \tag{2.10}$$

where δ_q^2 indicates quantization noise variance. For a sinusoidal input signal having an amplitude of A_m, an ideal A/D converter has a signal-to-noise ratio (SNR) of

$$10 \log \frac{P_S}{P_n} = 10 \log \frac{(A_m)^2/2}{\frac{1}{2}(V_{ref}/2^N)^2}, \tag{2.11}$$

where P_S and P_n denote signal and noise power, respectively. It is observed that the quantization SNR is a function of amplitude. The maximum SNR can be written as

$$\text{SNR}_{max} = 10 \log \frac{(V_{ref}/2)^2/2}{\frac{1}{2}(V_{ref}/2^N)^2} = 10 \log \frac{3}{2} 2^{2N} = 6.02N + 1.76 \text{ (dB)}. \tag{2.12}$$

2.2 Quantization

For instance, an ideal 16-bit A/D converter has a maximum SNR of about 97.8 dB. Quantization noise decreases by 6 dB for each additional bit.

In order to calculate the SNR of an A/D converter, a Matlab function snr.m is written and provided on the attached CD. It is assumed that the input signal is a sinewave whose frequency satisfies condition (2.6). The synopsis of this function appears next.

```
[SNR,SFDR,AmdB] = snr(x,fs,shg,dcRm);
%      -x   : samples of the signal
%      -fs  : sampling frequency (kHz)
%      -shg : show results on the screen;
%             0-> don't show; 1-> show
%      -dcRm: DC removal
%             0-> don't remove; 1-> remove
%
%  OUTPUTS:
%      -SNR : signal-to-noise ratio
%      -SFDR: spurious-free-dynamic-range
%      -AmdB: amplitude of sinewave (dB)
```

Figure 2-17 shows the SNR of an 8-bit A/D converter as a function of the input amplitude. The maximum occurs when the input sinewave amplitude is equal to one half of the full-scale voltage (scaled to 0 dB). Another Matlab function, named snr2vi.m, is provided to simulate and plot the quantization SNR. The function and description of the input and output arguments are as follows:

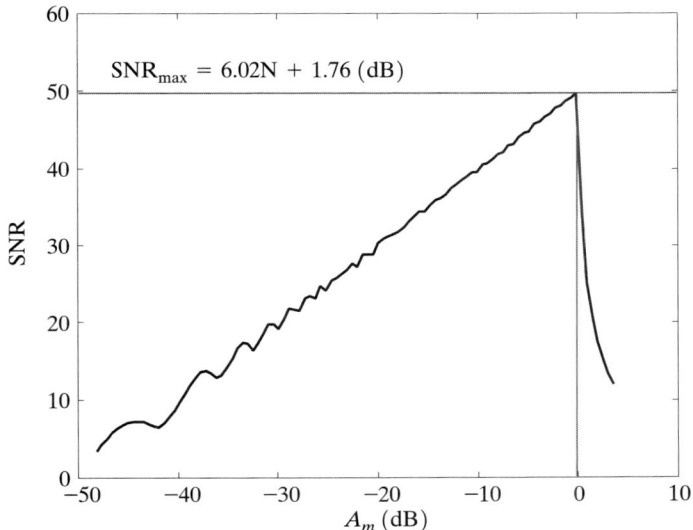

FIGURE 2-17 Signal-to-noise ratio of an ideal 8-bit A/D converter.

24 Chapter 2 Analog-to-Digital Signal Conversion

```
[SNR,AmdB] = snr2vi(fo,fs,N,nBit,noise,INL,shg);
%  INPUTS:
%      -fo   : input sine wave frequency (kHz)
%      -fs   : sampling frequency (kHz)
%      -N    : number of samples
%      -nBit : number of bits
%      -noise: electronic noise at different level (Gaussian noise)
%              noise (in dB) = 10*log ( Pn / Pgaus(+-1/2LSB) )
%              (1/2 LSB = 3*sigma)
%              Note: -inf means no noise
%      -INL  : Integral Nonlinearity (in +-LSB); vector or a max
%              INL. In case of max INL, vector is generated by using
%              uniformly distributed random variable.
%      -shg  : show results on the screen;
%              0-> don't show; 1-> show
%
%  OUTPUTS:
%      -SNR  : vector of SNR for different values of input amplitude
%      -AmdB : vector of input amplitude (dB)
```

To better understand the quantization effect, let us assume that the signal is zero mean Gaussian with $V_{ref} = K\sigma_x$, where σ_x denotes the standard deviation of the signal. By substituting for V_{ref} and σ_x in (2.11), we obtain

$$\text{SNR} = 10 \log \frac{\sigma_x^2}{\sigma_q^2} \cong 6N + 10.8 - 20 \log_{10} K. \qquad (2.13)$$

As an example, consider $K = 4$. The probability of signal samples falling in the $4\sigma_x$ range is 0.954. This means that out of 1000 samples, 954 samples fall in this range on average. In other words, 46 out of 1000 samples fall outside the indicated range and hence are represented by the maximum or minimum allowable value.

If the signal is scaled by α, the corresponding signal variance changes to $\alpha^2 \sigma_x^2$. Hence, the SNR changes to

$$\text{SNR} \cong 6N + 10.8 - 20 \log_{10} K + 20 \log_{10} \alpha. \qquad (2.14)$$

It is important to note that when we perform fractional arithmetic (discussed in Chapter 8), α is scaled to be less than 1, leading to a lower signal-to-noise ratio. This indicates that quantization noise should be kept in mind when scaling down the input signal. In other words, scaling down to achieve fractional representation cannot be done indefinitely, since, as a result, the signal would become buried in quantization noise.

The interested reader is referred to [3] for a more elaborate analysis of quantization noise. For example, for a linear time-invariant system, such as an FIR or an IIR filter, it can be shown that the noise variance σ_o^2 at the output of the system, caused by the input quantization noise, is given by [3]

$$\sigma_o^2 = \sigma_q^2 \sum_n h^2[n], \qquad (2.15)$$

where h denotes the unit sample response. For a first order system $y[n] = Ay[n-1] + x[n]$, it is easy to show that

$$\sigma_o^2 = \frac{\sigma_q^2}{1 - A^2}. \tag{2.16}$$

2.2.2 Sampling Time Jitter

In practice, the actual sampling time interval may not be exactly the same as the desired sampling time interval due to random jitters associated with clocks. (See Figure 2-18.) The deviation of sampling time interval from the desired interval adds errors to sampled values. The jitter error is considered to be an additive noise. Figure 2-19 shows the effect of time jitter on a sinusoidal signal.

For sinusoidal signals, the maximum allowable time jitter, which results in an error of less than one-half the LSB, is given by [6]

$$\Delta t_{max} = \frac{1}{\pi f_{in} 2^{N+1}}, \tag{2.17}$$

where N indicates the number of A/D bits and f_{in} is the input frequency. Figure 2-20 shows a plot of maximum jitter as a function of the input frequency and number of bits. For sinusoidal input signals, the root-mean-square (rms) of jitter noise is shown to be [6]

$$V_n \cong 2\pi A_m f_o T_j, \tag{2.18}$$

where A_m and f_o denote amplitude and frequency, respectively, and T_j is the rms value of time jitter Δt.

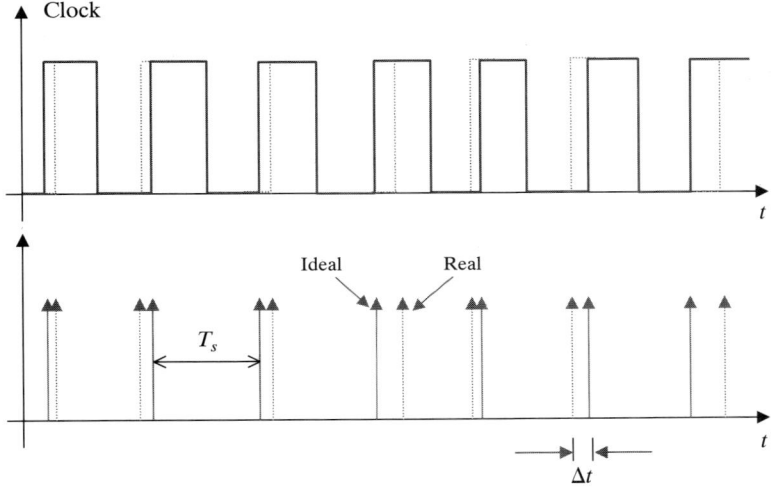

FIGURE 2-18 Sampling time jitter.

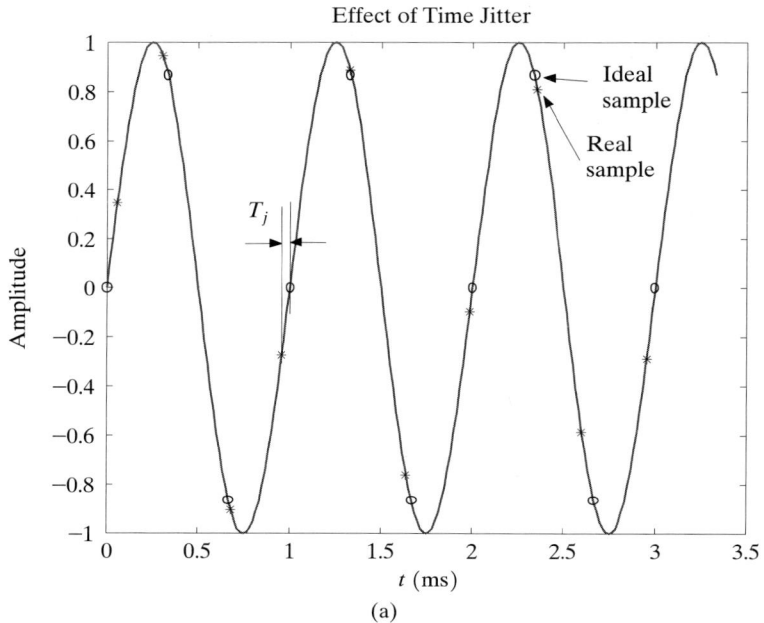
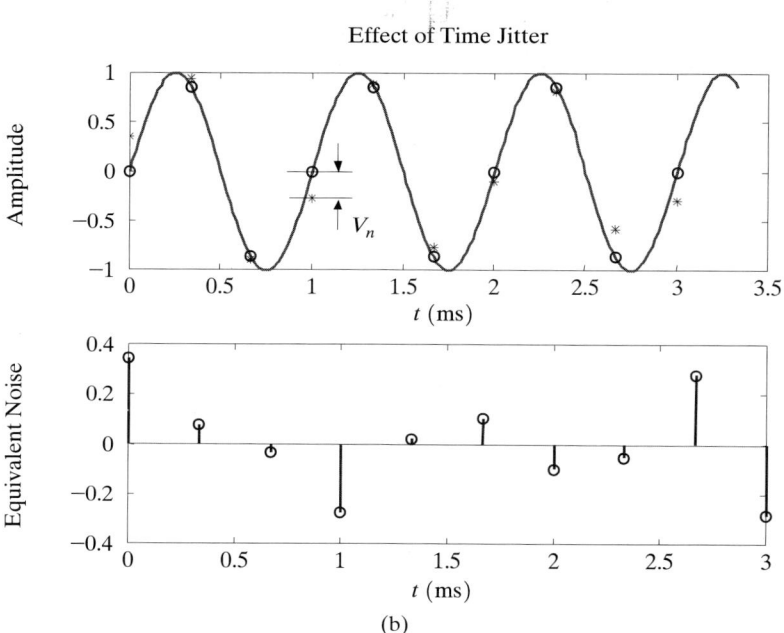

FIGURE 2-19 Effect of sampling time jitter: (a) sampling a sinewave and (b) illustration of equivalent noise of sampling time jitter.

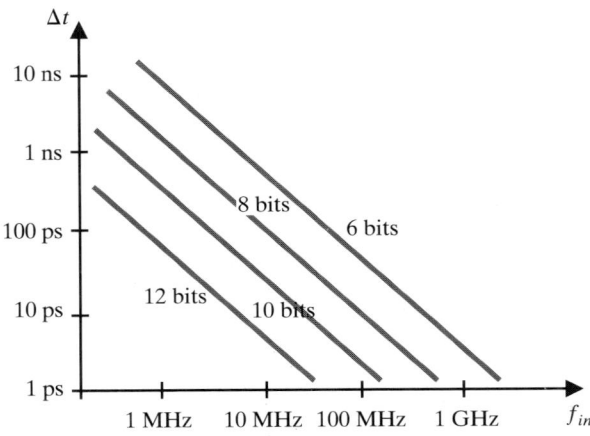

FIGURE 2-20 Maximum sampling jitter for an error of less than $\frac{1}{2}$ LSB.

The Matlab function `jitter.m` provided on the attached CD simulates the effect of jitter. The synopsis of this function is as follows:

```
[Vs,Vn,t] = jitter(fo,fs,N,Tj,phi,shg) ;
%   INPUTS:
%     -fo  : input signal frequency (kHz)
%     -fs  : sampling frequency (kHz)
%     -N   : number of samples
%     -Tj  : normalized (1/Ts) time jitter in RMS
%     -phi : initial phase of sinewave (degrees)
%     -shg : show graphics
%                     0-> No
%                     1-> Yes
%   OUTPUTS:
%     -Vs  : sampled signal
%     -Vn  : equivalent noise corrupting the signal
%     -t   : sampling time
```

For A/D signal conversion, one must compute the amount of time jitter that does not affect the A/D converter outcome. This depends on the resolution of the A/D converter. Figure 2-21 illustrates the power of jitter noise versus the rms value of time jitter for a sinusoid with $f_o = 0.5$ kHz and $f_s = 2$ kHz. The sampled signal is quantized by an 8-bit A/D converter. Figure 2-22 shows the corresponding SNR of the output signal versus the rms value of time jitter. It is seen that a jitter less than -53 dB is tolerable for this case.

The Matlab function `jitnoise.m` can be used to experiment with the tolerable amount of time jitter. The synopsis of this function is as follows:

```
[PndB,SNR,TjdB] = jitnoise(fo,fs,nBit,shg);
%   INPUTS:
%     -fo  : input signal frequency (kHz)
%     -fs  : sampling frequency (kHz)
```

28 Chapter 2 Analog-to-Digital Signal Conversion

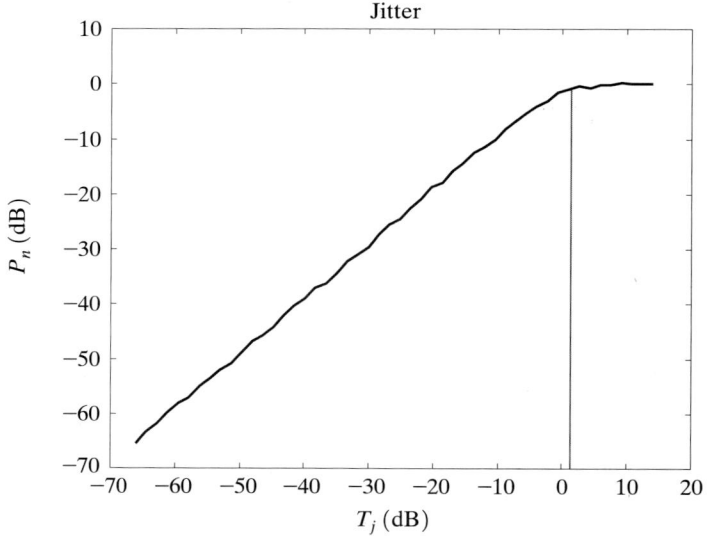

FIGURE 2-21 The equivalent noise caused by time jitter for a sinewave with $A_m = 1, f_o = 500$ Hz, and $f_s = 2$ kHz.

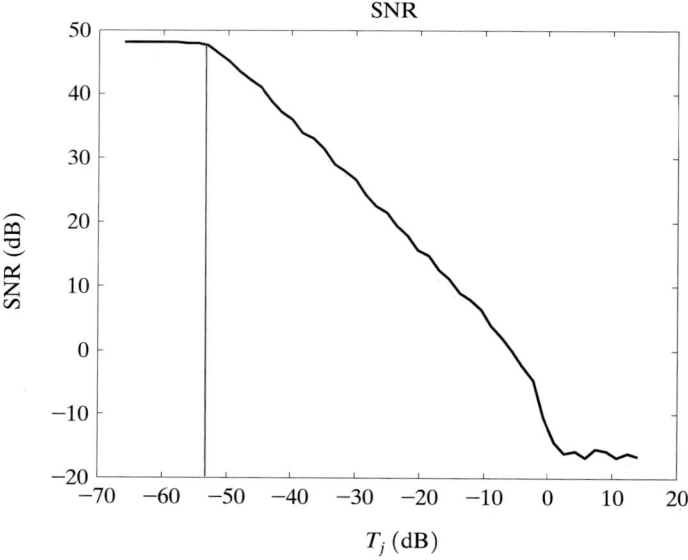

FIGURE 2-22 SNR of an 8-bit A/D converter for an input sinewave with $A_m = 1, f_o = 500$ Hz, and $f_s = 2$ kHz.

```
%       -nBit : number of bits (ideal case nBit = inf )
%       -shg  : show results on the screen;
%               0-> don't show; 1-> show
%
%   OUTPUTS:
%       -PndB : vector of noise power in dB
%       -SNR  : vector of SNR in dB
%       -TjdB : vector of time jitter RMS in dB
```

2.3 SIGNAL RECONSTRUCTION

So far, we have examined the forward process of sampling. At this point, it is also important to understand the inverse process of signal reconstruction from samples. According to the Nyquist theorem, an analog signal v_a can be reconstructed from its samples by using the following formula:

$$v_a(t) = \sum_{k=-\infty}^{\infty} v_a(kT_s) \left[\text{sinc}\left(\frac{t - kT_s}{T_s} \right) \right]. \tag{2.19}$$

One can see that the reconstruction is based on the interpolation of shifted sinc functions. Figure 2-23 illustrates the reconstruction of a sinewave from its samples. To experiment with sinc interpolation, a Matlab function named interpolsinc.m is written and provided on the attached CD. The synopsis of this function is as follows:

```
    [Vout,to] = interpolsinc(Vs,OSR,t,shg) ;
%
%
%   INPUTS:
%       - Vs : samples of signals ( must have an even number of samples)
%       - OSR: oversampling ratio
%       - t  : time vector or [tmin tmax]
%       - shg: show results on the screen;
%               0-> don't show; 1-> show
%
%   OUTPUTS:
%       - Vout: oversampled signal
%       - to  : time vector or [tmin tmax]
%
```

It is very difficult to generate sinc functions by electronic circuitry. That is why, in practice, an approximation of sinc function is used. Figure 2-24 shows an approximation of a sinc function by a pulse, which is easy to realize in electronic circuitry. In fact, the well-known sample-and-hold circuit performs this approximation. The final stage of a D/A converter is the sample-and-hold circuit. The transfer function of a D/A converter is

$$H(j\omega) = \frac{1}{j\omega} - \frac{1}{j\omega} e^{-j\omega T_s} = \frac{\sin(\omega T_s/2)}{\omega T_s/2} e^{-j\omega T_s/2} = \text{sinc}(f/f_s) e^{-j\pi f/f_s}. \tag{2.20}$$

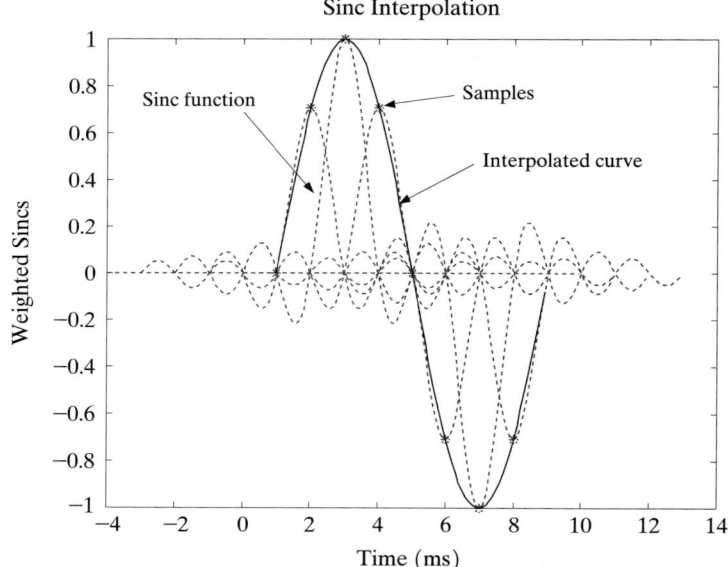

FIGURE 2-23 Reconstruction of an analog sinewave based on its samples $A_m = 1$, $f_s = 2$ Hz, and $f_s = 10$ kHz.

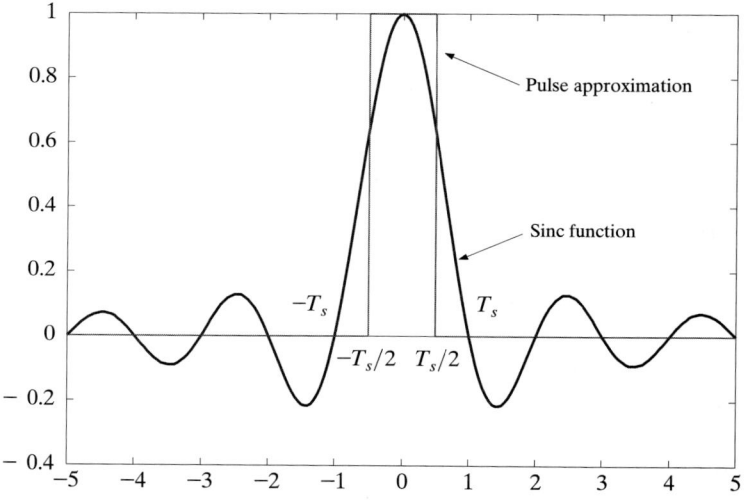

FIGURE 2-24 Approximation of a sinc function by a pulse.

2.3 Signal Reconstruction

Both the time and frequency domain response of a D/A converter are shown in Fig. 2-25.

Sample-and-hold circuits in D/A converters have two inherent nonidealities. First, as illustrated in Figure 2-26, the gain in the desired central band is not constant. It is possible to compensate for this nonideality by using an inverse filter as part of the DSP component. Another solution is to increase sampling frequency, which results in a narrower relative signal bandwidth. The second nonideality is caused by the presence of high-frequency replica of the signal spectrum, which can be removed by using a low-pass filter. These solutions are illustrated in Figure 2-27.

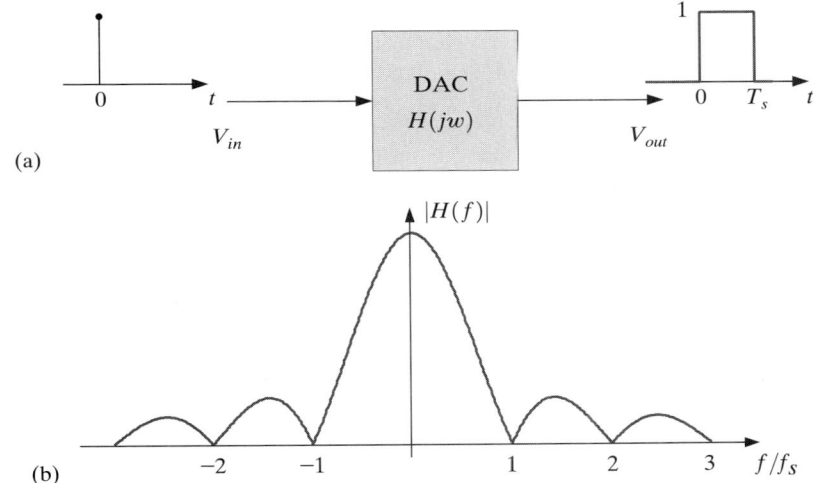

FIGURE 2-25 D/A converter: (a) time domain and (b) frequency response.

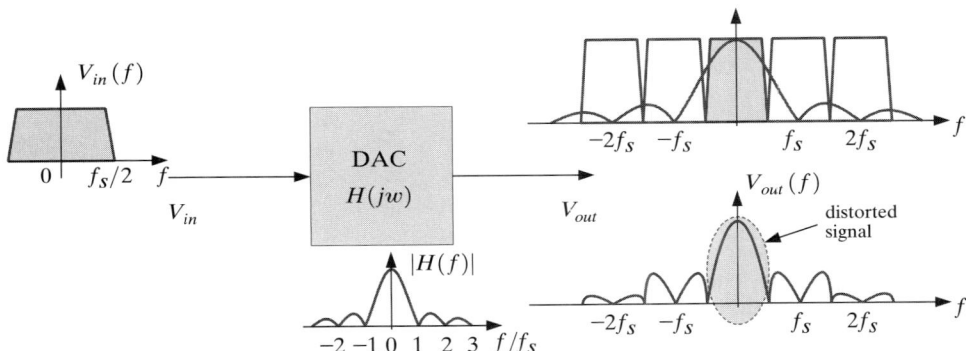

FIGURE 2-26 Nonidealities of a D/A converter.

32 Chapter 2 Analog-to-Digital Signal Conversion

To explore the functionality of D/A converters, a Matlab function named d2a.m is written and provided on the attached CD. The synopsis of this function appears next:

```
  [Vout,q,t] = d2a(Din,Nb,fs,Vref,INL,noise,shg) ;
%
%  INPUTS:
%      -Din : input digital code [0 , 2^Nb-1] ([]-> ramp input)
%       Note: if the input is not integer, the program assumes
%                zero 000..00 and Vref means 111...1
%      -Nb  : number of bits
%      -fs  : sampling frequency of sample and hold (kHz)
%      -Vref : reference voltage (Full Scale = 2^Nb - 1)
%      -INL : Integral Nonlinearity (in +-LSB); vector or a max INL
%             In case of max INL, vector is generated by
%             integrating a uniformly distributed random variable.
%      -noise: electronic noise at different levels (Gaussian noise)
%             noise (in dB) = 10*log( Pn / Pgaus(+-1/2LSB) )
%             (1/2 LSB = 3*sigma )
%       Note: -inf means no noise
%      -shg : show results on the screen;
%             0-> don't show; 1-> show
%
%  OUTPUTS:
%      -Vout : output signal of the A/D converter
%      -q   : quantization error in LSB
%      -t   : vector of time
%
```

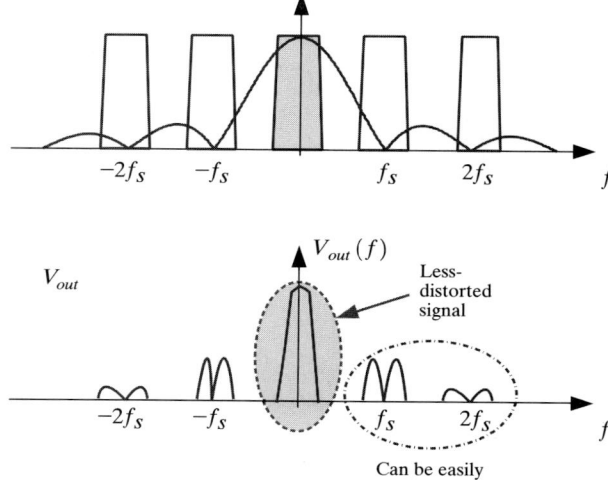

FIGURE 2-27 Reduction of frequency distortion of a D/A converter by increasing sampling frequency.

2.4 MATLAB TOOLBOX FOR DATA CONVERSION

The attached CD includes a Matlab toolbox for data converters. This toolbox is written for the purpose of providing readers with a hands-on experience in sampling and quantization as well as nonidealities of A/D and D/A converters. The programs are user-friendly and self-descriptive. In most cases, the program itself has predefined arguments. Hence, the input arguments are not necessary to be specified by the user. As an example, just by typing a2d on the command line, the help text dialog box and a static transfer characteristic of an A/D converter will appear. The user can also browse the toolbox by just typing helptool on the command line. A graphical user interface, as shown in Figure 2-28, will appear on the screen.

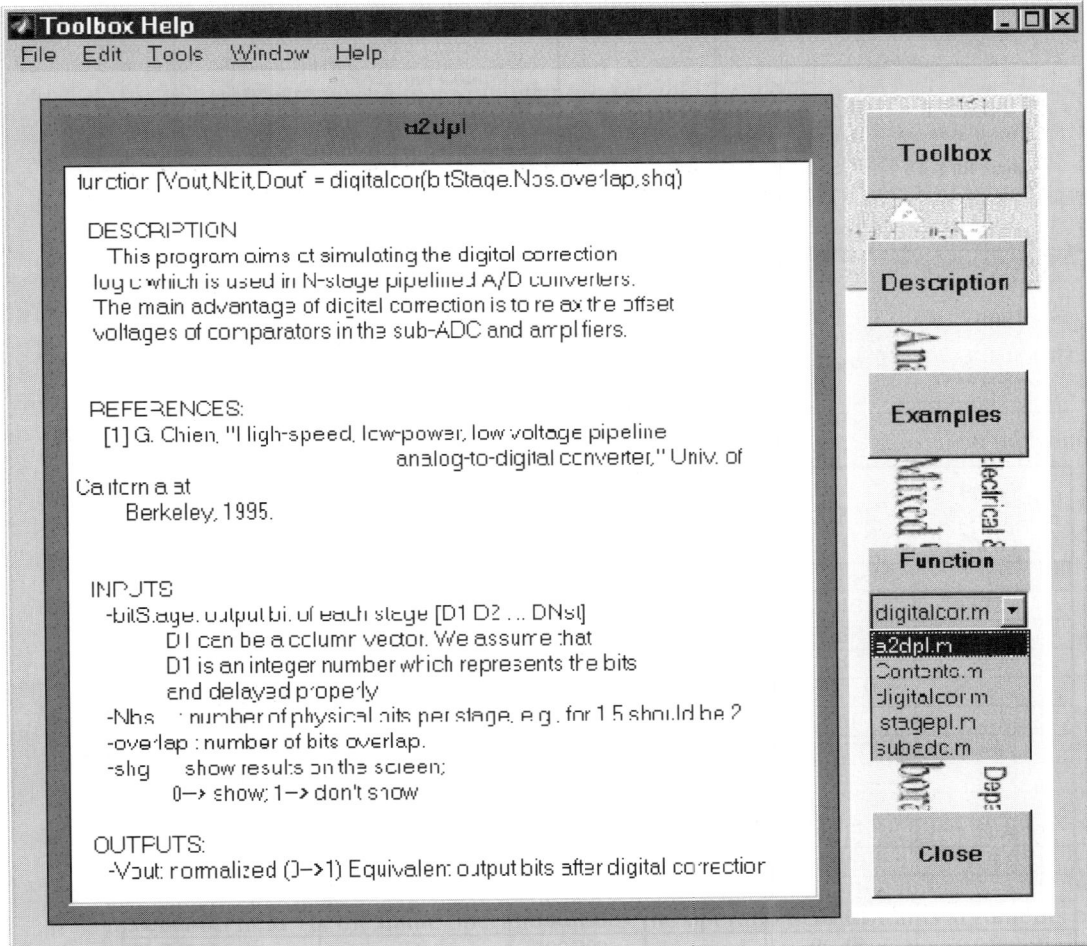

FIGURE 2-28 Matlab A/D toolbox browser.

CHAPTER 3

Data Converter Specifications

Two key specifications of data converters are sampling or conversion rate and bit resolution or precision. When designing DSP systems, a trade-off is often established between these two specifications. Different types of data converters offer DSP system designers a wide choice of sampling-rate and resolution specifications. It is important to realize that the product of each data converter is suitable for certain applications. This chapter is intended to provide DSP system designers with an understanding of data converters' specifications and limitations. The interested reader is referred to [4, 6, 20] for comprehensive descriptions of data converters.

3.1 SIGNAL CONDITIONING

Normally, an analog signal generated by a sensor cannot be connected directly to A/D converters. It must first go through an analog signal processing stage known as *signal conditioning*. Likewise, the digital output of an A/D converter, which may be parallel or serial, must meet certain voltage level and impedance requirements before it can be fed into a DSP. As illustrated in Figure 3-1, signal conditioning consists of three components to meet these objectives: (a) linearization of analog signals generated by a sensor, (b) amplification of analog signals so that they fall within the A/D input voltage range, and (c) limiting bandwidth of analog signals to meet the sampling rate capability of the A/D converter.

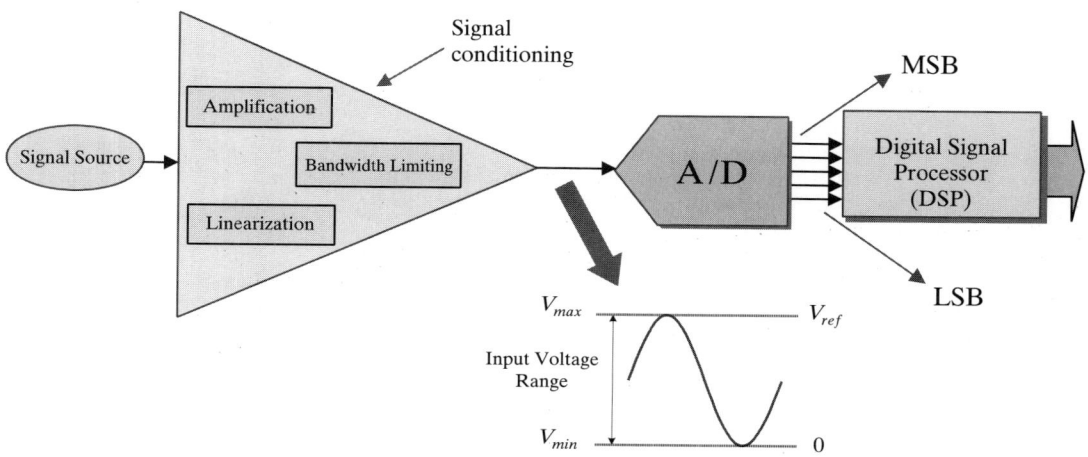

FIGURE 3-1 Signal conditioning in DSP systems.

3.1.1 Linearization and Amplification

In many situations, the analog signal from a sensor is not exactly linear with respect to the quantity being measured. Figure 3-2 shows an example corresponding to the characteristic response of a pressure sensor [7]. Hence, to ensure a linear response, the signal must pass through an amplifier with an inverse transfer characteristic. Alternatively, such a correction can be carried out by an appropriate mapping software residing on the DSP chip.

For situations in which the analog signal of a source is not large enough to fall within the full input signal range of an A/D converter, amplification is needed. As an example, the circuit shown in Figure 3-3 can be used to amplify the output voltage of the above pressure sensor to a level suitable for an A/D converter.

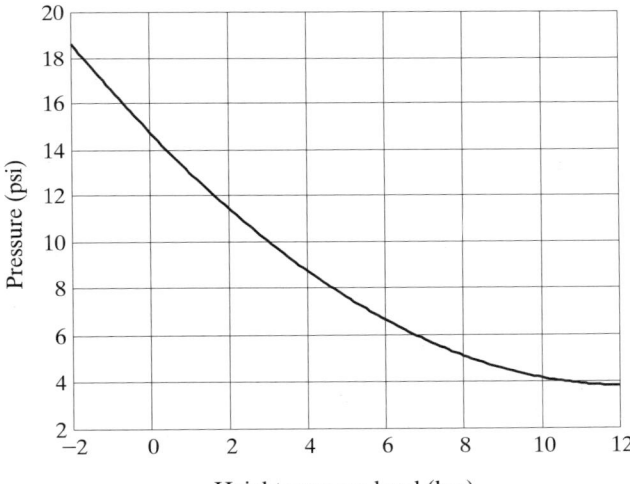

FIGURE 3-2 Nonlinearity in static characteristic of a pressure sensor.

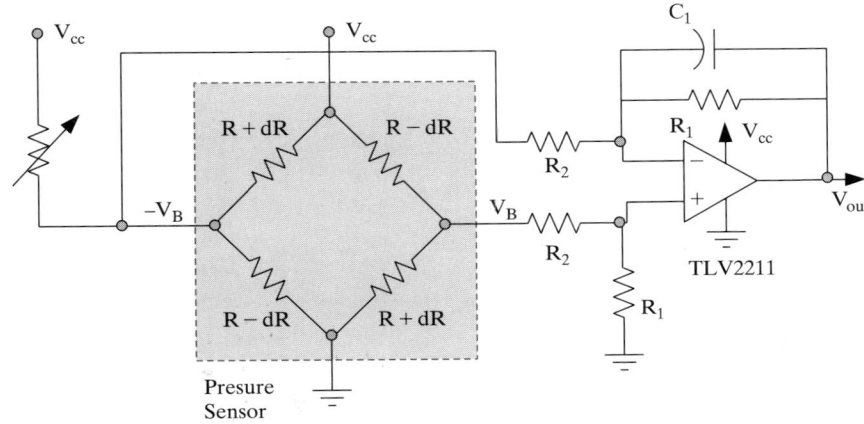

FIGURE 3-3 Signal conditioning circuit for a pressure sensor.[†]

3.1.2 Limiting Bandwidth

According to the Nyquist criterion, the analog signal bandwidth must be less than or equal to one-half the sampling frequency for no aliasing to occur. Therefore, normally in DSP systems, an analog front-end lowpass filter is used to limit the signal bandwidth. Such a filter removes high frequencies (noise and interference) that are present in most analog signals. In Figure 3-3, the lowpass filtering is done as part of the amplification stage.

Care must be taken in designing the lowpass filter. The order of the filter depends on how close the cutoff frequency is to the Nyquist limit, or one-half the sampling frequency $(f_s/2)$. A higher order filter is required as the cutoff frequency gets closer to $f_s/2$, since the filter slope needs to be higher. The maximum variation of the lowpass filter transfer function is related to the resolution of the A/D converter by [20]

$$\Delta \leq \frac{1}{2^N}, \qquad (3.1)$$

where N is the number of resolution bits. For a first-order filter, the transfer function is given by

$$H(f) = \frac{1}{\sqrt{1 + (f/f_o)^2}} \angle \arctan(f/f_o), \qquad (3.2)$$

where f_o denotes -3 dB cutoff frequency. Table 3-1 lists the amplitude drop and excess phase of this filter for several normalized frequencies. For instance, the signal bandwidth must be less than one-tenth of f_o for an 8-bit A/D converter in order to satisfy (3.1). It is also worth mentioning that the architecture of an A/D converter can influence the lowpass filter requirements. For example, for a sigma-delta A/D converter, the requirements are relaxed because of the oversampling involved.

TABLE 3-1 First order filter response.

f/f_o	$A(f)\%$	$A(f)$dB	$\theta(f)$(degrees)
0.05	−0.12	−0.01	2.9
0.1	−0.5	−0.14	5.7
0.2	−1.9	−0.17	11.3
0.3	−4.2	−0.37	16.7
0.4	−7.2	−0.65	21.8
0.5	−10.6	−0.97	26.6
0.6	−14.3	−1.34	30.9
0.7	−18.1	−1.73	35.0
0.8	−21.9	−2.15	38.7
0.9	−25.7	−2.58	42.0
1	−29.3	−3.0	45.0

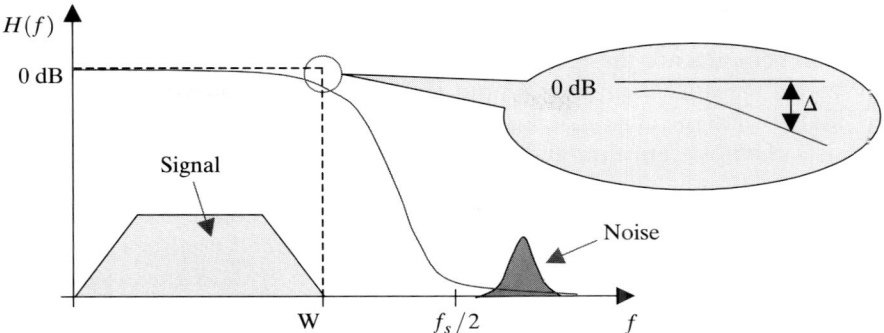

FIGURE 3-4 Analog lowpass filter design.

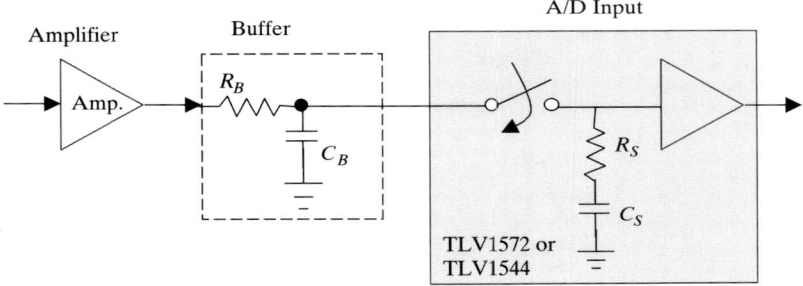

FIGURE 3-5 Output buffer before A/D converter.

3.1.3 Output Buffer

To ensure a high-resolution response, it is recommended that one use an RC buffer, as indicated in Figure 3-5, even though the amplifier output voltage may be close to the supply voltage. Capacitor C_B stores the energy from the amplifier. Current is flowing out of the amplifier during the entire sample-and-hold time. When the sampling switch is closed, the energy will be transferred from the buffer capacitor C_B to the sampling capacitor C_s. For a resolution of N bits, the buffer capacitance must be about 2^N times larger than C_s. Also, the buffer must possess enough bandwidth to ensure no change in the signal frequency components.

3.2 SAMPLE AND HOLD

Signals must be sampled and held before being quantized. Thus, the first stage of an A/D converter comprises a so-called sample-and-hold (S&H) or a track-and-hold (T&H) circuit. In high-speed applications, the term *track-and-hold* is usually used instead of *sample-and-hold*.

3.2.1 Track-and-Hold Signal Distortion

T&H circuits are the most important part of a high-speed, high-resolution A/D converter, because samples of input signals must be taken and held with a resolution higher than the conversion resolution. Figure 3-6 shows a simple T&H circuit. It consists of a hold capacitor and an electronic switch, which is usually an NMOS or CMOS transistor.

In T&H circuits, switches exhibit a number of nonidealities: clock jitter, charge injection, clock feedthrough, nonlinear resistance, and nonlinear stray capacitance [6]. These switch nonidealities result in a distortion of the signal, as shown in Figure 3-7. For instance, a signal-to-noise-and-distortion (SNDR) ratio of 80 dB is required in digital radio, meaning that the amplitudes of distortion harmonics must be 10,000 times less than that of the fundamental. This is very difficult to achieve.

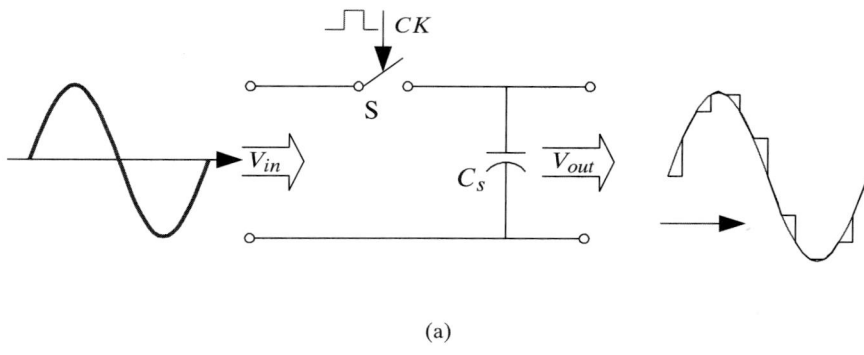

(a)

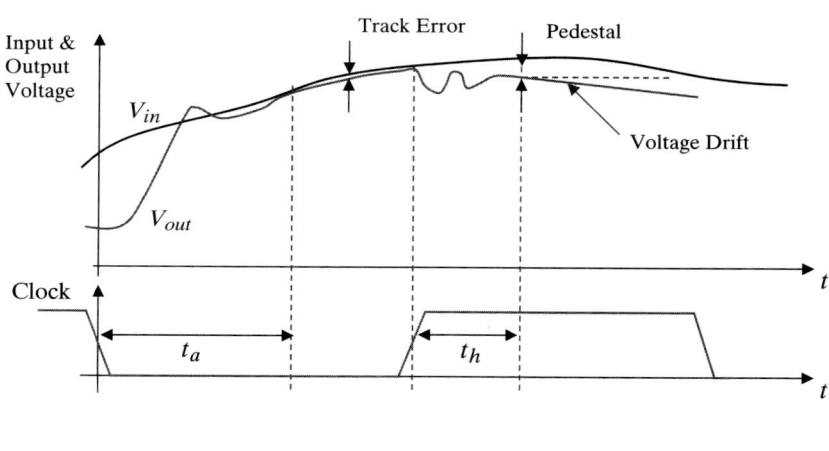

(b)

FIGURE 3-6 (a) Simple Track and Hold circuit, and (b) input/output voltages.

3.2 Sample and Hold

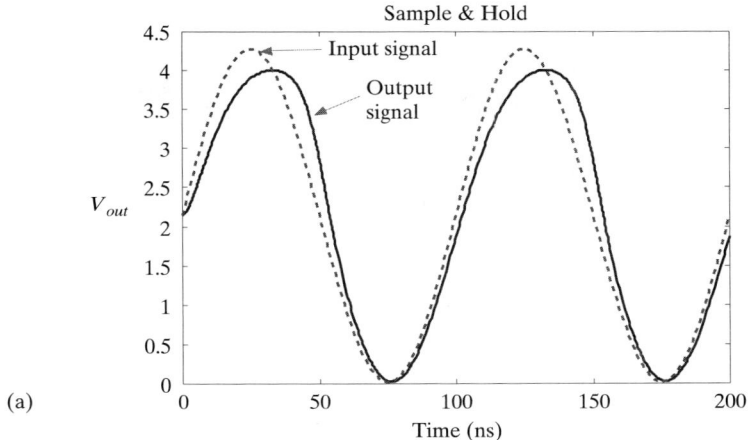

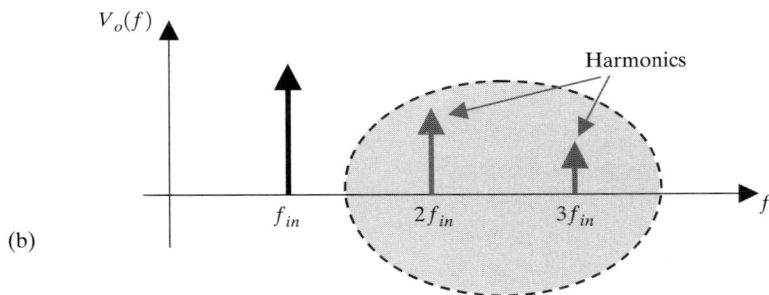

FIGURE 3-7 Distortion in T&H circuit: (a) input signal, and (b) distorted signal before sampling.

A Matlab function shn1.m is written and provided on the attached CD to simulate a simple T&H circuit. The input/output arguments of the function are as follows:

```
[SNDR,SFDR,t,Vo] = shn1(fo,Cs,WoL,AmPrct,shg);
%
%   INPUTS:
%       -fo      : frequency of input sinewave
%       -Cs      : sample & hold capacitor
%       -WoL     : W/L of the switch
%       -AmPrct  : Signal amplitude in percentage of full-scale (Vref/2)
%       -shg     : show results on the screen;
%                  0-> don't show; 1-> show
%
%   OUTPUTS:
%       -SNDR: signal-to-noise ratio
%       -SFDR: spurious-free-dynamic-range
%       -t   : vector of time
%       -Vo  : output voltage
%
```

One way to reduce the T&H nonidealities is by using differential signals, as illustrated in Figure 3-8. That is why in some high-performance A/D and D/A converters, the input and output signals are differential. Figure 3-9 shows the SNDR of a single-end output and a differential-output A/D converter as a function of the input signal amplitude. As can be seen from this figure, the differential output generates a higher SNDR as compared to the single-end output.

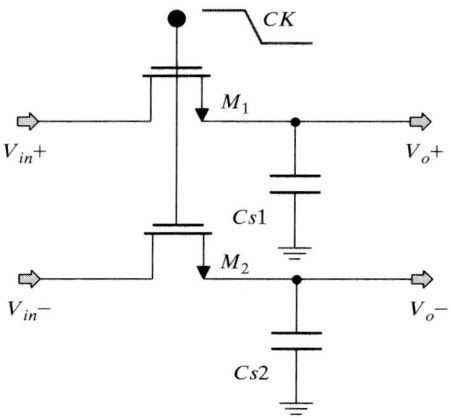

FIGURE 3-8 Differential T&H circuit to reduce nonidealities.

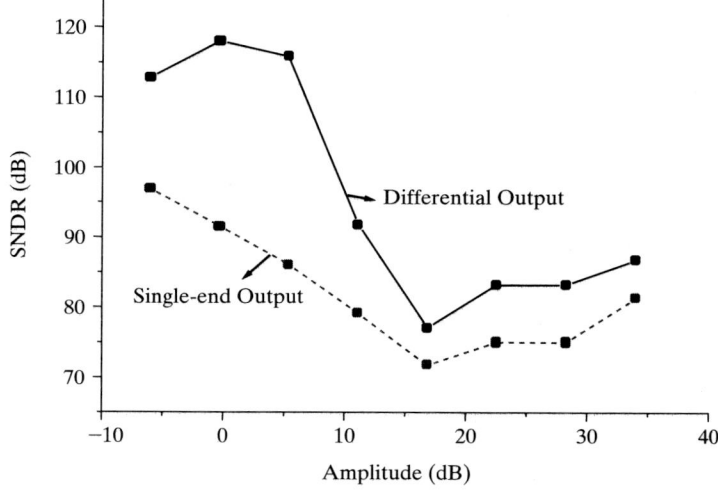

FIGURE 3-9 SNDR of differential and single-end T&H circuits; $f_s = 32$ MHz, $f_o = 3/512*f_s$, 0 dB corresponds to amplitude of 20 mV.

3.2.2 Input Impedance of A/D Converters

In most cases, the input circuit of A/D converters can be approximated by a first-order RC circuit, as shown in Figure 3-10. This circuit, together with the output impedance of the signal source (or amplification stage), acts as a lowpass filter for the signal. As an example, the frequency response of the input circuit of a 10-bit, 20-MSPS A/D converter (TI TLC876) is shown in Figure 3-11.

It should be noted that the output impedance of the source must be designed in such a way that the total distortion is less than 0.5 LSB of the A/D converter. (See Table 3-1.) The minimum tracking time (switch on) for the T&H circuit to achieve this required accuracy is calculated to be [6]

$$t_{min} = (R_S + R_{in})C_{in}(N + 1)\ln(2). \quad (3.3)$$

In high-frequency applications, the stray parasitic inductance of connection wires cannot be neglected. As a result, the first-order approximation of the input equivalent circuit is not sufficient, and the second-order circuit shown in Figure 3-12 is considered.

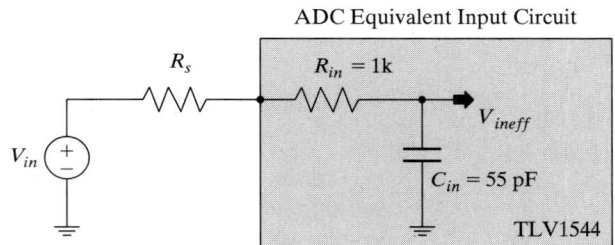

FIGURE 3-10 Input impedance of A/D converters.

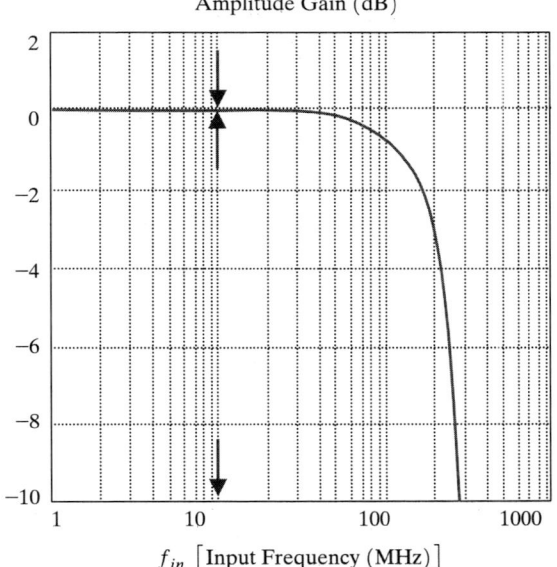

FIGURE 3-11 Frequency response of input circuit of a 10-bit, 20 MSPS A/D converter (TI TLC876).[†]

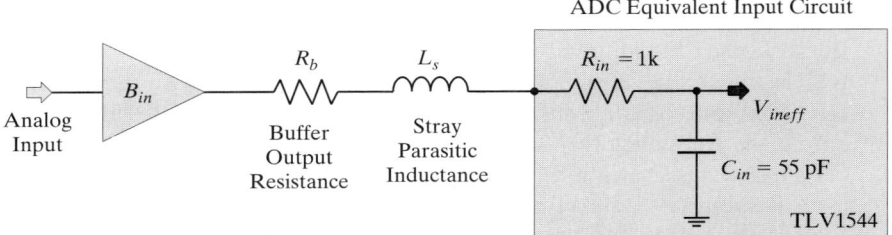

FIGURE 3-12 Second-order equivalent input circuit of an A/D converter.

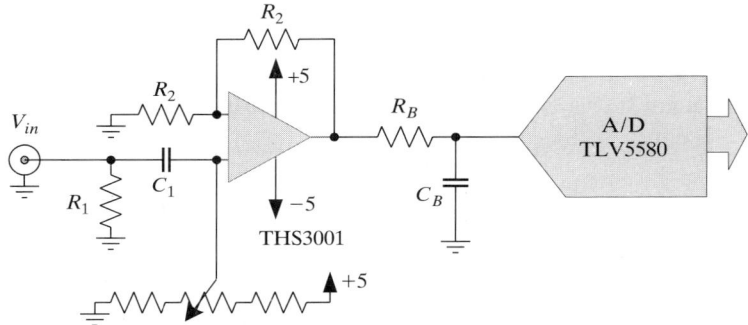

FIGURE 3-13 Driving a high-speed A/D converter (TI TLV5580).[†]

The response of this second-order filter is optimized (no underdamped or overdamped response) when we have [6]

$$R_{in} + R_B = 2\sqrt{\frac{L_S}{C_{in}}}. \tag{3.4}$$

In order to drive the input capacitance of an A/D converter, one may use an input buffer. Figure 3-13 shows an example circuit for driving the single-pipeline, switched-capacitor A/D converter (TI TLV5580, 8-bit, 80 MSPS). This circuit takes care of the DC biasing of the input signal and provides a low source impedance. A high-speed opamp (e.g., THS3001) is used to level-shift the input signal. The noninverting amplifier provides a gain of 2 for both DC and AC. A simple lowpass filter is placed on the output of the amplifier. The input bandwidth of the A/D converter is extremely high (750 MHz), suitable for use in the IF section of communication receivers.

3.3 PERFORMANCE METRICS OF A/D CONVERTERS

There exists no universal A/D converter that can be used for all applications. However, it is possible to select an optimum A/D converter for a specific application based on the required specifications. Therefore, it is important for DSP system designers to be familiar

3.3 Performance Metrics of A/D Converters

with the performance metrics of A/D converters. In the subsections that follow, we focus on the definitions and physical meanings of various performance metrics of A/D converters, which are usually specified in manufacturers' data sheets.

3.3.1 A/D Static Metrics

Figure 3-14 shows an ideal input/output static characteristic of a 3-bit A/D converter. The input analog voltage must be between V_{ref-} and V_{ref+} in order to ensure no saturation of the output digital words. In practice, the input/output characteristic deviates from the ideal staircase characteristic. The deviation also depends on the input frequency (dynamic behavior). The static input/output characteristic is measured by setting a DC level at the input and by obtaining the output digital word. Basically, this characteristic is represented by a set of threshold voltages corresponding to different output digital words. The transition levels are random because of electronic noise in the circuitry. Figure 3-15 shows a realization of the actual static input/output characteristic of a 3-bit A/D converter. It is seen that the output digital words oscillate around transition levels.

The interested reader can use the Matlab function a2d.m on the attached CD to visualize the effects of electronic noise on the input/output characteristic. For example, one can use the following command to generate Figure 3-15:

```
[Vout,q,t,Dout] = a2d([],3,[0 7],7,-13,0,1);
```

Figure 3-16 illustrates the indeterministic input/output characteristic of an A/D converter along with the cumulative (probability) density function (CDF) of transition levels. In practice, transition levels are defined as the levels for which the probability of obtaining consecutive words is 50%. Most manufacturers' data sheets use this definition of transition levels.

According to transition voltage levels, the following performance metrics are defined: differential nonlinearity (DNL), integral nonlinearity (INL), and monotonicity. Figure 3-17 illustrates a realistic static input/output characteristic of an A/D converter.

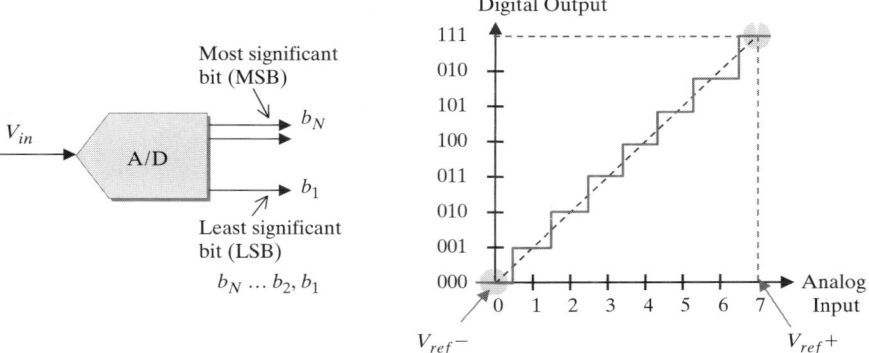

FIGURE 3-14 An ideal A/D converter: (a) symbolic representation, and (b) input/output static transfer function.

44 Chapter 3 Data Converter Specifications

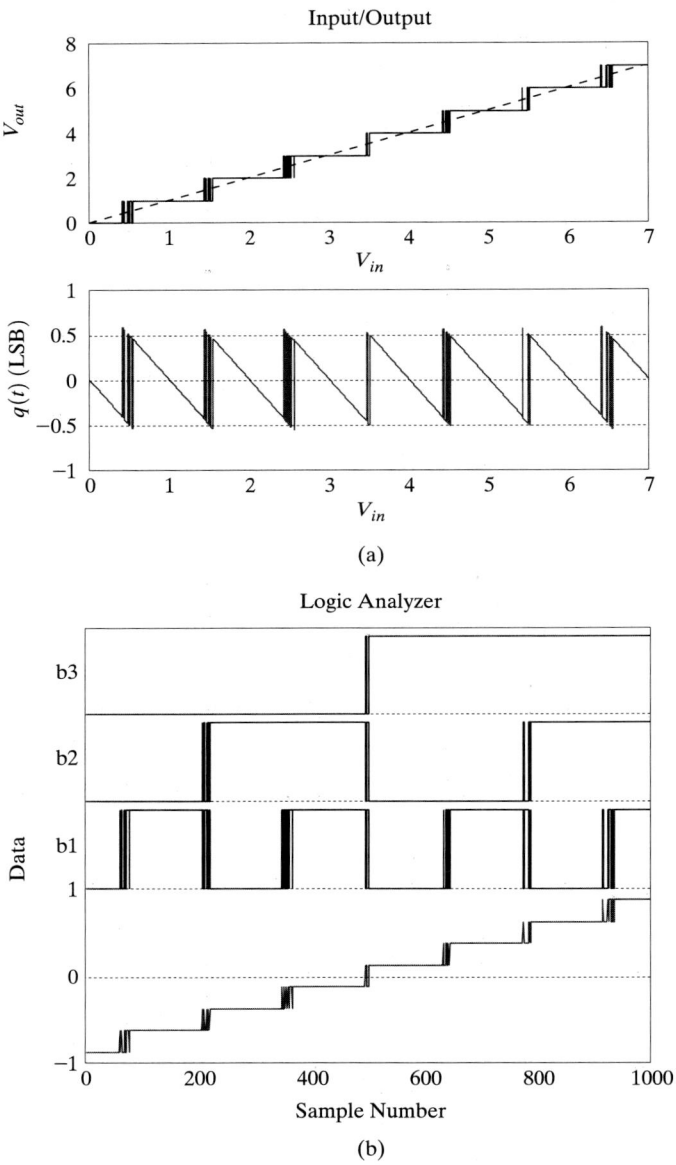

FIGURE 3-15 Characteristic of an A/D converter: (a) realizations of indeterministic input/output characteristic and quantization error, and (b) output bits.

3.3 Performance Metrics of A/D Converters 45

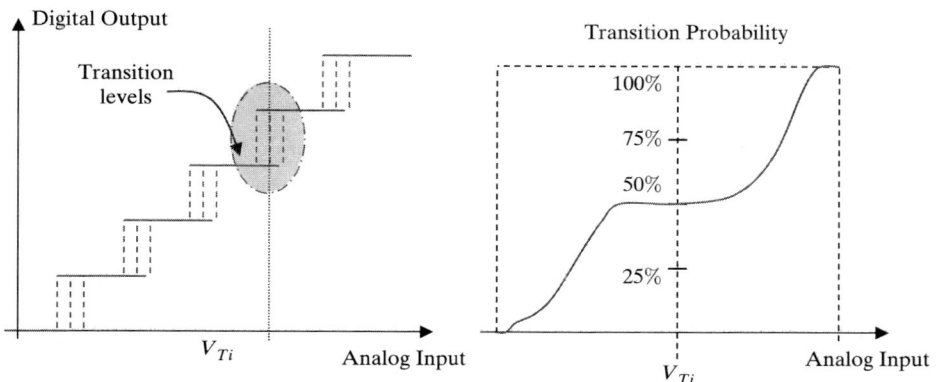

FIGURE 3-16 Realistic A/D converter: (a) indeterministic input/output characteristic, and (b) probability of transition levels.

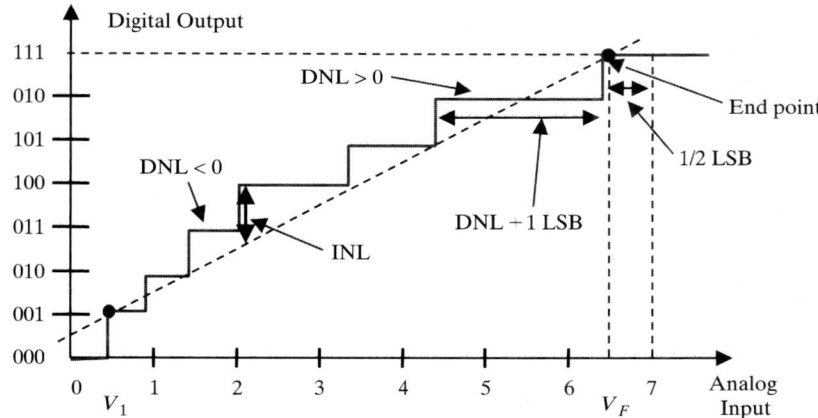

FIGURE 3-17 Realistic static input/output characteristic of an A/D converter.

Differential Nonlinearity: Ideally, the input range of an N-bit A/D converter is equally divided into 2^N small steps, and an analog input change of $V_{ref}/2^N$ causes a 1-LSB change in the output digital word. Differential nonlinearity (DNL) is a measure of the deviation of the actual input voltage step from the ideal voltage step for 1 LSB (see Figure 3-17), and is defined as

$$\text{DNL}_i = \frac{V_{i+1} - V_i}{\text{LSB}_{nom}} - 1 \, (\text{LSB}), i = 1, \ldots, 2^N - 2, \quad (3.5)$$

where LSB_{nom} is

$$\text{LSB}_{nom} = \frac{V_F - V_1}{2^N - 2}. \quad (3.6)$$

The largest positive and negative numbers are usually used to reflect the static performance of an A/D converter.

Integral Nonlinearity: The overall linearity of an A/D converter can be specified in terms of integral nonlinearity (INL), which is a measure of the deviation of the actual output digital word from the straight line which passes through the end points (see Figure 3-17), and is defined as

$$\text{INL}_i = \frac{V_i - [(i-1)\text{LSB}_{nom} + V_1]}{\text{LSB}_{nom}} \text{ (LSB)}, i = 1, \ldots, 2^N - 1;$$

$$\text{INL}_1 = \text{INL}_{2^N-1} = 0. \tag{3.7}$$

The largest positive and negative numbers are usually used to reflect the static performance of an A/D converter. INL can also be considered to be the integral of the DNL curve. Most data-converter manufacturing companies employ the best-fit-line method to determine INL. The best-fit line is obtained by least-square curve fitting. This method, illustrated in Figure 3-18, results in less INL than the standard method.

The Matlab function inlgen.m on the attached CD can be used to generate and visualize the effect of INL on the output of an A/D converter. Figure 3-19 shows a typical INL and DNL outcome of this Matlab function for an 8-bit A/D converter with a maximum INL of 1 LSB. The synopsis of this function is as follows:

```
        [INL] = inlgen (Nb,INLmax,type,shg)
%
%       INPUTS:
%          - Nb       : number of bits
%          - INLmax   : maximum of abs ( INL ) (LSB)
%          - type     : type of INL
%                       0-> uncorrelated
%                       1-> correlated
%                       2-> correlated & best-fit
%                       3-> parabolic & best-fit
%          - shg      : show results on the screen;
%                       0-> don't show; 1-> show
%
%       OUTPUTS:
%          - INL: integral nonlinearity (+-1 LSB)
%
```

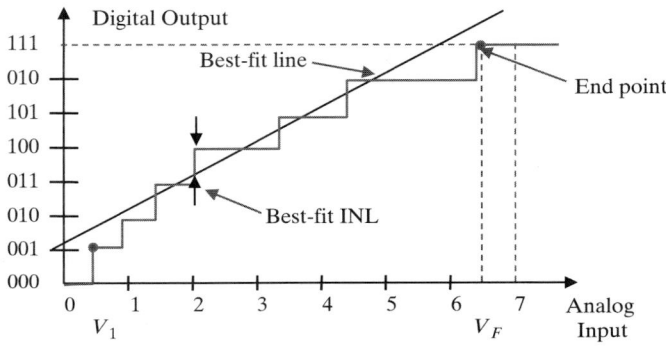

FIGURE 3-18 INL definition based on best-fit method.

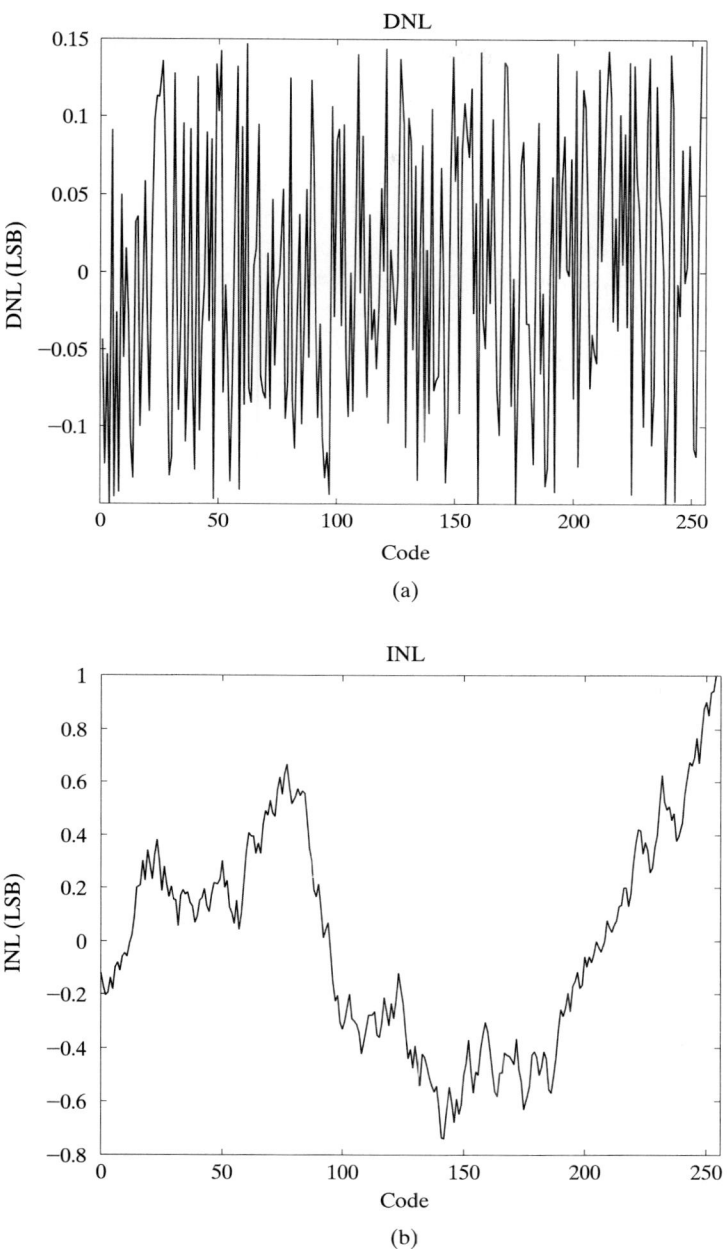

FIGURE 3-19 Static nonlinearity of an 8-bit A/D converter with $INL_{max} = 1$ LSB: (a) DNL, and (b) INL.

48 Chapter 3 Data Converter Specifications

Monotonicity: The output digital word of an A/D converter should increase over its full range as the input analog signal increases. This is referred to as *monotonicity*. Monotonicity is critical in many applications, in particular in digital control, because a decrease in the A/D transfer characteristic could create oscillations (limit cycles) in the system. A DNL less than −1 LSB results in a nonmonotonic A/D converter. Figure 3-20 illustrates a nonmonotonic static input/output characteristic of an A/D converter.

Offset: The input analog signal should be 0 V for the output $D = 0$. However, there exists an offset, even though the output is zero. The offset is defined in LSB as the difference between the nominal and actual offset points, as shown in Figure 3-21(a). The offset voltage of an A/D converter does not pose a major problem, because it can easily be compensated for, in either the analog or the digital domain.

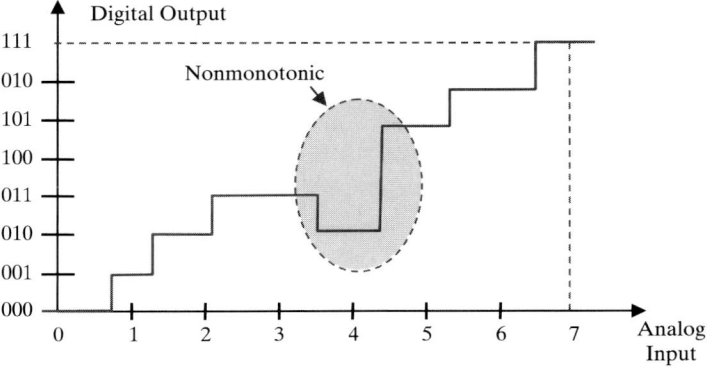

FIGURE 3-20 Nonmonotonic static transfer function of an A/D converter.

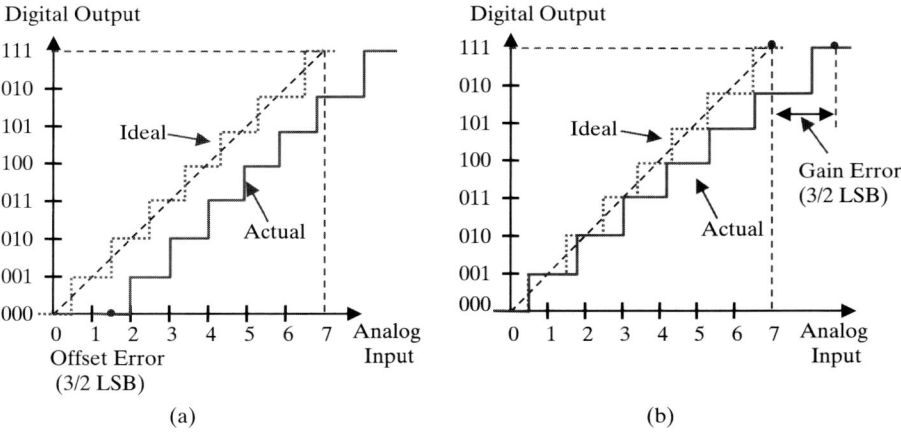

FIGURE 3-21 Static performance metrics of an A/D converter: (a) offset and (b) gain error.

Gain Error: A gain error exists if the slope of the best-fit line through the transfer characteristic curve is different from the slope of the best-fit line for the ideal case, as shown in Figure 3-21(b). Gain error is defined as the difference between the ideal and actual slopes. Gain error acts as a scaling factor which can be corrected in digital domain.

Latency: This specification indicates the total time from the moment the input analog signal is sampled to the time the digital output is valid. Latency should be taken into account when designing DSP systems.

3.3.2 A/D Dynamic Metrics

The resolution of A/D converters is a function of the frequency of the input signal. At high frequencies, the performance starts degrading because of a number of frequency dependent imperfections. A number of performance metrics are defined to describe the dynamic behavior of A/D converters. The importance of these performance metrics depends on the application of interest.

Signal-to-Noise Ratio (SNR): The ratio of signal power to noise power at the output of an A/D converter is defined as SNR. SNR is usually measured for a sinusoidal input signal and is a function of its amplitude and frequency. The theoretical maximum of SNR is only a function of the number of bits, N. That is,

$$\text{SNR}_{max} = 6.02N + 1.76 \text{ dB}. \quad (3.8)$$

Signal-to-Noise + Distortion Ratio (SNDR): The ratio of signal power to the sum of noise and harmonics power at the output of an A/D converter is defined as SNDR. Again, SNDR is measured for a sinusoidal input signal. Since it incorporates the nonlinearity information of an A/D converter, in practice, SNDR is more significant than SNR. SNDR is a particular nuisance in communication systems due to the undesirable effects of harmonics distortion.

In order to see the effect of INL on SNDR, one can use the following Matlab functions to generate a sinewave and to obtain the spectrum of the corresponding digital output:

```
[Vs , t] = sinesample( 31/512*4 , 4 , 512 , 0 , 0 , 1 );
[Vout, q, t, Dout] = a2d( Vs , 8 , t , 2 , -inf, 1, 1 );
```

Figure 3-22 shows the spectrum of the digital output of an 8-bit A/D converter for a unity amplitude sinusoidal input, and the maximum INL of 1 LSB.

Dynamic Range (DR): Dynamic range is another useful performance metric. Dynamic range is the ratio of the power of the full-scale input to the power of a sinusoidal input for which SNR is equal to 0 dB.

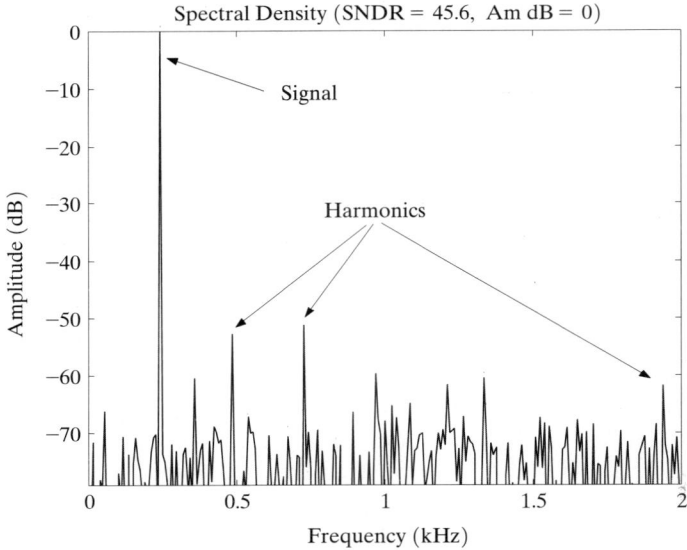

FIGURE 3-22 Spectrum of the output code for a sinewave input and an 8-bit A/D converter with INLmax = 1 LSB.

Effective Number of Bits (ENOB): This performance metric is defined based on the maximum SNDR and the theoretical value of SNR,

$$\text{ENOB} = \frac{\text{SNDR}_{max} - 1.76 + 20 \log\left(\dfrac{\text{Full-scale input amplitude}}{\text{Actual input amplitude}}\right)}{6.02}. \qquad (3.9)$$

For the full-scale input signal situation, (3.9) can be rewritten as

$$\text{ENOB} = \frac{\text{SNDR}_{max} - 1.76}{6.02}. \qquad (3.10)$$

Spurious-Free-Dynamic-Range (SFDR): SFDR is defined as the difference between the maximum signal component and the largest distortion component in dB. SFDR is an important dynamic specification for high-frequency applications, such as wireless communications. Figure 3-23 illustrates the definition of SFDR based on the spectrum of the output for a full-scale sinusoidal input. The spurious-free-dynamic range can also be used as a useful dynamic range, because it is related to the minimum detectable amplitude of the input signal resulting in no distortion.

It is well known that the maximum SNDR for a full-scale sinewave quantized by an ideal A/D converter is equal to $6N + 1.76$ dB. In contrast, the SFDR expression is rather involved. In [6], it is derived to be

$$\text{SFDR}_{max} = 9N - 6 \text{ dBc}, \qquad (3.11)$$

3.3 Performance Metrics of A/D Converters

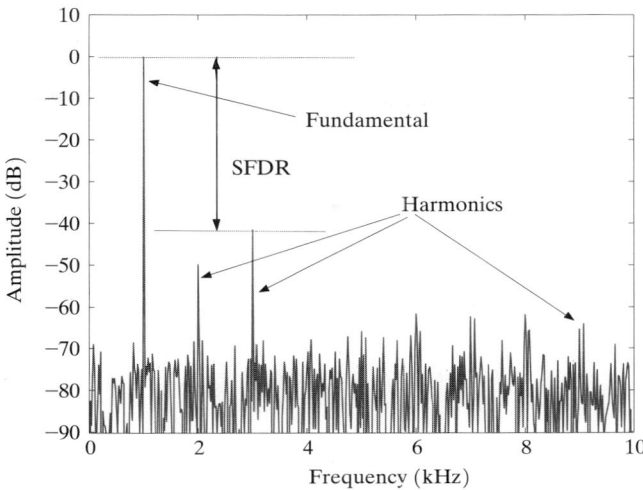

FIGURE 3-23 Definition of SFDR, based on the spectrum of the output code.

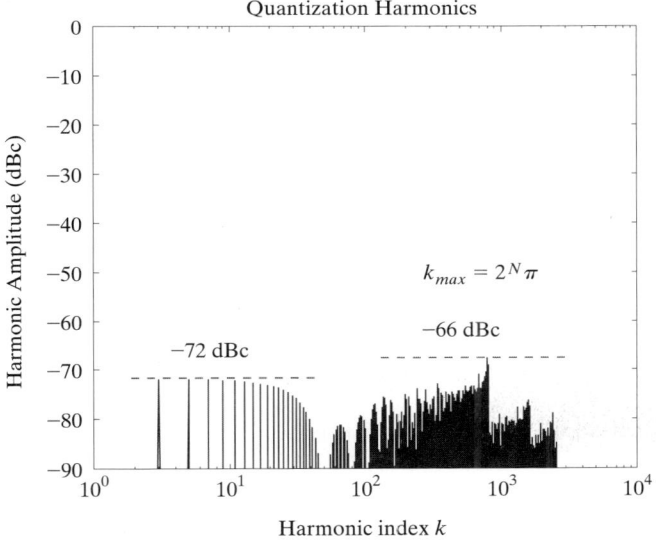

FIGURE 3-24 Output harmonics of an ideal 8-bit A/D converter.

where dBc stands for dB with respect to either the carrier or the fundamental component. Figure 3-24 shows the harmonic components of an ideal 8-bit A/D converter for a full-scale sinewave. In general, SFDR is a function of both the input frequency and the trend of the INL curve. Figure 3-25 shows the spectral density of an ideal 8-bit A/D-converter output. It is seen that the actual SFDR is 13 dB less than the predicted value, which is not negligible in wireless communications applications.

52 Chapter 3 Data Converter Specifications

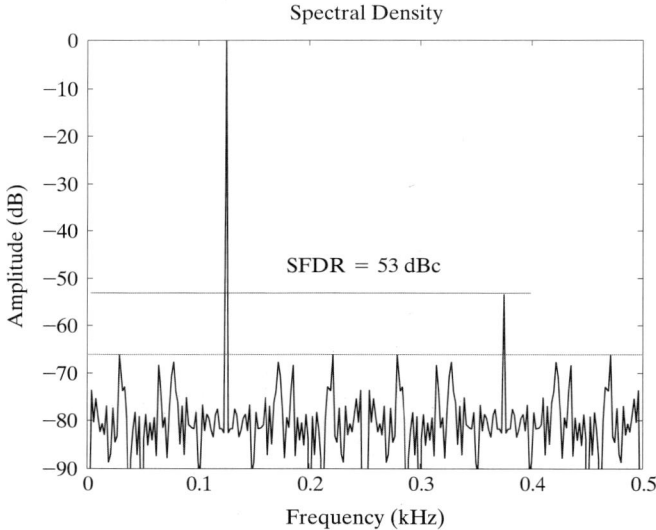

FIGURE 3-25 An example of SFDR which is less than the theoretical prediction, $f_o = 125$ Hz, $f_s = 1$ kHz, $N_s = 512$, $N = 8$ bits.

A Matlab function, named sfdr2f.m, is provided on the attached CD to plot SFDR versus frequency. The effects of INL can easily be understood from such a plot. The synopsis of this function is as follows:

```
        [SFDR,SNR,fo] = sfdr2f(fs,Nb,noise,INL,Ns,phi,shg)
%
%       INPUTS:
%           -fs   : sampling frequency (kHz)
%           -Nb   : number of bits
%           -noise: electronic noise at different levels (Gaussian noise)
%                   noise (in dB) = 10*log( Pn / Pgaus(+-1/2LSB) )
%                   (1/2 LSB = 3*sigma )
%                   Note: -inf means no noise
%           -INL  : Integral Nonlinearity (in +-LSB); vector or a max INL.
%                   In case of max INL, vector is generated by using
%                   uniformly distributed random variable.
%           -Ns   : number of samples (must be 2^M)
%           -phi  : phase of sinewave (degree) ( [] means random phase )
%           -shg  : show results on the screen;
%                   0-> don't show; 1-> show
%
%       OUTPUTS:
%           - SFDR: vector of SFDR
%           - SNR : SNR
%           - fo  : vector of input frequency
```

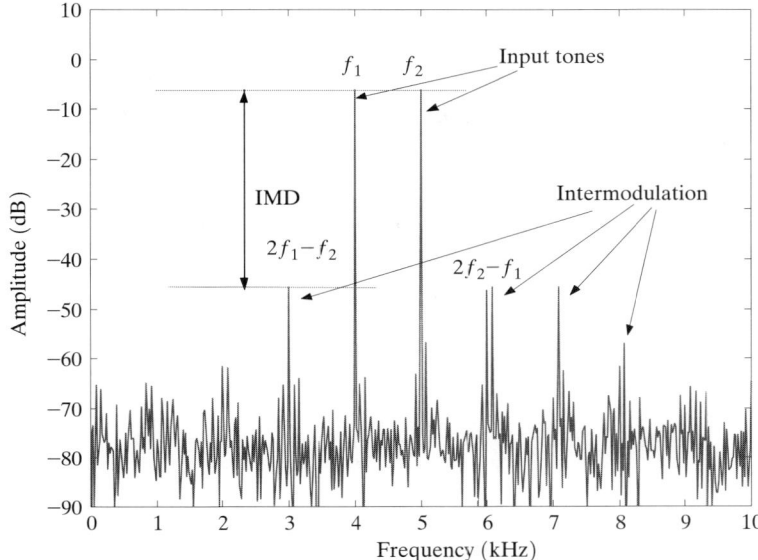

FIGURE 3-26 Definition of IMD, based on the spectrum of the output code.

Intermodulation Distortion (IMD): This is a measure of how much one frequency modulates other frequencies. IMD can be calculated as indicated in Figure 3-26. The more linear an A/D converter is, the lower the intermodulation products become. IMD quantifies how strongly close band interferers can appear in the signal bandwidth.

3.4 PERFORMANCE METRICS OF D/A CONVERTERS

D/A converters are the gates for connecting the digital domain (DSP) to the analog world. Figure 3-27 depicts the general usage of a D/A converter. The output of a D/A converter can be current or voltage. In high-speed applications, the analog output is usually current. In either case, we need both an output buffer and an amplifier, since the output impedance of D/A converters is not suitable for direct connection to a transducer or an instrumentation device. Figure 3-28 shows two different output buffers for the output voltage and current modes of D/A converters.

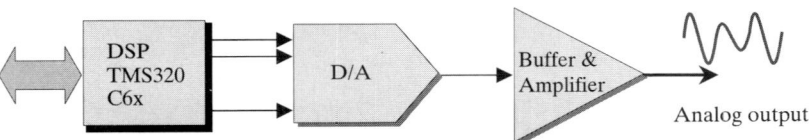

FIGURE 3-27 General usage of a D/A converter.

54 Chapter 3 Data Converter Specifications

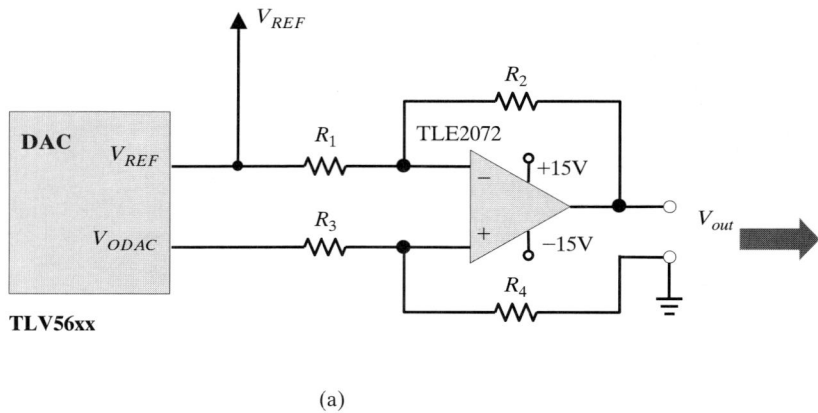

(a)

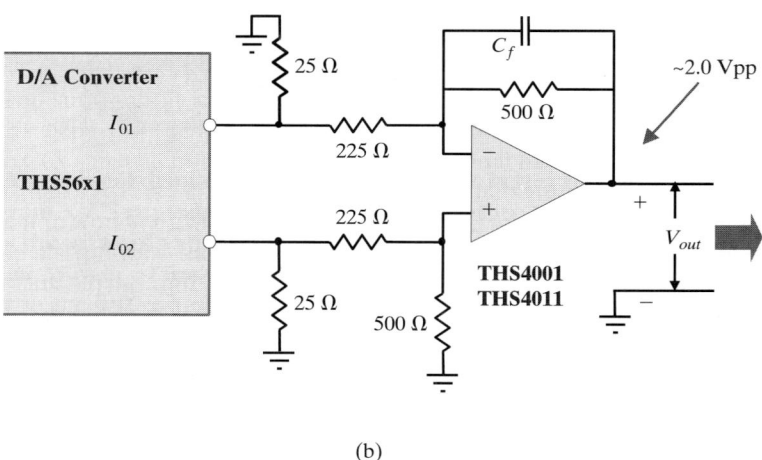

(b)

FIGURE 3-28 Output buffer of D/A converter: (a) voltage output, and (b) current output.[†]

3.4.1 D/A Static Metrics

Input/Output Characteristic: The ideal analog output of a D/A converter is expressed by

$$V_{out}(D_i) = V_{ref}\left(\frac{b_N}{2} + \frac{b_{N-1}}{2^2} + \ldots + \frac{b_1}{2^N}\right) \quad (3.12)$$

where $\{D_i, i = 0, 1, \ldots, 2^N - 1\}$ denotes a digital word, as shown in Figure 3-29. The reference point is usually considered to be the midpoint of the full-scale range $-V_{ref}/2$ to $V_{ref}/2$. As an example, the static input/output characteristic of an ideal 4-bit D/A converter is shown in Figure 3-30.

3.4 Performance Metrics of D/A Converters 55

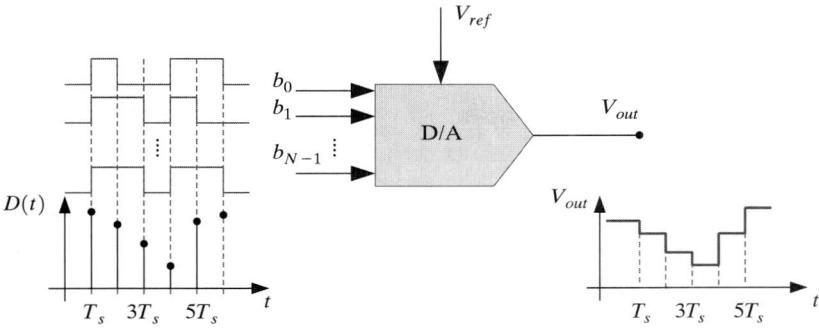

FIGURE 3-29 Input and output signals of a D/A converter.

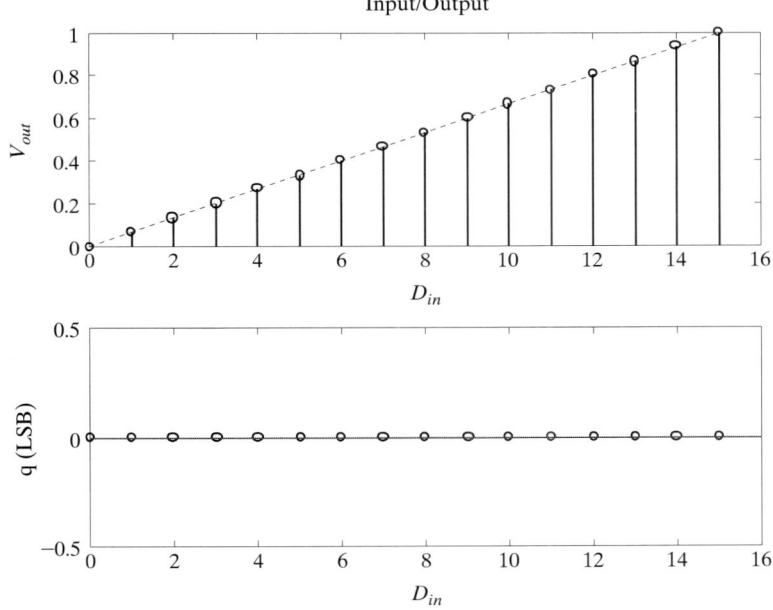

FIGURE 3-30 Static input/output characteristic of an ideal 4-bit D/A converter.

Differential Nonlinearity (DNL): The output range of an N-bit D/A converter is equally divided into 2^N small steps, and a 1-LSB change in the input digital word produces a change of $V_{ref}/2^N$ in the analog output. Differential nonlinearity (DNL) is a measure of the deviation of the actual D/A converter output voltage step from the ideal voltage step for 1 LSB (refer to Figure 3-31) and is defined as

$$\text{DNL}_i = \frac{V_{out}(D_{i+1}) - V_{out}(D_i) - V_{ref}/2^N}{V_{ref}/2^N}, \text{ for } i = 0, 1, \ldots, 2^N - 2. \quad (3.13)$$

The largest positive and negative numbers are usually used to reflect the static performance of a D/A converter.

Integral Nonlinearity (INL): The overall linearity of a D/A converter can be specified in terms of integral nonlinearity (INL). INL is a measure of the deviation of the actual D/A converter output voltage from the ideal line. It is defined as

$$\text{INL}_i = \frac{V_{out}(D_i) - i \times V_{ref}/2^N}{V_{ref}/2^N}, \text{ for } i = 0, 1, \ldots, 2^N - 1. \quad (3.14)$$

The largest positive and negative numbers are usually used to reflect the static performance of a D/A converter. The Matlab function d2a.m can be used to better understand the effects of INL on the output analog signal.

Offset: The analog output should be 0 V for $D = 0$. However, as indicated in Figure 3-32, there exists an offset voltage similar to the offset voltage of an operational amplifier.

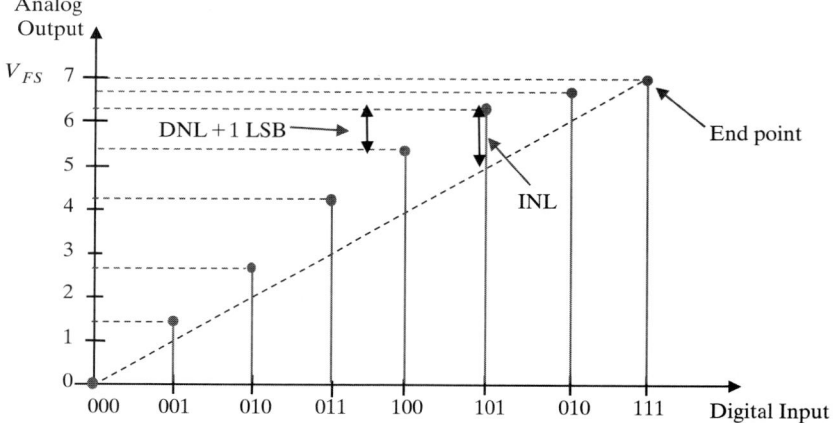

FIGURE 3-31 Realistic input/output characteristic of a D/A converter.

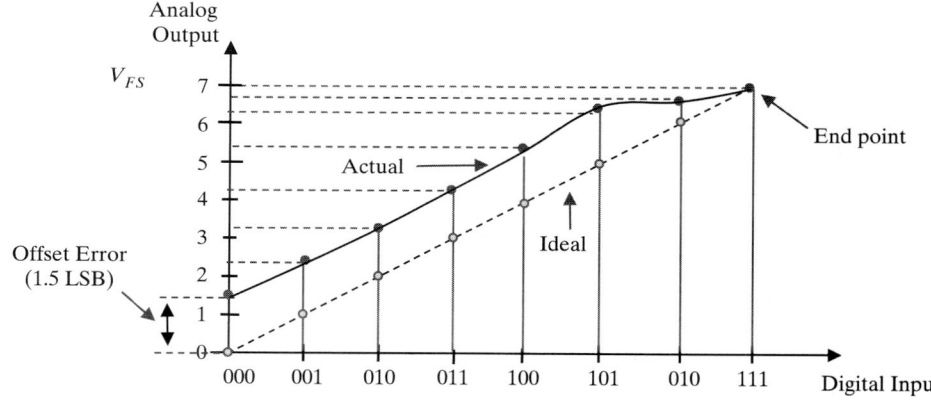

FIGURE 3-32 Offset of static input/output characteristic of a D/A converter.

3.4 Performance Metrics of D/A Converters

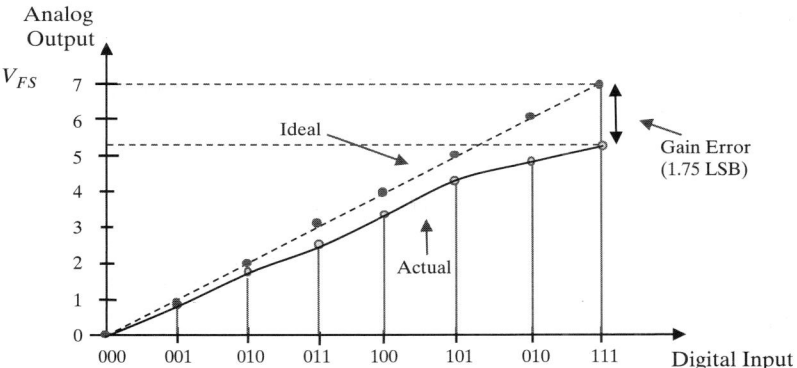

FIGURE 3-33 Gain error of static input/output characteristic of a D/A converter.

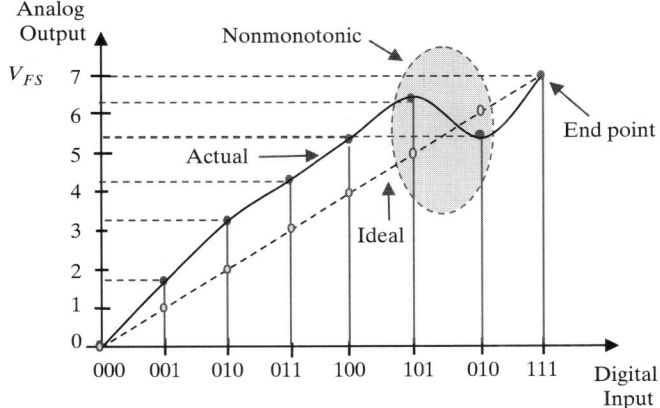

FIGURE 3-34 Monotonicity of a D/A converter.

Gain Error: A gain error exists if the slope of the best-fit line through the transfer curve is different from the slope of the best-fit line in the ideal case. This error is illustrated in Figure 3-33.

Monotonicity: The D/A converter output should increase over its full range as the digital input word to the D/A converter increases. DNL must be greater than −1 LSB in order for the D/A converter to be monotonic. Monotonicity is of importance in most applications, particularly in digital control. Figure 3-34 illustrates a nonmonotonic characteristic of a D/A converter.

3.4.2 D/A Dynamic Metrics

Settling Time: As shown in Figure 3-35, the time required for the output to experience transition and settle down within an error limit is called *settling time*.

58 Chapter 3 Data Converter Specifications

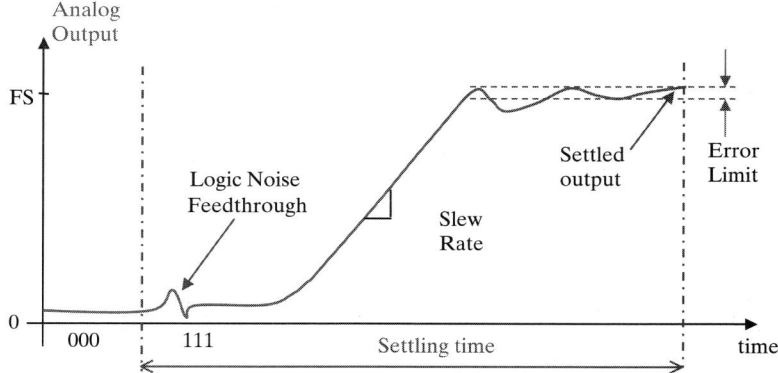

FIGURE 3-35 Time domain response of a D/A converter.

Latency: The total delay to obtain a valid output after the input word changes is called *latency*.

Glitch Area: The maximum area under the glitch of the output when the input word changes is called *glitch area*. Figure 3-36 illustrates how a glitch is produced when the delays are associated with different switches.

Signal-to-Noise + Distortion Ratio (SNDR): Resolution is the term used to describe the minimum signal level that a D/A converter can resolve. The fundamental limit of a D/A converter is governed by quantization noise. If the input digital word is N bits, the minimum step that a D/A converter can resolve is $V_{ref}/2^N$. If output voltages are

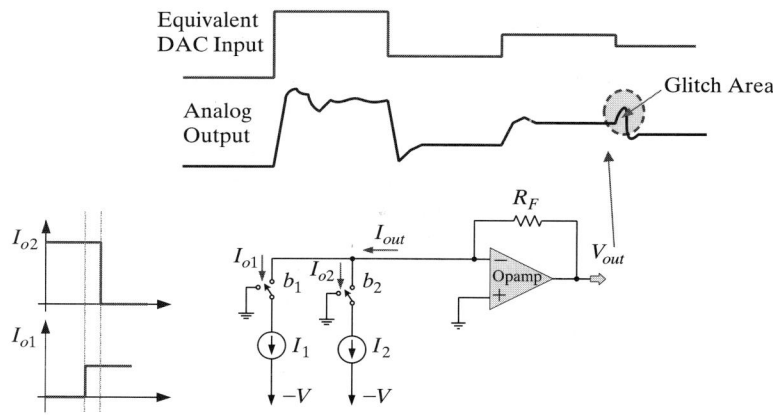

FIGURE 3-36 Output glitch in D/A converters.

reproduced with this minimum step of uncertainty, an ideal D/A converter should have a minimum signal-to-noise ratio of

$$\text{SNR} = 10 \log \frac{(V_{ref}/2)^2/2}{\frac{1}{12}(V_{ref}/2^N)^2} = 10 \log \frac{3}{2} 2^{2N} = 6.02N + 1.76 \text{ (dB)}. \quad \textbf{(3.15)}$$

For example, an ideal 16-bit D/A converter has an SNR of about 97.8 dB. The spectrum of quantization noise is evenly distributed over up to one-half of the sampling frequency. As a result, this inband quantization noise decreases by 6 dB when the oversampling ratio is doubled. Normally, the resolution of a digital-to-analog converter (DAC) is characterized in terms of SNR, but SNR accounts for only uncorrelated noise. The performance is better represented by SNDR, which is the ratio of signal power to the total inband noise, including harmonics distortion. Figure 3-37 shows the output spectrum of an 8-bit D/A converter with a maximum INL of 1 LSB.

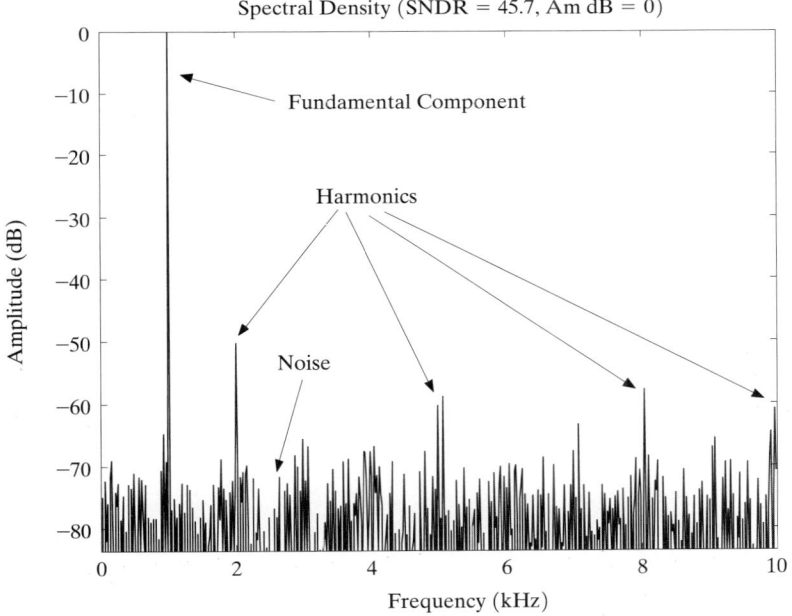

FIGURE 3-37 Output spectrum of an 8-bit D/A converter with a maximum INL of 1 LSB.

CHAPTER 4

Architectures of Data Converters

This chapter presents different architectures of analog-to-digital and digital-to-analog data converters. Considering that there are numerous data converter products on the market, with a wide variety of conversion rate and resolution, much attention must be paid to the selection of an A/D or D/A converter for a specific application. To make a proper selection, DSP system designers must be familiar with different data converter architectures, as well as their advantages and disadvantages. The major A/D architectures are flash, pipeline, folding, interleaving, successive-approximation register (SAR), and sigma-delta, and the major D/A architectures are resistor ladder, current steering, charge redistribution, and sigma-delta.

4.1 A/D ARCHITECTURES

Figure 4-1 shows how different A/D architectures compare in terms of sampling rate and resolution—the two most important attributes of data converters. As can be seen from this figure, the flash architecture is a good candidate for high-speed applications, and the sigma-delta architecture is a good candidate for high-resolution applications.

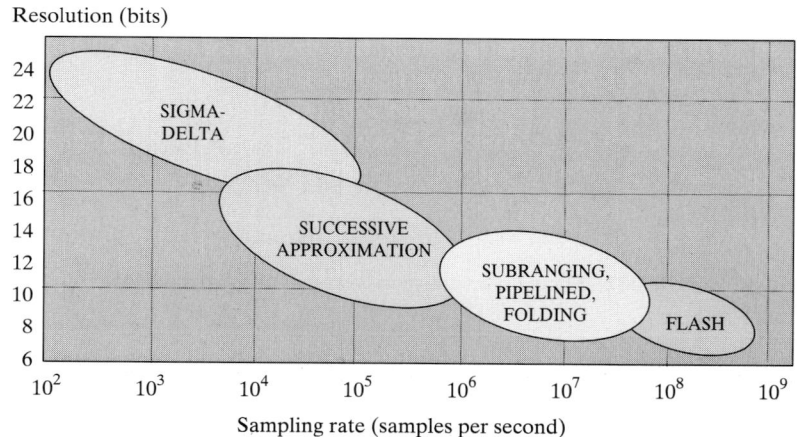

FIGURE 4-1 Different A/D conversion techniques vs. sampling rate and resolution.

4.1.1 Flash

In the flash architecture, the input signal is concurrently compared to all possible quantization levels allowing very high conversion rates. Figure 4-2 illustrates the architecture of a flash A/D converter, where 2^N precise comparators, along with a voltage divider, are used to generate 2^N reference voltages. The outputs of comparators are then passed through a thermometer decoder to produce an N-bit digital word. Ideally, as depicted in Figure 4-3, a flash A/D does not require a track-and-hold circuit for data conversion. However, due to the delay in the signal path and comparator circuit, a track-and-hold circuit is often used at its input.

The randomness associated with the offset voltages of comparators limits the resolution of flash A/D converters. Their sampling rate (or speed) is also limited by the switching speed of comparators and by the propagation delay of the thermometer decoder. The number of comparators grows exponentially with the number of bits. As a result, flash A/D converters are usually designed for a resolution of 10 bits or less, but for high sampling rates (e.g., 40 MSPS). Their main disadvantages include requiring a large silicon area (high cost), having high power consumption, and a high input capacitance.

4.1.2 Subranging

Subranging means resolving an analog signal in several steps, from MSB to LSB bits. Various subranging architectures have been introduced to overcome the limitations of the flash architecture. One of the simplest subranging architectures is the two-step architecture shown in Figure 4-4. The idea is to distribute the quantization load between two

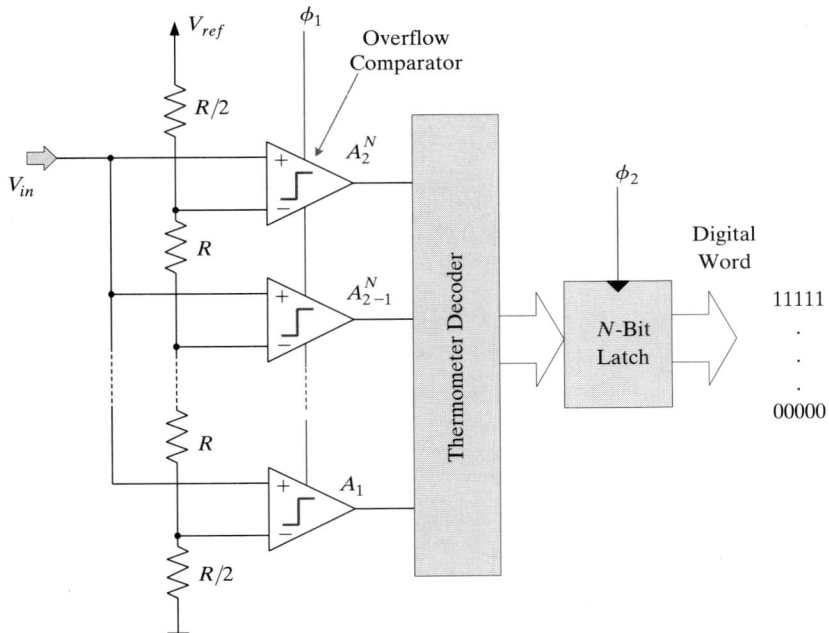

FIGURE 4-2 Conceptual circuit of flash A/D conversion.

62 Chapter 4 Architectures of Data Converters

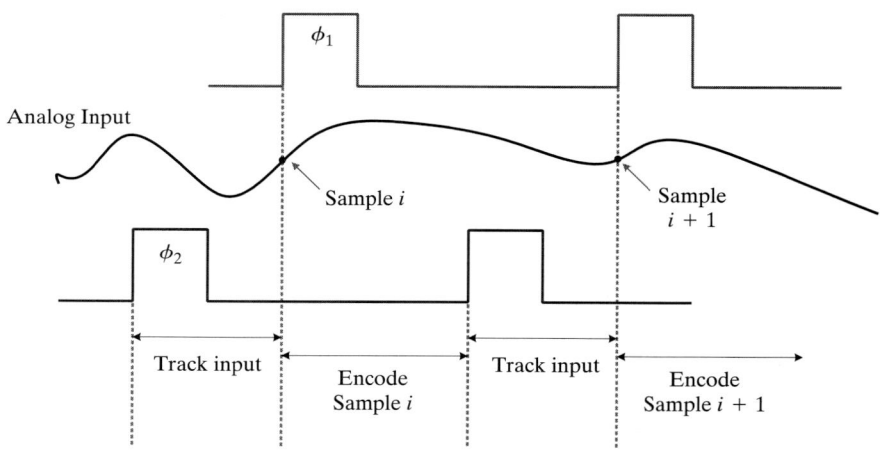

FIGURE 4-3 Different clock phases in flash A/D converters.

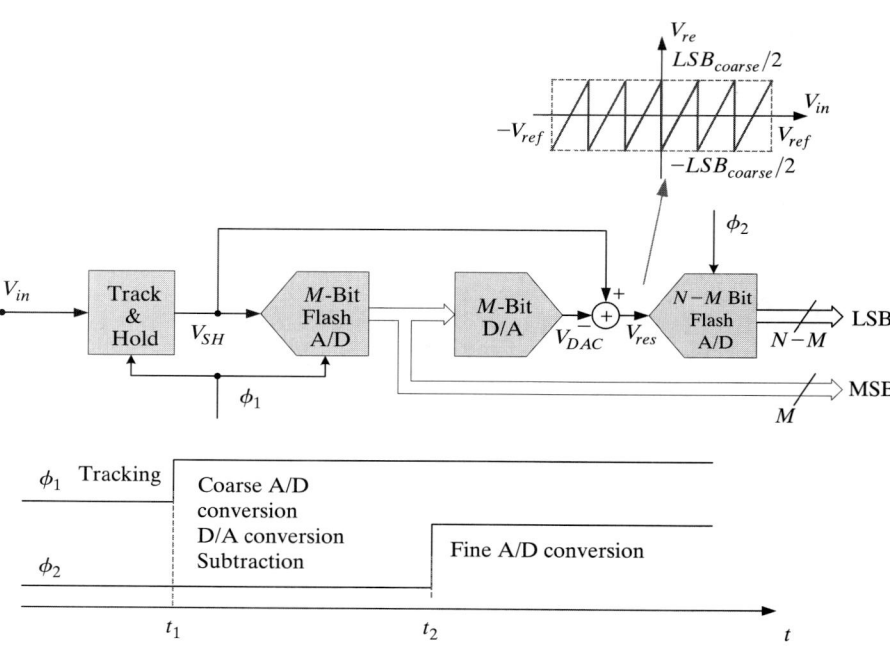

FIGURE 4-4 Simple subranging A/D architecture.

A/D converters in order to reduce chip area and power dissipation. This architecture consists of a coarse-resolution flash A/D converter, an interstage D/A converter, a subtractor, and a fine-resolution A/D converter. First, M most significant bits are obtained by the coarse converter. Then, the fine-resolution converter is used to obtain $N - M$ less significant bits, based on the difference between the input signal and its coarse estimate.

The primary advantage of the two-step subranging architecture is that power dissipation, circuit complexity, input capacitance, and chip area are proportional to $2^{N/2}$, as compared with 2^N when using the flash architecture. Two quantizations are required here before a new input sample is acquired. As a result, the conversion rate is half that achieved by the flash architecture. However, due to the delay in the D/A converter and subtractor, the two-step architecture is more than 50% slower than the flash architecture. Normally, a front-end track-and-hold circuit is used to minimize the sensitivity to clock and signal dispersions.

4.1.3 Pipelined

The idea of distributing the quantization load can be further extended to a multistage architecture. The general configuration of a multistage pipelined architecture is shown in Figure 4-5. Each stage performs the following tasks: (a) samples the analog signal originating from the previous stage and, using a flash A/D converter, quantizes it to k bits, (b) subtracts the quantized signal from its input analog signal, and (c) amplifies the difference. The output words of all stages are combined and digitally corrected to get the final N-bit digital word.

The performance of a pipelined A/D converter is limited by the nonlinearity of its sub-A/D and sub-D/A converters and by the gain error of interstage track-and-hold amplifiers. The fact that the pipelined architecture is constructed by cascading a number of identical pipelined stages is considered to be an advantage, since its hardware cost increases in a linear fashion with resolution. Furthermore, due to its concurrent processing, the throughput of the pipelined architecture is at a rate equal to the analog signal sampling

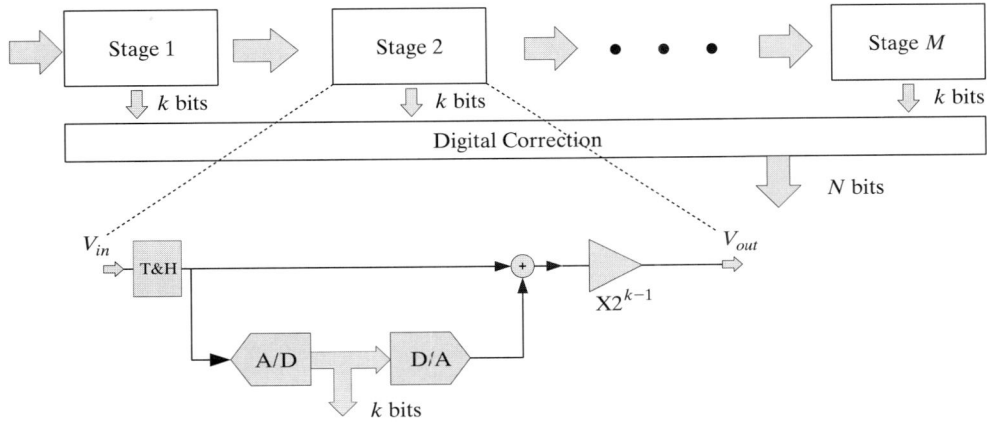

FIGURE 4-5 Pipelined A/D converter architecture.

TABLE 4-1 TI flash and pipelined A/D converter products.

Part Number	Bits	Sampling Rate [Msps]	Power Supply [V]	SNR	SFDR	Power Consumption [mW]
8-bit Analog-to-Digital Converters						
THS0842	8	40	3	44 dB	N/A	275
TLV5580	8	80	3	45 dB	48 dB	165
THS8083	8	80	3	40 dB	N/A	1280
10-bit Analog-to-Digital Converters						
HS1030	10	30	3 - 5	57 dB	60 dB	87
THS1050	10	50	5	60 dB	73 dB	500
THS1060	10	60	5	60 dB	73 dB	600
12-bit Analog-to-Digital Converters						
THS1240	12	40	5	66 d B	73 dB	500
THS1252	12	52	5	67 dB	70 dB	N/A

rate. Fewer numbers of bits per stage eases the requirements for the sub-A/D comparators, increasing the inherent speed of each stage. The major disadvantage of the pipelined architecture is latency. The full digital outcome is delayed in the pipeline by the number of samples multiplied by the time it takes to pass through successive stages.

Table 4-1 provides a list of TI flash and pipelined A/D converters at the time of writing.

4.1.4 Folding

The limitations of the two-step subranging architecture can be overcome by generating a residue (or folding signal) in parallel with a coarse A/D converter, as shown in Figure 4-6. The main advantage of the folding architecture is that it does not require a track-and-hold circuit, subtractor, or sub-D/A converter. The folding signal is generated by a folding circuit. This architecture is suitable for 8-to-10-bit resolution applications requiring very high sampling rates (e.g., 150 MSPS).

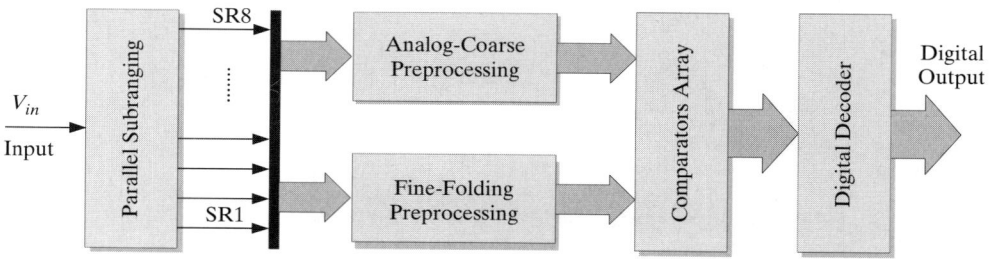

FIGURE 4-6 Block diagram of folding and interpolating A/D converter.

4.1.5 Successive Approximation

Successive-approximation-register (SAR) A/D converters produce N bits in N consecutive cycles, from MSB to LSB. Figure 4-7 shows a block diagram of the successive-approximation architecture. In this architecture, N clock cycles are needed for N sequential comparisons. A conversion time of 1 μs and a resolution up to 16 bits are achievable.

A number of TI SAR A/D converters are listed in Table 4-2. It can be seen that the maximum sampling rate is 1.25 MSPS. This type of A/D converter is quite popular and is suitable for both low-speed and low-power applications.

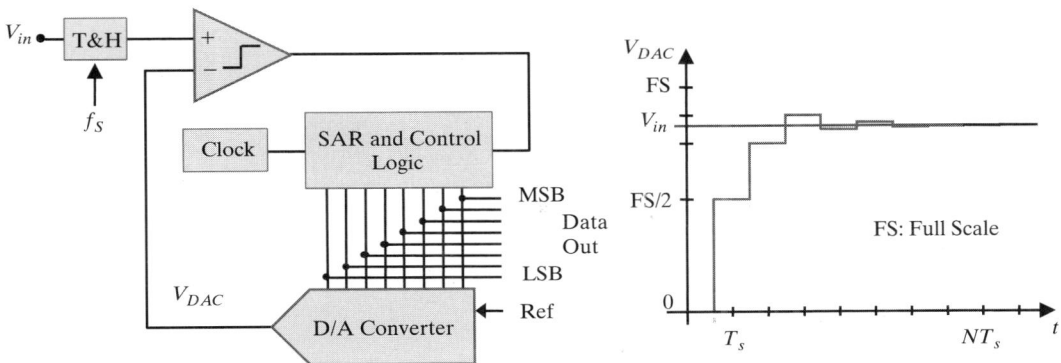

FIGURE 4-7 Block diagram of successive-approximation A/D converter.

TABLE 4-2 TI successive-approximation A/D converter products.

Part Number	Bits	Sampling Rate [Ksps]	Power Supply [V]	SNDR	Power Consumption [mW]
10-bit Analog-to-Digital Converters					
TLC1518	10	400	5	72 dB	22
TLV1508	10	200	3-5	72 dB	2.7
TLV1505	10	200	3-5	72 dB	2.7
12-bit Analog-to-Digital Converters					
TLV2541	12	200	3-5	84 dB	2.3
TLC2551	12	400	5	84 dB	15
TLC2578	12	200	5	86 dB	N/A
14-bit Analog-to-Digital Converters					
TLC3544	14	200	5	86 dB	N/A
TLC3541	14	100/200	5	90 dB	2
16-bit Analog-to-Digital Converters					
TLC4542	16	100/200	5	96 dB	2
TLC4545	16	100/200	5	96 dB	2

66 Chapter 4 Architectures of Data Converters

4.1.6 Interleaved

By placing several A/D converters in parallel (interleaved architecture), it is possible to achieve very high conversion speeds. Figure 4-8(a) shows an interleaved A/D converter, consisting of M N-bit A/D converters. This architecture requires a special clock generator to provide M nonoverlapping clocks, as shown in Figure 4-8(b). A multiplexer is used to route the digital output of each converter at the proper time.

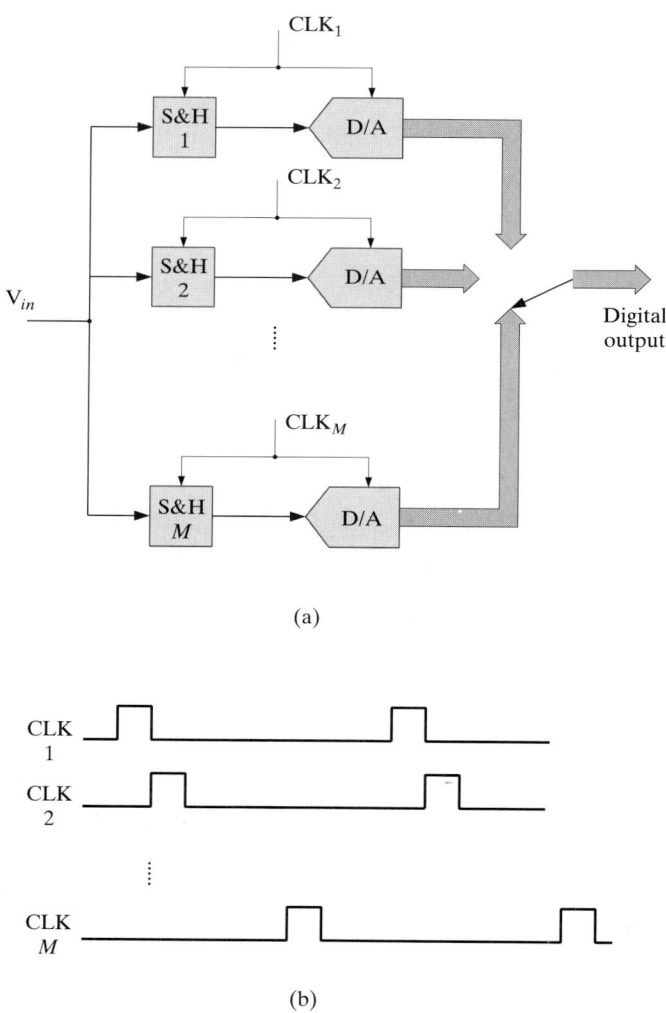

FIGURE 4-8 (a) Block diagram of an interleaved A/D converter and (b) nonoverlapping clock signals.

4.1.7 Sigma-Delta

A/D converters can be grouped into two categories: Nyquist rate and oversampling. In the Nyquist rate category, sampling frequency is determined by the Nyquist rate. For example, flash A/D converters belong to this category. In the oversampling category, sampling rate is much higher (e.g., 20 to 50 times) than the Nyquist rate. Sigma-delta converters belong to the oversampling category, and provide the most promising converter for high-resolution applications because of their tolerance to manufacturing process imperfections. Figure 4-9 illustrates the difference between the Nyquist rate and oversampling A/D converters. The oversampling converters increase the effective number of bits by using a digital filter.

As depicted in Figure 4-10, the quantization-noise power, expressed as

$$n_q^2 = \frac{(\Delta = 1\ \text{LSB})^2}{12}, \quad (4.1)$$

is independent of the sampling frequency. The oversampling ratio (OSR) is defined as

$$\text{OSR} = \frac{f_s}{2W}, \quad (4.2)$$

where W denotes the signal bandwidth. Given that the spectrum of noise is uniformly distributed over 0 to $f_s/2$, and OSR is greater than unity, one can decrease the noise power by using a lowpass digital filter. It is shown that the general expression of SNR is given by [6]

$$\text{SNR} = 6.02N + 1.76\ (\text{dB}) + 10\log(\text{OSR})\ \text{dB}. \quad (4.3)$$

A major advantage of oversampling A/D converters is that they allow the specifications of the input antialiasing filter to be relaxed, resulting in a lower order filter, which in turn lowers the implementation cost. This is illustrated in Figure 4-11.

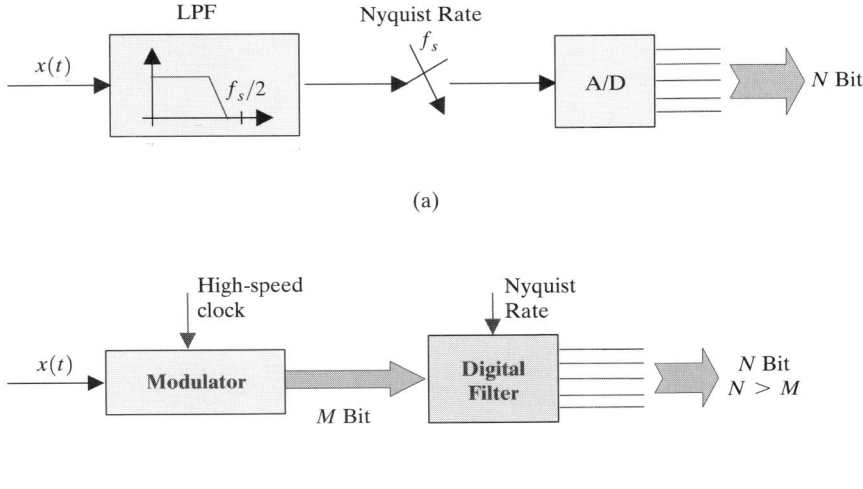

FIGURE 4-9 Two basic categories of A/D converters: (a) Nyquist rate, and (b) oversampling.

FIGURE 4-10 Oversampling technique: (a) effect of oversampling on signal and quantization-noise spectra and (b) signal and noise spectra after filtering.

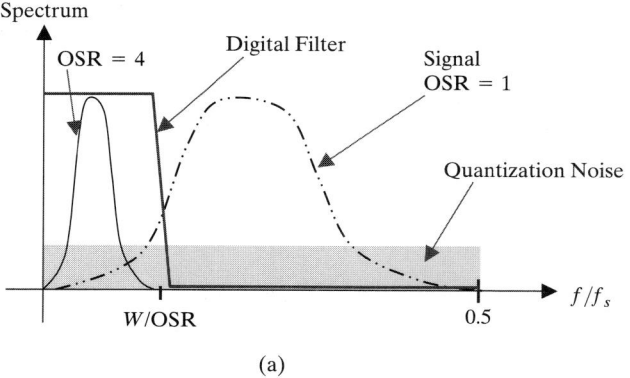

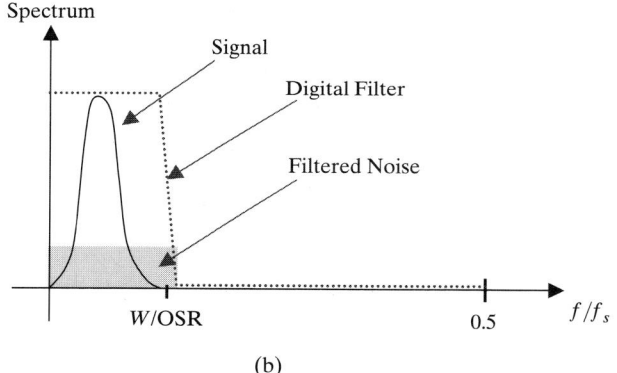

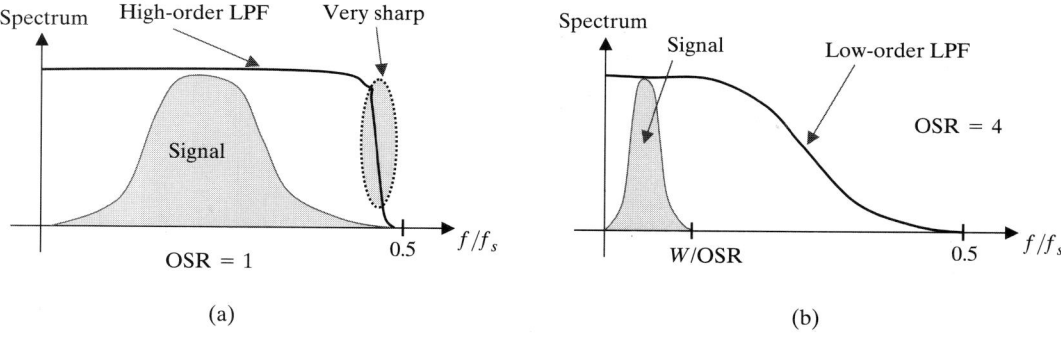

FIGURE 4-11 Specification of antialiasing input filter: (a) Nyquist rate and (b) oversampling.

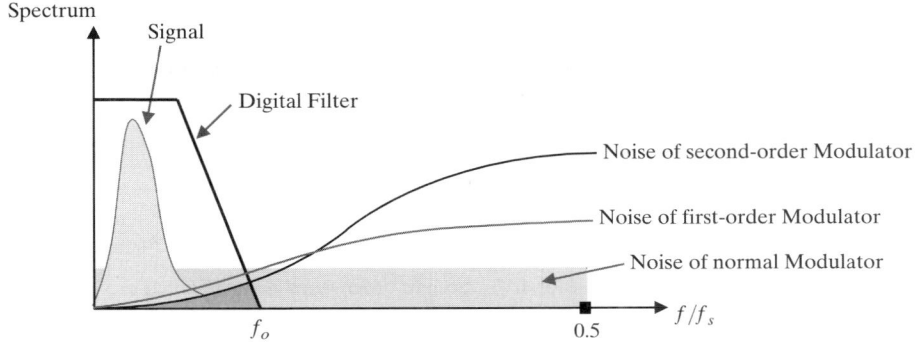

FIGURE 4-12 Noise shaping in sigma-delta A/D converters.

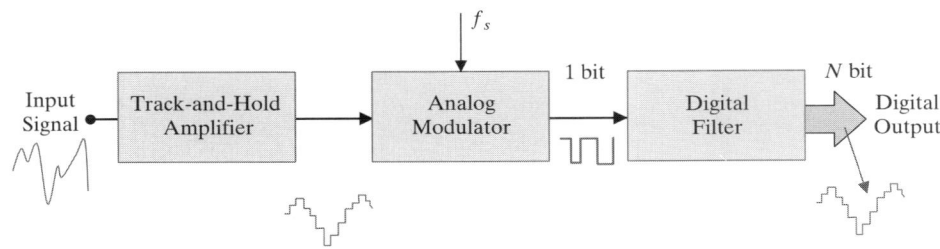

FIGURE 4-13 Block diagram of sigma-delta A/D converters.

A common thrust in all sigma-delta architectures is the alteration of the quantization noise spectrum in such a way that little noise occurs in the signal frequency band. The out-of-band noise is then filtered out by a digital filter. The total power of signal and noise is constant. A modulator is therefore used to push noise to higher frequencies. The noise shaping process is illustrated in Figure 4-12.

As shown in Figure 4-13, the sigma-delta converter comprises (a) a track-and-hold, (b) a modulator, and (c) a digital decimation filter. As mentioned, the modulator shapes the quantization noise to push a significant percentage of it out of the band of interest. The modulator can be implemented by employing continuous-time or sampled-data filters. Figure 4-14 provides an example of a simple modulator consisting of two integrators followed by a 1-bit quantizer and a 1-bit D/A converter feedback loop. The advantage of using a 1-bit D/A converter is that it is extremely linear. The number of integrators in the loop determines the order of the sigma-delta modulator. The higher the modulator order, the greater is the amount of quantization noise that gets pushed out of the band of interest. However, higher order modulators tend to be more difficult to stabilize. The order of sigma-delta modulators is usually less than 5. Sigma-delta converters are well suited for high-resolution applications (for example, between 14 bits and 24 bits). A commercial example of a sigma-delta converter is the TI TLC320AD75 with 20-bit resolution at 48 kSPS.

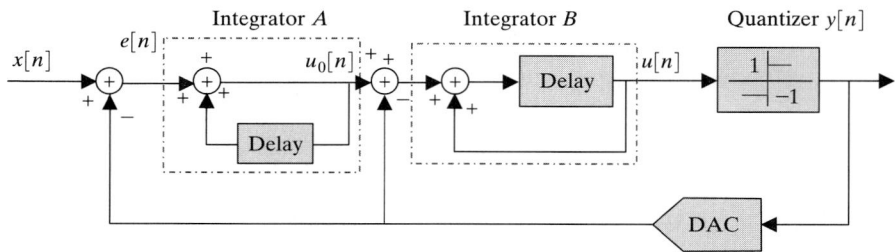

FIGURE 4-14 Second-order sigma-delta modulator.

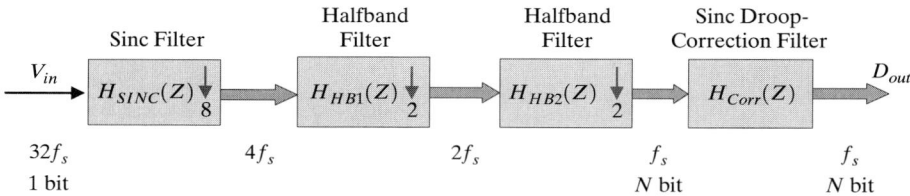

FIGURE 4-15 Implementation of decimation filter.

The 1-bit quantizer output must be filtered in order to reject the out-of-band noise. The digital decimation filter can be implemented in DSP or by a dedicated digital circuitry. The overall order of the decimation filter is made lower by dividing it into multiple stages via multirate filtering, as shown in Figure 4-15. The major advantages of the sinc filter appearing in Figure 4-15 are its simple digital circuitry and high-speed operation. The optimum order of the sinc filter is the order of the sigma-delta modulator (L) plus 1. The optimum transfer function is [6]

$$H_{opt}(f) = H_{opt}(z)\bigg|_{z=e^{j2\pi fT}} = \left[\frac{\text{sinc}(\pi f MT)}{\text{sinc}(\pi f T)}\right]^{L+1}. \quad (4.4)$$

In the half-band filters appearing in Figure 4-15, the passband and stopband ripples are the same, and the cutoff frequencies are symmetrical around $f_s/4$. Half-band filters belong to a family of filters whose impulse responses $h[n]$ are zero for all even values of n, except $n = 0$. Therefore, these filters can be implemented with one half of the coefficients.

4.2 D/A ARCHITECTURES

A D/A converter converts a sequence of bits into analog levels. Different architectures can perform this conversion with different sampling speeds and resolution. DSP systems have different requirements ranging from very high speed (>40 MHz), low-resolution (<10 bits), such as video systems, to moderately high speed (1–10 MHz), high-resolution (12–16 bits), such as DSL systems, and to low speed (<50 kHz), very high-resolution (>16 bits), such as high-quality audio systems.

4.2.1 Resistor Ladder

In the resistor ladder architecture, the reference voltage is divided into $2^N - 1$ sub-voltages (see Figure 4-16.) The difference between two subsequent voltages is exactly 1 LSB. A network of switches selects the output voltage from the resistor string, based on the digital input word. In order to avoid loading the resistor ladder, a buffer is used at the output of the D/A converter. The output buffer is usually integrated as part of the D/A converter. The major advantage of this architecture is its simplicity. It also possesses a monotonic input/output transfer characteristic. Its disadvantages are (a) it requires $2^N - 1$ matched resistors and (b) it needs an output buffer, which degrades its overall speed. High-resolution resistor ladder D/A converters require a large number of both resistors and switches. This problem can be overcome by decomposing the architecture into a coarse and a fine section. This allows reducing the number of devices to approximately $2^{N/2}$.

In order to have a rough estimate of the conversion time, one can approximate the signal path by a first-order circuit, as shown in Figure 4-17. Each switch is considered to be a series resistor along with a parasitic capacitor to ground. Using the open-circuit, time-constant method, the time constant of this first-order circuit is

$$\tau \cong \sum_i \tau_i = RC + 2RC + \ldots + nRC = \frac{n(n+1)}{2} RC \approx \frac{n^2}{2} RC. \quad (4.5)$$

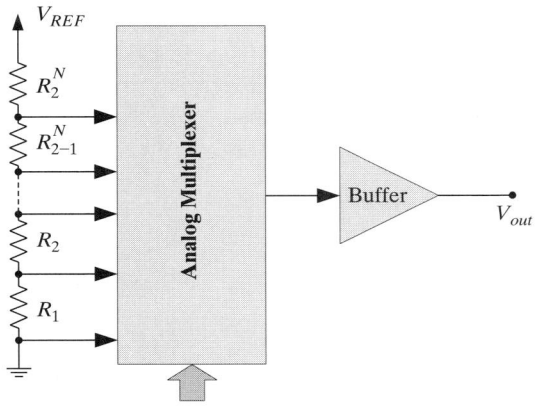

FIGURE 4-16 Basic block diagram of resistor ladder D/A converters.

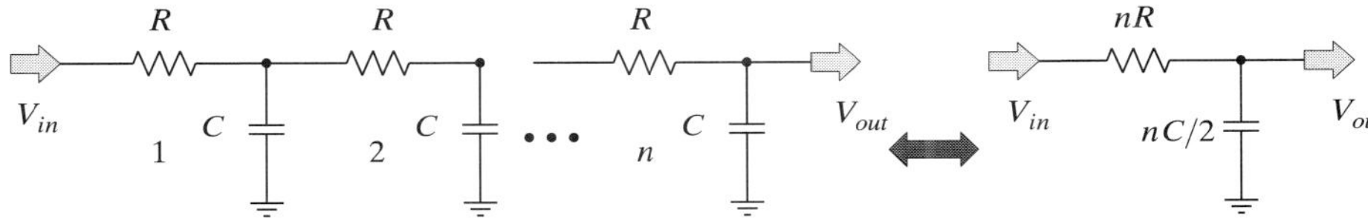

FIGURE 4-17 First-order approximation of signal path in resistor ladder D/A converters.

72 Chapter 4 Architectures of Data Converters

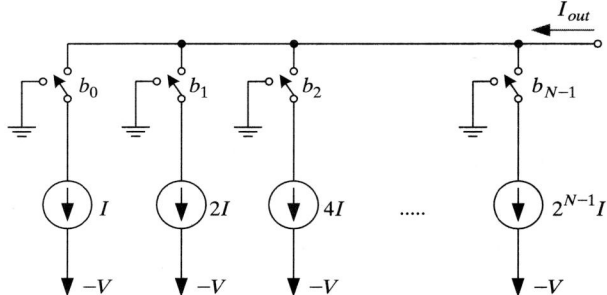

FIGURE 4-18 Architecture of a current steering D/A converter.

4.2.2 Current Steering

In a current steering D/A converter, an array of identical current sources is used to generate an output current corresponding to the digital input word, as shown in Figure 4-18. The major advantage of the current steering converter is that it can drive a resistor load directly and can convert the output current to voltage without using any output buffer. Similar to the resistor string architecture, some of the current steering architectures also guarantee monotonicity. In addition, they allow a higher conversion rate. Their disadvantage is the static power dissipation in their current sources.

4.2.3 Charge Redistribution

Another type of D/A converter is charge redistribution, which uses an array of identical capacitors. A simple circuit of the charge redistribution architecture is shown in Figure 4-19. Its main disadvantage is the linear capacitor requirement, which demands a considerable chip area. The performance of this architecture becomes limited at high sampling rates because it is required to have a highly linear output buffer.

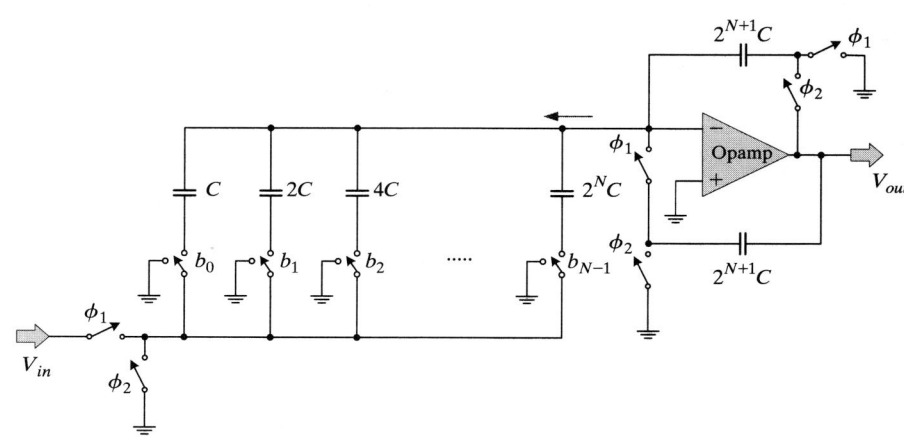

FIGURE 4-19 Charge redistribution D/A converter.

4.2.4 Sigma-Delta

The aforementioned architectures are not suitable for high-resolution (>14 bits), D/A data conversion. Sigma-delta converters can achieve high resolution at the expense of lower effective sampling rates—that is, high oversampling ratios. The sigma-delta architecture is not sensitive to the imperfections of analog components. Figure 4-20 shows the basic block diagram of a sigma-delta D/A converter, as well as the signal spectra at various points. A sigma-delta D/A converter consists of a digital interpolation filter, a noise-shaping filter, a 1-bit D/A converter, and an output lowpass filter (reconstruction filter). The interested reader is referred to [4] for a comprehensive description of sigma-delta data converters.

The function of the interpolation filter is to generate more samples between consecutive samples (oversampling). This increases the correlation among samples. Figure 4-21 illustrates the operation of the interpolation filter. The criteria in the design of an

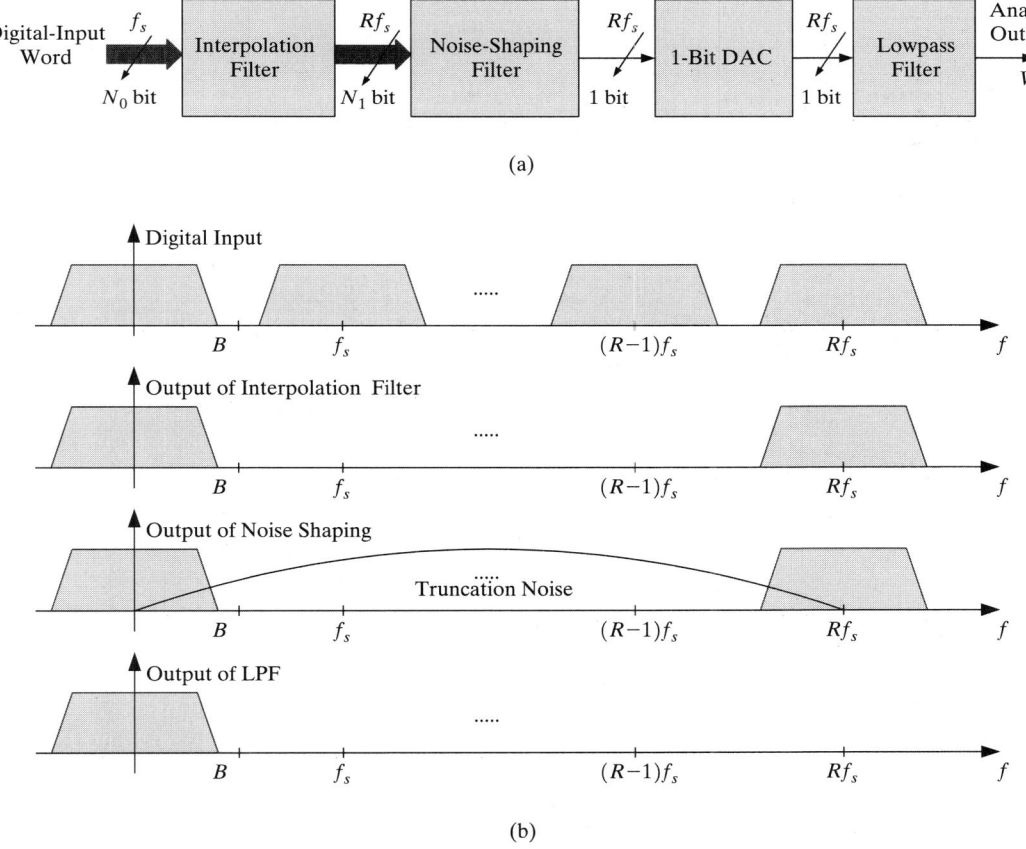

FIGURE 4-20 Sigma-delta D/A converter: (a) basic block diagram and (b) spectrum of signals at different points.

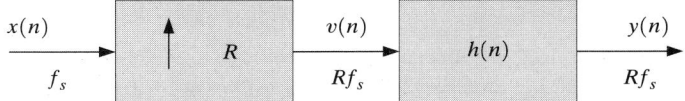

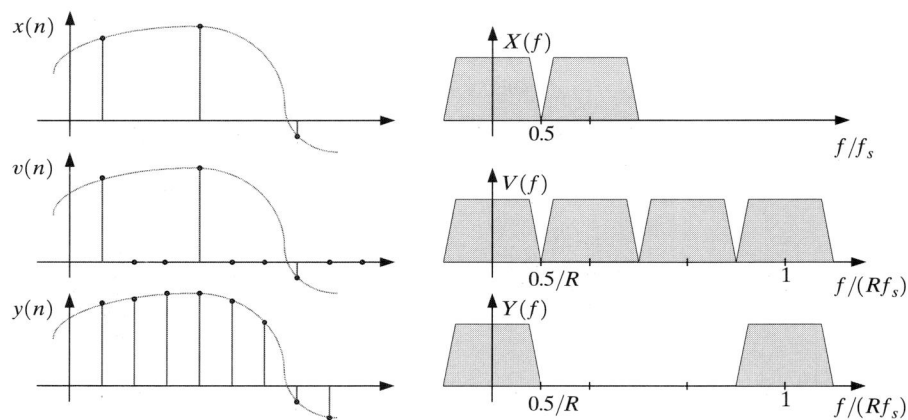

FIGURE 4-21 Interpolation filter: (a) block diagram and (b) different operations.

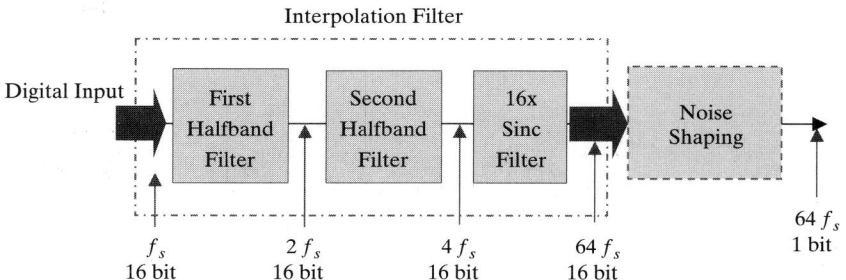

FIGURE 4-22 Implementation of interpolation filter.

interpolation filter include power consumption efficiency, chip area, and circuit complexity. In practice, a multistage interpolation is typically used, as illustrated in Figure 4-22.

After increasing data rates by using the interpolation filter (refer to Figure 4-23), a sigma-delta modulator is used to generate a 1-bit data stream. The modulator pushes the quantization noise to the frequencies out of the band of interest. The output of the

4.2 D/A Architectures

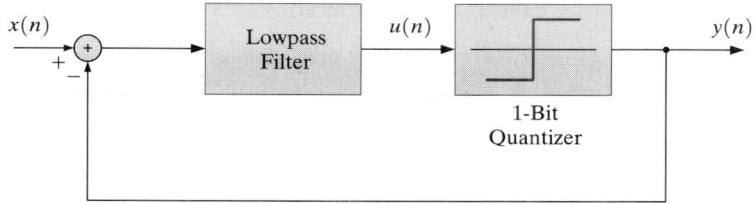

FIGURE 4-23 Sigma-delta noise shaping.

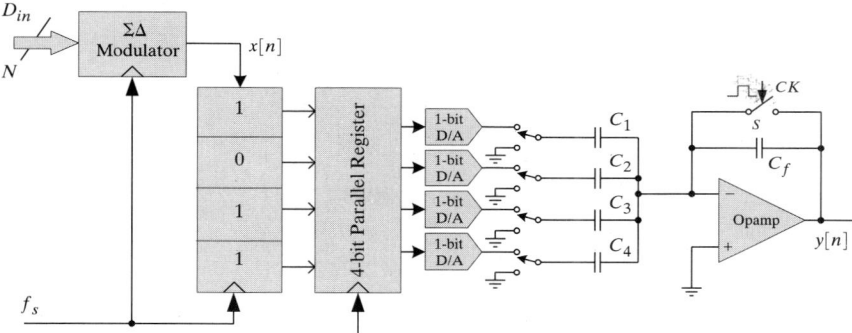

FIGURE 4-24 A switched-capacitor comb filter.

modulator is then applied to a decimation filter. Figure 4-24 shows a third-order decimate-by-2 switched capacitor comb filter. The output voltage is determined by [6]

$$y[n] = \sum_{i=0}^{3} \frac{C_i}{C_f} V_{ref} x_i. \qquad (4.6)$$

An example of a 16-bit sigma-delta D/A converter for digital audio, and with a 44.1-kHz sampling rate, is shown in Figure 4-25. Here, a simple zero order hold register is used to provide the sinc interpolation.

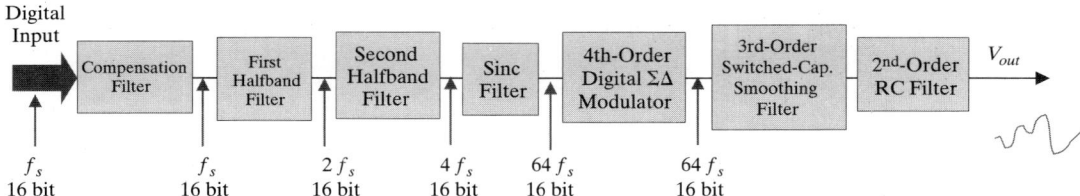

FIGURE 4-25 A 16-bit sigma-delta D/A converter for digital audio.

4.3 SELECTION OF DATA CONVERTERS FOR DSP SYSTEMS

In this section, a number of tables are provided to help DSP system designers select an appropriate TI data converter for a specific application of interest. Figure 4-26 provides some of the high-speed A/D and D/A data converter products manufactured by TI. Table 4-3 lists the interfacing requirements for connecting various TI A/D converter products to different families of TI DSP processors. Table 4-4 lists the interfacing requirements for connecting various TI D/A converter products to different families of TI DSP processors. Finally, Table 4-5 provides a list of possible data converters for telecommunication applications. The information in these tables have been extracted from *TI Seminar Series on DSP/Analog Technologies*.

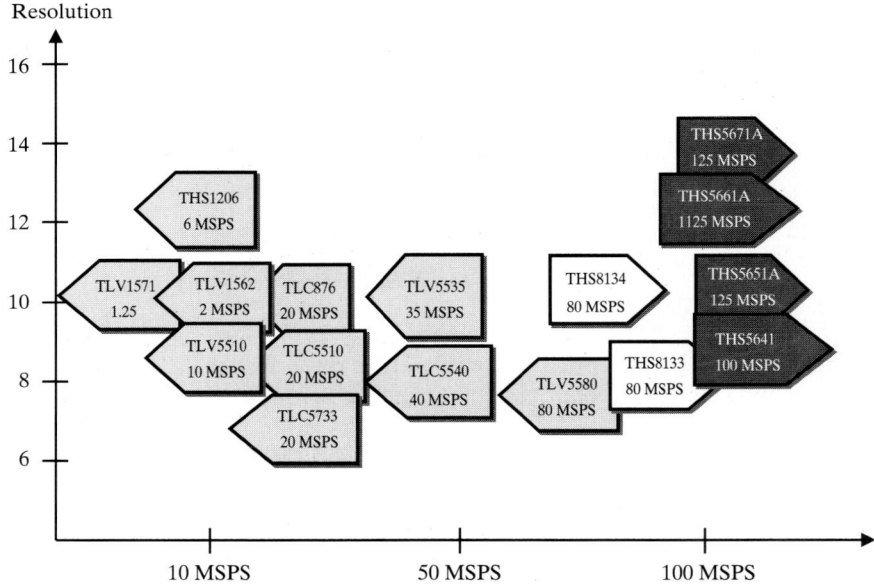

FIGURE 4-26 TI high-speed data converters.[†]

4.3 Selection of Data Converters for DSP Systems

TABLE 4-3 TI A/D converter for the TMS320 DSP family.[†]

ADC	C203	C206	C209	C24x	C3x	C4x	C54x	C6x
TLC540				■				
TLC541				■				
TLC542				■				
TLC545				■				
TLC546				■				
TLC548				■				
TLC549				■				
TLC876	■	P	■		P	P	■	P
TLC1542				■				
TLC1549				■				
TLC1550	P	P	P	■	P	P	P	●
TLC2543				■				
TLC5510	■			P	P	P	■	
TLV 0831								
TLV1543				■				
TLV1544	■	■	■	■	■		■	■
TLV1570	■		■	■		■	■	■
TLV1572	■	■	■	■	■		■	■
TLV2543				■				
TLV5510	■			P	P	P	■	P

Legend:	
	Needs glue logic to work with DSP
■	Glueless interface, 100% compatible
	Glueless interface, some timings are critical
	Simple glue logic, like clock inverter
P	Parallel interface
P	Parallel interface

TABLE 4-4 TI D/A converter for the TMS320 DSP family.[†]

D/A	C203	C206	C209	C24x	C3x	C4x	C54x	C6x
TL5632								
TLC5615								
TLV5617								
TLV5618								
TLC5628								
TLC7225	P	P	P	P	P	P	P	
TLC7226	P	P	P	P	P	P	P	
TLC7524	P	P	P	P	P	P	P	
TLC7628	P	P	P	P	P	P	P	
TLV5604								
TLV5613	P	P	P	P	P	P	P	P
TLV5616								
TLV5619	P	P	P	P	P	P	P	P
TLV5628								
TLV5633	P	P	P	P	P	P	P	P
TLV5636								
TLV5637								
TLV5638								
TLV5639	P	P	P	P	P	P	P	P

Legend:	
	Needs glue logic to work with DSP
	Glueless interface, 100% compatible
P	Parallel interface. Glue logic needed
	Parallel interface, Timing circuit and glue logic needed
P	Simple glue logic, like clock inverter
	Parallel interface
P	Parallel interface, DSP bus access needed

4.3 Selection of Data Converters for DSP Systems

TABLE 4-5 TI Data Converters for Telecom Applications.[†]

Application	Year	Supply	Technology	Bits	fs / OSR	Type	Area (mm^2)	Power (mW)
Speech	1980	5V	5 μ NMOS	8	8 kHz	Suc. Appro A/D		50
	1985	5V	2.4 μ CMOS	12	1 MHz / 256x	2nd-order ΣΔ A/D / ΣΔ D/A	3.5 / 2	20
	1990	+5V	1.2 μ CMOS	13	2 MHz / 512x	2nd-order ΣΔ A/D / ΣΔ D/A	2 / 1	6 / 4
	1995	+5V	0.7 μ CMOS	14	2 MHz / 512x	2nd-order ΣΔ A/D / ΣΔ D/A	1.5 / 2	5 / 3
ISDN	1987	+5V	2 μ CMOS	10	16 MHz / 128x	2nd-order ΣΔ A/D / ΣΔ D/A	21	5
	1996	+3V	0.5 μ CMOS	10 / 1.2	16 MHz / 128x	4th-order ΣΔ A/D / 6th-order ΣΔ D/A	2 / 1.5	35 / 10
GSM	1990	+5V	μ CMOS	8	270 kHz / 1x	Suc. Appr. A/D / Binary Weig D/A	11	0
	1993	+5V	0.7 μ CMOS	8	270 kHz / 1x	Suc. Appr. A/D / Binary Weig D/A	11	0
	1995	+3V	0.5 μ CMOS	13 / 8	6.5 MHz / 24x	4th-order ΣΔ A/D / Bin.-Weig. D/A	1.5 / 0.4	14 / 3
ADSL	1993	+5V	0.7 μ CMOS	12	53 MHz / 24x	4th-order ΣΔ A/D / 6th-order ΣΔ D/A	9 / 7	850 / 700
	1997	+3V	0.5 μ CMOS	12	8.8 MHz / 4x	Pipelined A/D / Switched-I D/A	5 / 2	120 / 30
VDSL	1998	+3V	0.35 μ CMOS	12	40 MHz / 1x	Pipelined A/D / Switched-I D/A	5 / 2	250 / 60

CHAPTER 5

TMS320C6000 Architecture

The choice of a DSP processor to implement an algorithm in real-time is application dependent. There are many factors that influence this choice. These factors include cost, performance, power consumption, ease-of-use, time-to-market, and integration/interfacing capabilities.

The family of TMS320C6000 processors, manufactured by Texas Instruments, are built to deliver speed. They are designed for MIPS (million instructions per second) intensive applications, such as 3G wireless, DSL cable modems, and digital imaging. Table 5-1 provides a list of currently available fixed-point and floating-point C6x processors (at the time of this writing). As can be seen from the table, instruction cycle time, speed, power consumption, memory, peripherals, packaging, and cost specifications vary for different products in this family. For example, the fixed-point C6201 version can operate at 200MHz (5 ns cycle time), delivering a peak performance of 1600 MIPS or 400 million multiplies and accumulates (MACs) per second. The floating-point C6701 version can operate at 167MHz (6 ns cycle time), delivering a peak performance of 1000 million floating-point operations per second (MFLOPS). Figure 5-1 illustrates the processing power of C62x and C67x by showing a speed benchmarking comparison with some other common DSP processors.

Figure 5-2 shows the block diagrams of the generic C6x architecture together with the C6201 and C6211 architectures. The CPU consists of 8 functional units divided into two data paths labeled A and B. Each side has a so-called .M unit (used for the multiplication operation), a .L unit (used for logical and arithmetic operations), a .S unit (used for branch, bit manipulation and arithmetic operations) and a .D unit (used for loading, storing and arithmetic operations). Many instructions such as ADD can be executed by more than one unit. There are sixteen 32-bit registers associated with each side. Interaction with the CPU must be done through these registers. A listing of the C6x instructions, as divided by the four functional units, appears in Appendix A (Quick Reference Guide). These instructions are fully discussed in the *TI TMS320C6x CPU and Instruction Set Reference Guide* [14].

As shown in Figure 5-3, the internal buses consist of a 32-bit program address bus, a 256-bit program data bus accommodating eight 32-bit instructions, two 32-bit data address buses (DA1 and DA2), two 32-bit (64-bit for the floating-point version) load data buses (LD1 and LD2), and two 32-bit (64-bit for floating-point version) store data buses (ST1 and ST2). In addition, there are a 32-bit DMA data and a 32-bit DMA address bus. The off-chip, or external, memory is accessed through a 26- or 29-bit address bus, depending on the product version, and a 32-bit data bus.

The peripherals on a typical C6x processor include EMIF (External Memory Interface), DMA, Boot Loader, McBSP (Multichannel Buffered Serial Port), HPI (Host Port Interface), Timer, and Power Down unit. EMIF provides the necessary timing for

TABLE 5-1 (a) Fixed-point TMS320C62x, (b) floating-point TMS320C67x DSP product specifications, (prices represent year 2000 pricing).†

(a)

Device	RAM Data	RAM Prog	Mc BSP	DMA	Parallel	Timers	MHz	Cycle (ns)	MIPS	Typical Activity CPU Power (mA/MPS)	Total internal Power(W) (Full Device Speed)	Packaging (BGA-ball grid array)	$US/1KU	$US/25KU
TMS3206201-200	64KB	64KB	2	4	HPI/16	2	200	5	1600	0.15	1.3	352BGA, 35/27mm	91.27	72.58
TMS3206202-200	128KB	256KB	3	4	Exp. Bus/32	2	200	5	1600	0.15	1.7	352BGA, 27mm	109.54	87.10
TMS3206202-250	128KB	256KB	3	4	Exp. Bus/32	2	250	4	2000	0.15	2.1	384BGA, 18mm 352BGA, 27mm	159.41	116.21
TMS3206203-250	512KB	384KB	3	4	Exp. Bus/32	2	250	4	2000	0.07	1.1	384BGA, 18mm 352BGA, 27mm	191.29	139.45
TMS3206203-300	512KB	384KB	3	4	Exp. Bus/32	2	300	3.3	2400	0.07	1.3	3848GA, 18mm 352BGA, 27mm	248.67	181.28
TMS3206204-200	64KB	64KB	2	4	Exp. Bus/32	2	200	5	1600	0.07	0.8	384BGA, 18mm 384BGA, 18mm	43.38	31.63
TMS3206205-200	64KB	64KB	2	4	PCI/32	2	200	5	1600	0.07	0.8	306BGA, 16mm	58.20	42.43
TMS3206211-150	4KB/4KB	64KB	2	16	HPI/16	2	150	6.7	1200	0.15	0.9	256BGA, 27mm	33.05	25.00

(b)

Device	RAM Data	RAM Prog	Mc BSP	DMA	Parallel	Timers	MHz	Cycle (ns)	MIPS	Typical Activity CPU Power (mA/MPS)	Total internal Power(W) (Full Device Speed)	Packaging (BGA-ball grid array)	$US/1KU	$US/25KU
TMS3206701-150	64KB	64KB	2	4	HPI/16	2	150	6.7	900M	0.22	1.3	352BGA, 35mm	109.54	87.10
TMS3206701-167	64KB	64KB	2	4	HPI/16	2	167	6	1G	0.22	1.4	352BGA, 27mm	171.68	125.15
TMS3206711-100	4KB/4KB	64KB	2	16	HPI/16	2	100	10	600M	0.22	0.8	256BGA, 27mm	31.36	21.59
TMS3206711-150	4KB/4KB	64KB	2	16	HPI/16	2	150	6.7	900M	0.22	1.1	256BGA, 27mm	44.07	37.79

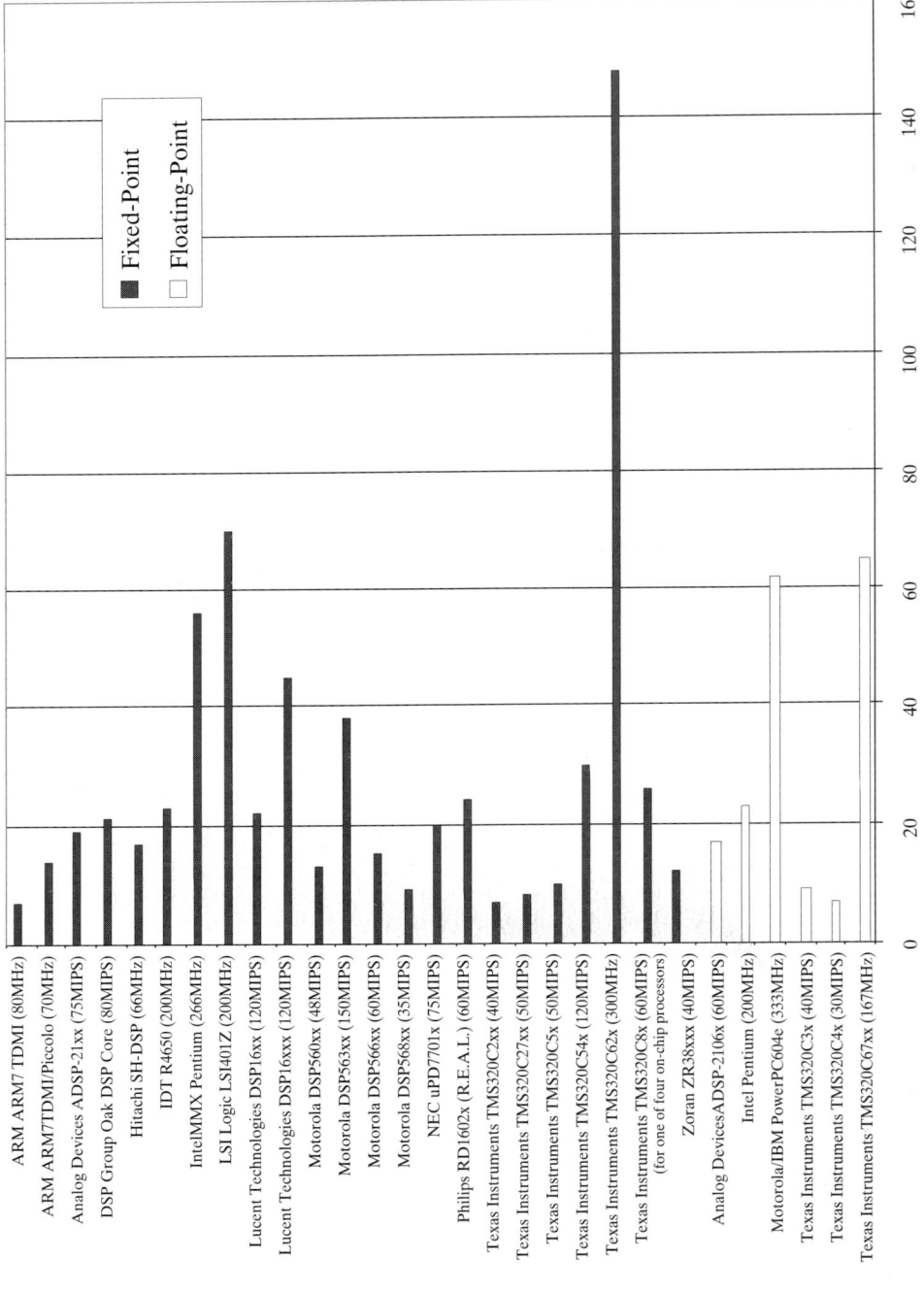

FIGURE 5-1 BDTImark(TM) DSP Speed Metric benchmark, by Berkeley Design Technology, Inc.[1]

[1] The BDTImark is a summary measure of DSP speed, distilled from a suite of DSP benchmarks developed and independently verified by Berkeley Design Technology, Inc. A higher BDTImark score indicates a faster processor. For a complete description of the BDTImark and underlying benchmarking methodology, as well as additional BDTImark scores, refer to http://www.bdti.com. (c) 2000 Berkeley Design Technology, Inc.

Introduction 83

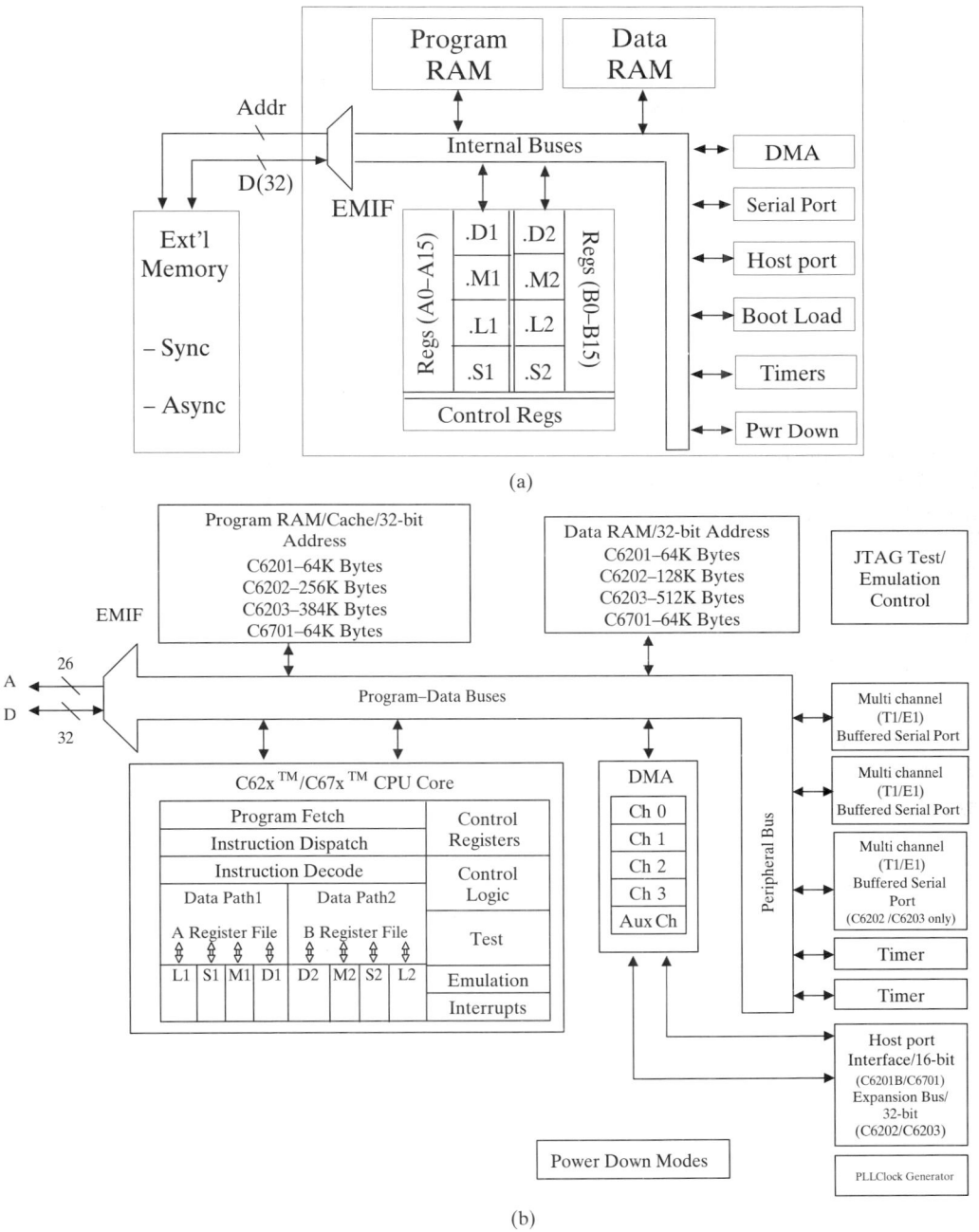

FIGURE 5-2 (a) Generic C6x architecture, (b) C6201 & C701 architecture, and (c) C6211 & C6711 architecture.† *(continues on next page)*

84 Chapter 5 TMS320C6000 Architecture

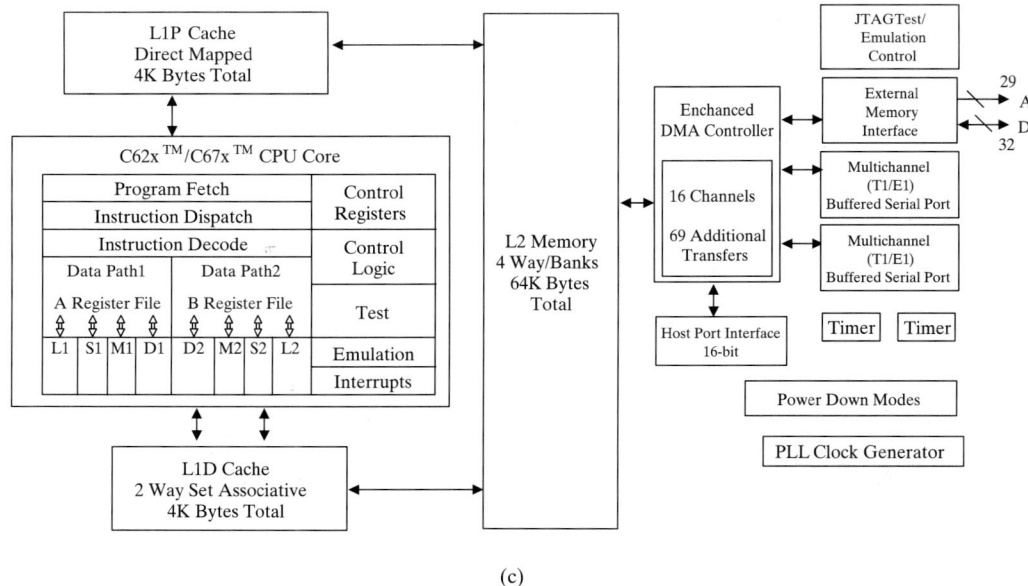

(c)

FIGURE 5-2 *(continued)*

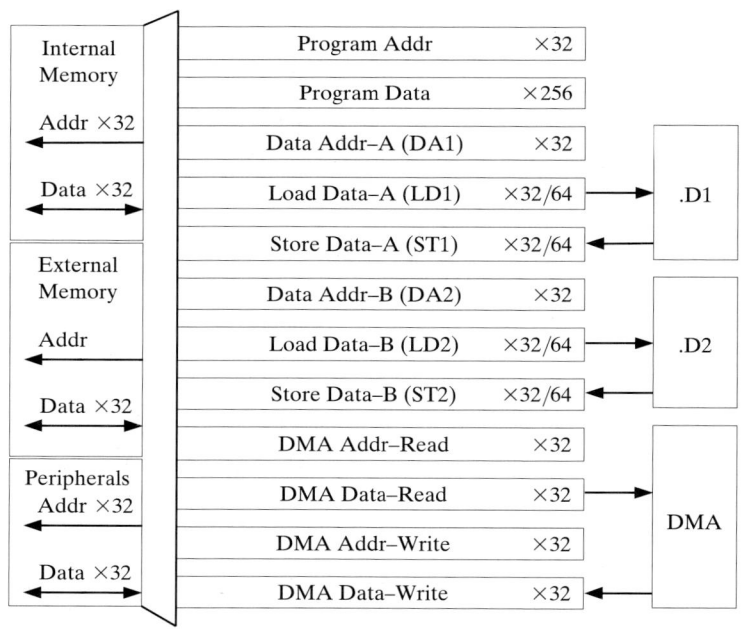

FIGURE 5-3 C6x internal buses.

accessing external memory. DMA allows the movement of data from one place in memory to another place without interfering with the CPU operation. Boot Loader boots the loading of code from off-chip memory or HPI to internal memory. McBSP provides a high-speed multichannel serial communication link. HPI allows a host to access internal memory. Timer provides two 32-bit counters. Power Down unit is used to save power for durations when the CPU is inactive.

5.1 CPU OPERATION (DOT-PRODUCT EXAMPLE)

As shown in Figure 5-2, the C6x CPU is divided into two data paths–data path A (or 1), and data path B (or 2). An effective way to understand the CPU operation is by going through an example. Figure 5-4 shows the assembly code for a 40-point dot product y between two vectors **a** and **x**, $y = \sum_{n=1}^{40} a_n \times x_n$. This code appears in the TI Technical Training Notes on TMS320C6x DSP [10]. At this point, it is worth mentioning that the assembler is not case sensitive (i.e., instructions and registers can be written in lower or uppercase).

The registers assigned to a_n, x_n, loop count, product, y, &a[n] (address of a_n), &x[n] (address of x_n), and &y[n] (address of y_n) are shown in Figure 5-5. In this example, only the A-side functional units and registers are used.

A loop is created by the instructions indicated by ●'s. First, a loop counter is set up by using the move constant instruction MVK. This instruction uses the .S1 unit to place the constant 40 in register A2. The beginning of the loop is indicated by the label loop

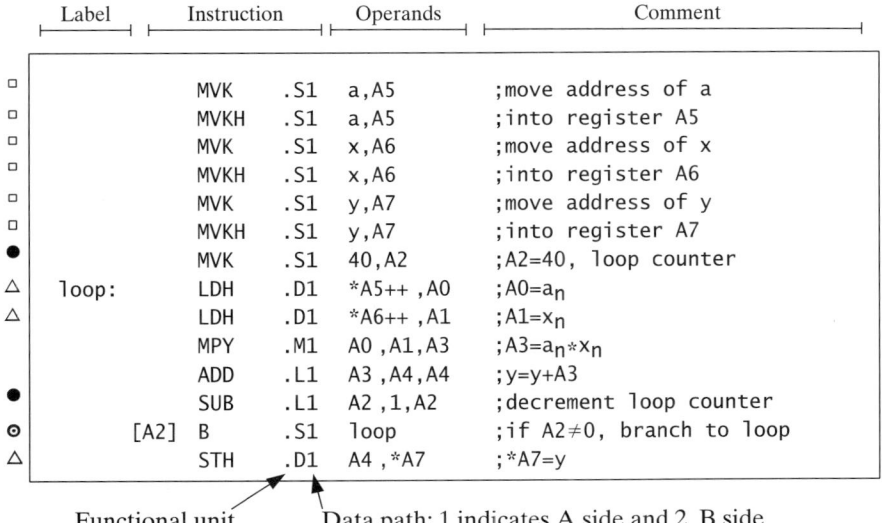

FIGURE 5-4 Dot-product assembly code.

FIGURE 5-5 A-side registers.

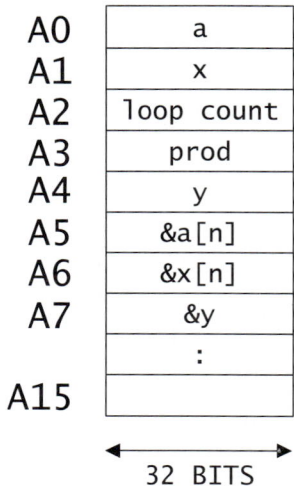

and the end by a subtract instruction SUB, to decrement the loop counter, followed by a branch instruction B to return to loop.

The subtraction is performed by the .L1 unit and branching by the .S1 unit. The brackets as part of the branch instruction indicate that this is a conditional instruction. All the C6x's instructions can be made conditional based on a zero or nonzero value in one of the registers A1, A2, B0, B1, and B2. The syntax [A2] means "execute the instruction if A2 ≠ 0", and [!A2] means "execute the instruction if A2 = 0". As a result of these instructions, the loop is repeated 40 times.

Considering that the interaction with the functional units is done through the A-side registers, these registers must be set up in order to start the loop. The instructions labeled by □'s indicate the necessary instructions for doing so. MVK and MVKH are used to load the addresses of a_n, x_n, and y into registers A5, A6, and A7, respectively. These instructions must be executed in the order indicated to load the lower 16 bits of the full 32-bit address first, followed by the upper 16 bits. These registers are used as pointers to load a_n and x_n into the A0 and A1 registers, respectively, and store y from the A4 register (instructions labeled by ▲'s). The C programming language notation * indicates that a register is being used as a pointer. Depending on the datatype, any of the following loading instructions can be used: bytes (8-bit) LDB, halfwords (16-bit) LDH, or words (32-bit) LDW. Here, the data are assumed to be halfwords. The loading–storing is done by the .D1 unit, since .D units are the only units capable of interacting with data memory.

Note that the pointers A5 and A6 need to be postincremented (C notation), so that they point to the next values for the next iteration of the loop. When registers are used as pointers, there are several ways to perform pointer arithmetic. These include pre- and postincrement/decrement options by some displacement amount, where the pointer is modified before or after it is used (e.g., *++A1[disp] and *A1++[disp]). In addition, a pre-offset option can be executed with no modification of the pointer (e.g., *+A1[disp]). Displacement within brackets specifies the number of data elements (depending on the datatype), whereas displacement in parentheses specifies the number of bytes. These pointer offset options are listed in Figure 5-6 together with some examples.

Syntax	Description	Pointer Modified
*R	Pointer	No
*+R[disp]	+Pre-offset	No
*-R[disp]	−Pre-offset	No
*++R[disp]	Pre–increment	Yes
*--R[disp]	Pre–decrement	Yes
*R++[disp]	Post–increment	Yes
*R--[disp]	Post–decrement	Yes

[disp] specifies # elements–size in W, H, or B
(disp) specifies # bytes

(a)

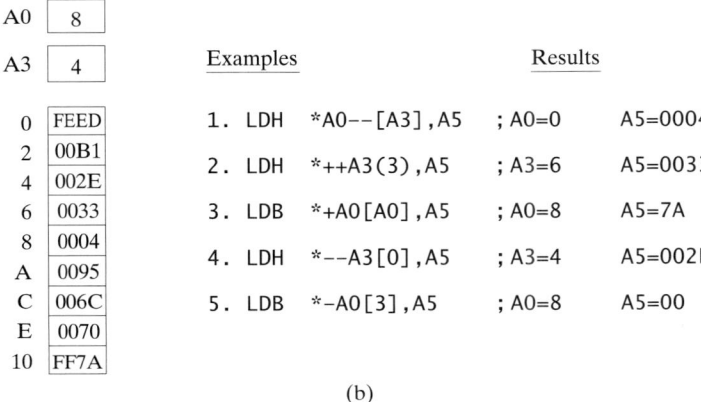

(b)

FIGURE 5-6 (a) Pointer offsets, (b) pointer examples (*note*: Instructions are independent, not sequential).[†]

Finally, the instructions MPY and ADD within the loop perform the dot-product operation. The instruction MPY is executed by the .M1 unit and ADD by the .L1 unit. It should be mentioned that the above code as is will not run properly on the C6x because of its pipelined CPU, which is discussed next.

5.2 PIPELINED CPU

In general, it takes several steps to perform an instruction. Basically, these steps are fetching, decoding, and execution. If they are completed serially, not all of the resources on the processor, such as multiple buses or functional units, are fully utilized. In order to increase throughput, DSP CPUs are designed to be *pipelined*. This means that the foregoing steps are carried out simultaneously. Figure 5-7 illustrates the difference in processing time for three instructions executed on a serial or nonpipelined and a pipelined

88 Chapter 5 TMS320C6000 Architecture

	Clock Cycles								
CPU Type	1	2	3	4	5	6	7	8	9
Non–Pipelined	F_1	D_1	E_1	F_2	D_2	E_2	F_3	D_3	E_3
Pipelined	F_1	D_1 F_2	E_1 D_2 F_3	E_2 D_3	E_3				

F_x = fetching of instruction x
D_x = decoding of instruction x
E_x = execution of instruction x

FIGURE 5-7 Pipelined vs. nonpipelined CPU.[†]

CPU. As can be seen, a pipelined CPU requires fewer clock cycles to complete the same number of instructions.

On the C6x processor, fetching consists of four phases, each requiring a clock cycle. These include generate fetch address (denoted by F1), send address to memory F2, wait for data F3, and read opcode from memory F4. Decoding consists of two phases, each requiring a clock cycle. These are dispatching to appropriate functional units (denoted by D1), and decoding D2. Due to the delays associated with the instructions multiply (MPY–1 delay), load (LDx–4 delays), and branch (B–5 delays), the execution step may consist of up to six phases, E1 through E6, accommodating a maximum of five delays. Hence, as shown in Figure 5-8, the F step consists of four, the D step of two, and the E step of six substeps, or phases.

When the outcome of an instruction is used by the next instruction, an appropriate number of NOPs (no operation or delay) must be added after multiply (one NOP), load (four NOPs/or NOP 4), and branch (five NOPs/or NOP 5) instructions in order to allow the pipeline to operate properly. Therefore, for the above example to run on the C6x, NOPs, as shown in Figure 5-9, should be added after the instructions MPY, LDH, and B.

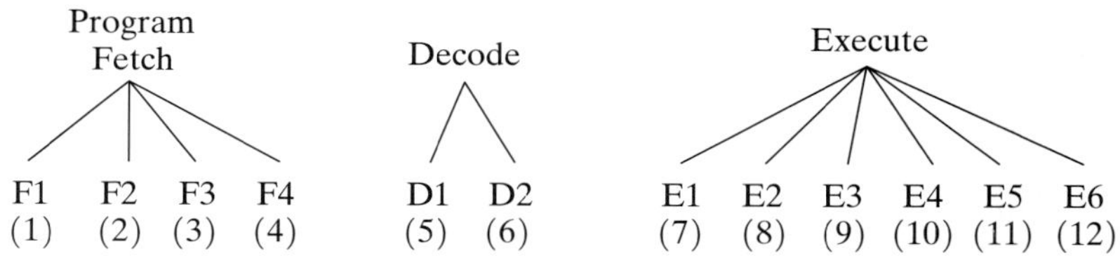

```
              Program
              Fetch                  Decode              Execute

         F1  F2  F3  F4           D1    D2         E1   E2   E3   E4   E5   E6
         (1) (2) (3) (4)          (5)   (6)        (7)  (8)  (9)  (10) (11) (12)
```

FIGURE 5-8 Stages of the pipeline.

FIGURE 5-9 Pipelined code with NOPs inserted.

```
              MVK     .S1     40, A2
       loop:  LDH     .S1     *A5++, A0
              LDH     .S1     *A6++, A1
              NOP             4
              MPY     .M1     A0, A1, A3
              NOP
              ADD     .L1     A3, A4, A4
              SUB     .L1     A2, 1, A2
       [A2]   B       .S1     loop
              NOP             5
              STH     .D1     A4, *A7
```

Figure 5-10 illustrates an example of a pipeline situation that requires adding an NOP. The plus signs indicate the number of substeps or latencies required for the instruction to be completed. In this example, it is assumed that the addition operation is done before one of its operands is made available from the previous multiply operation, hence the need for adding an NOP after the MPY. Later on, it will be seen that, as part of code optimization, NOPs can be reduced or removed leading to an improvement in efficiency.

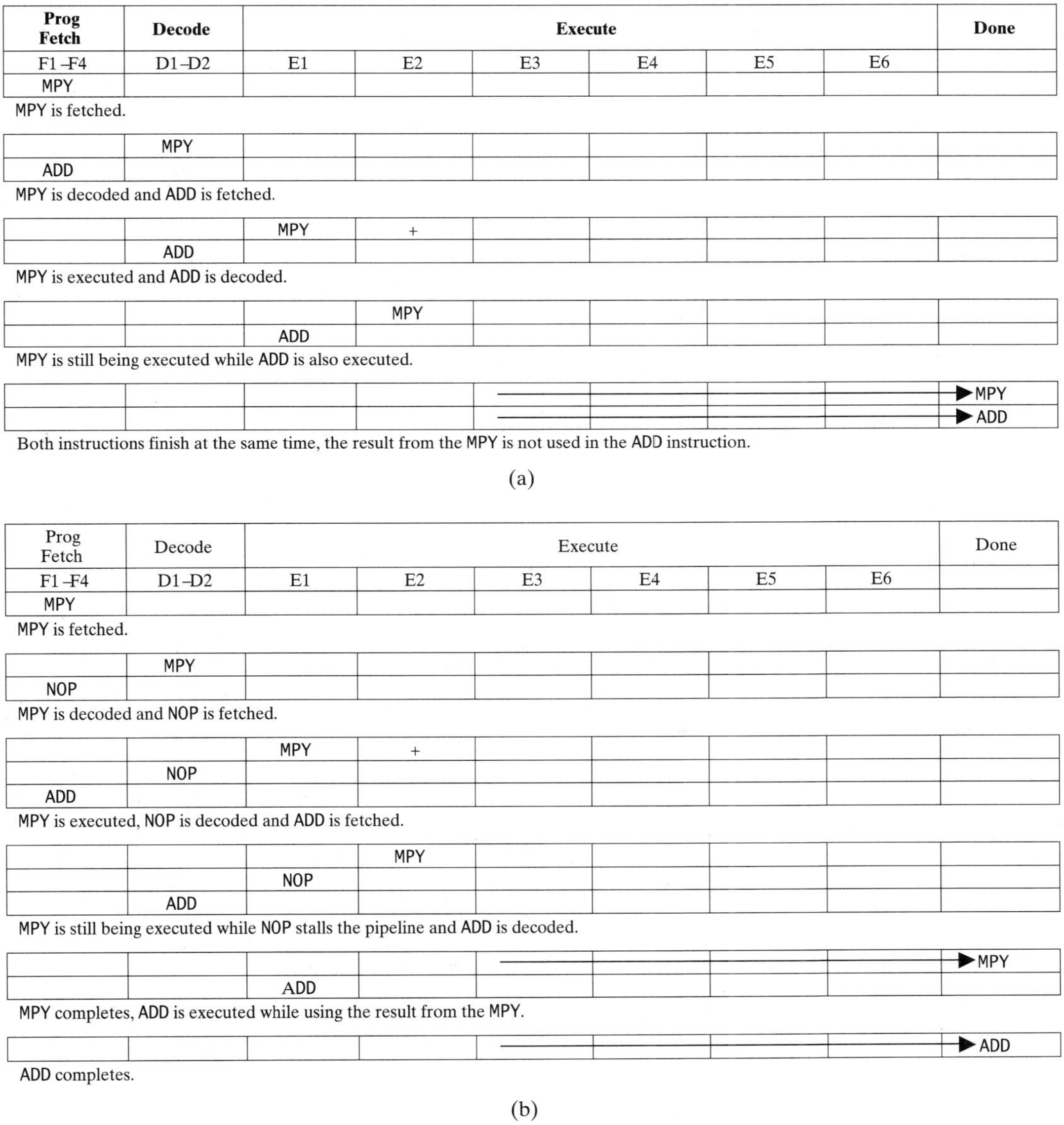

Prog Fetch	Decode	Execute						Done
F1–F4	D1–D2	E1	E2	E3	E4	E5	E6	
MPY								

MPY is fetched.

	MPY							
ADD								

MPY is decoded and ADD is fetched.

		MPY	+					
	ADD							

MPY is executed and ADD is decoded.

			MPY					
		ADD						

MPY is still being executed while ADD is also executed.

								▶ MPY
								▶ ADD

Both instructions finish at the same time, the result from the MPY is not used in the ADD instruction.

(a)

Prog Fetch	Decode	Execute						Done
F1–F4	D1–D2	E1	E2	E3	E4	E5	E6	
MPY								

MPY is fetched.

	MPY							
NOP								

MPY is decoded and NOP is fetched.

		MPY	+					
	NOP							
ADD								

MPY is executed, NOP is decoded and ADD is fetched.

			MPY					
		NOP						
	ADD							

MPY is still being executed while NOP stalls the pipeline and ADD is decoded.

								▶ MPY
		ADD						

MPY completes, ADD is executed while using the result from the MPY.

								▶ ADD

ADD completes.

(b)

FIGURE 5-10 (a) Multiply then add and (b) need for NOP insertion.

5.3 VELOCITI

The C6x architecture is based on the Very Large Instruction Word (VLIW) architecture. In such an architecture, several instructions are captured and processed simultaneously. This is referred to as a Fetch Packet (FP). (See Figure 5-11.)

The C6x uses VLIW, allowing eight instructions to be captured simultaneously from on-chip memory onto its 256-bit wide program data bus. The original VLIW architecture has been modified on the C6x to allow several so-called Execute Packets (EP) to be included within the same Fetch Packet, as shown in Figure 5-12. An EP constitutes a group of parallel instructions. Parallel instructions are indicated by double pipe symbols (||), and, as the name implies, they are executed together, or in parallel. Instructions within an EP move together through every stage of the pipeline. This VLIW modification, introduced by TI, is called VelociTI. Compared with VLIW, VelociTI reduces code size and increases performance when instructions reside off-chip.

5.4 C64X DSP

The C64x is a newly released DSP core, as part of the C6x family, with higher MIPS power operating at higher clock rates. The initial release of this core operates in the range of 600–800 MHz clock rates, giving a processing power of 4800–6400 MIPS. The

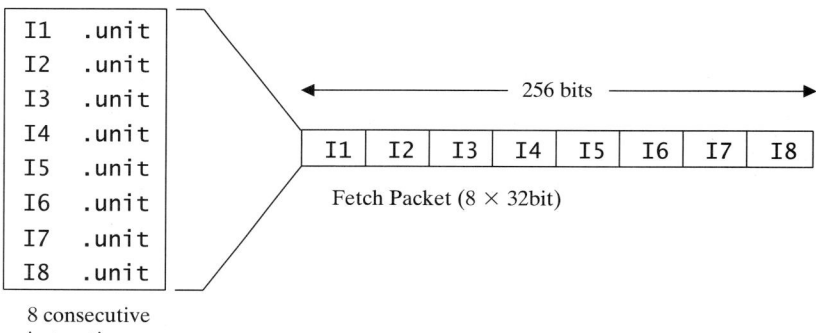

FIGURE 5-11 C6x fetch packet: C6x fetches eight 32-bit instructions every cycle.

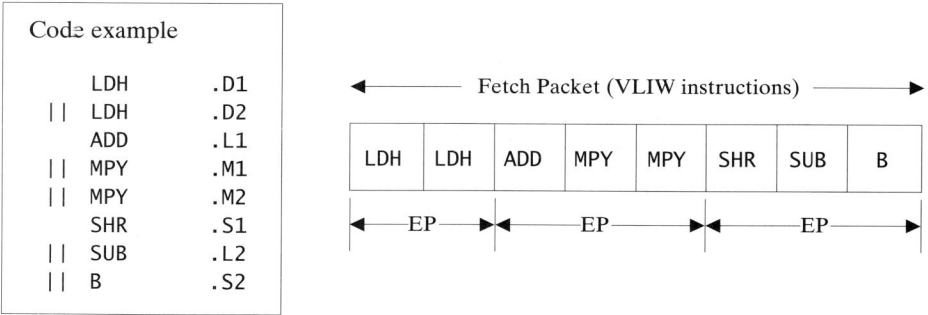

FIGURE 5-12 A fetch packet containing three execute packets.

5.4 C64x DSP

clock rate is expected to increase to 1.1 GHz and higher, leading to a processing rate of 8800+MIPS. The TI website http://www.ti.com/ provides the C64x speedups obtained over the C62x for various wireless communication and digital imaging algorithms. Such speedups are achieved due to many enhancements, some of which are mentioned here.

Per CPU data path, the number of registers is increased from 16 to 32, A0–A31 and B0–B31. These registers support packed datatypes, allowing storage and manipulation of four 8-bit or two 16-bit values within a single 32-bit register.

Although the C64x is code compatible with the C62x, (i.e., all the C62x instructions run on the C64x), the C64x can run additional instructions on packed datatypes, boosting parallelism. For example, the new instruction MPYU4 performs four, or quad, 8-bit multiplications, or the instruction MPY2 performs two, or dual, 16-bit multiplications in a single instruction cycle on an .M unit. This packed-data processing capability is illustrated in Figure 5-13. Table 5-2 provides a listing of the C64x packed-data instructions.

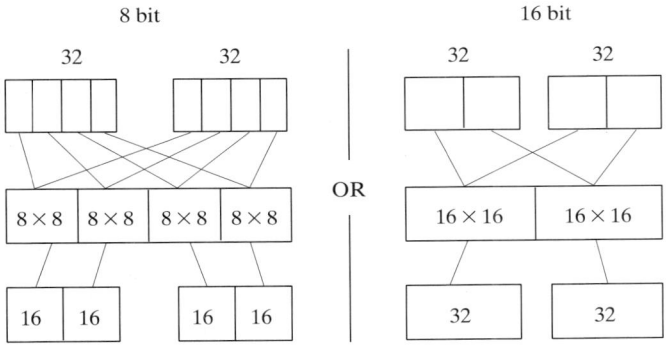

FIGURE 5-13 C64x packed-data processing capability.†

TABLE 5-2 A listing of C64x packed-data instructions.†

Operations	Quad 8-bit	Quad 16-bit
Multiply	X	X
Multiply with Saturation		X
Addition/Subtraction	X	X
Addition with Saturation	X	X
Absolute Value		X
Subtract with Absolute Value	X	
Compare	X	X
Shift		X
Data pack/unpack	X	X
Data pack with Saturation	X	X
Dot Product with Optional Negate	X	X
Min/Max/Average	X	X
Bit expansion (Mask generation)	X	X

TABLE 5-3 C64x special-purpose instructions.†

Instruction	Description	Example Application
BITC4	Bit count	Machine vision
GMPY4	Galois Field MPY	Reed–Solomon support
SHFL	Bit interleaving	Convolution encoder
DEAL	Bit de-interleaving	Cable modem
SWAP4	Byte swap	Mixed Multiprocessor support
XPNDx	Bit expansion	Graphics
MPYHIx, MPYLIx	Extended precision 16x32 MPYs	Audio
AVGx	Quad 8-bit Dual 16-bit average	Motion compensation
SUBABS4	Quad 8-bit Absolute of differences	Motion estimation
SSHVL, SSHVR	Signed variable shift	GSM

Additional hardware has been added to each functional unit on the C64x so that it may perform 10 special-purpose instructions in order to accelerate key functions encountered in wireless and digital imaging applications. For example, the instruction GMPY4 allows four 8-bit Galois-field multiplications in a single instruction as part of Reed–Solomon decoding. Table 5-3 provides a list of these special-purpose instructions.

In addition, the efficiency of each functional unit on the C64x has been improved, leading to a greater orthogonality, or generality, of operations. For example, the .D unit can perform 32-bit logical operation just as the .S and .L units, or the .M unit can perform shift and rotate operations just as the .S unit. The C64x .S unit is capable of performing additional branching instructions, such as branch positive BPOS. Furthermore, the C64x allows multiple units on one side to read the same crosspath source from the other side.

The C64x supports 64-bit loads and stores with a single instruction. There are four 32-bit paths for loading data to the registers. LD1a and LD1b are the load paths for 32 LSBs (least significant bits) on side A and B, respectively, and LD2a and LD2b for 32 MSBs (most significant bits). Similarly, ST1a, ST1b, ST2a, ST2b are the 32-bit store paths for storing data from memory. The C64x also allows nonaligned loads and stores, meaning that loading and storing of words and double words can be done on any byte boundary by using nonaligned load and store instructions. Figure 5-14 provides a comparison between the data paths of the C62x and C64x CPUs.

Finally, similar to the C6211, the C64x contains a two-level cache, allowing it to make better use of the CPU speed when interacting with off-chip memory that has a lower speed.

5.4 C64x DSP 93

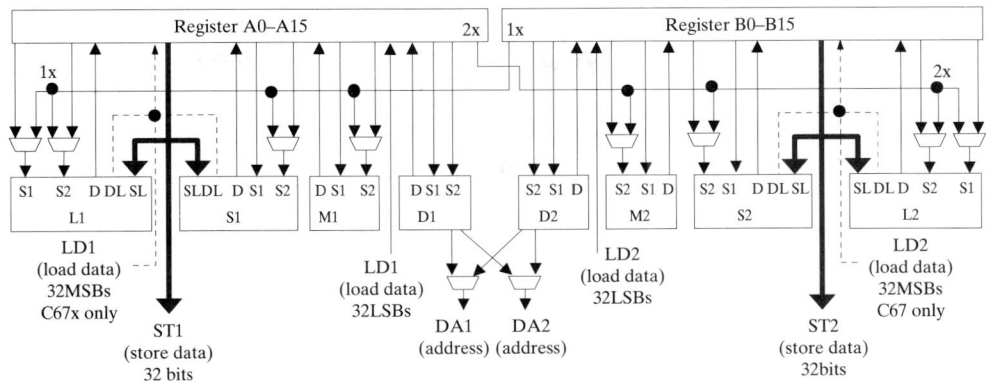

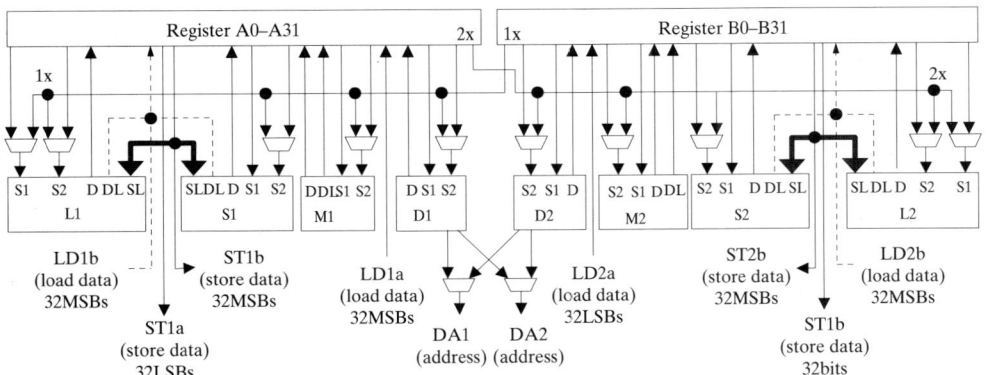

FIGURE 5-14 Data paths of C62x and C64x (S1 = source1, S2 = source2, D = destination, SL = source long, DL = destination long).†

CHAPTER 6

Software Tools

Programming most DSP processors can be done either in C or assembly. Although writing programs in C would require less effort, the efficiency achieved is normally less than that of programs written in assembly. Efficiency means having as few instructions or as few instruction cycles as possible by making maximum use of the resources on the chip.

In practice, one starts with C coding to analyze the behavior and functionality of an algorithm. Then, if the required processing rate is not met by using the C compiler optimizer, the time-consuming portions of the C code are identified and converted into assembly, or the entire code is rewritten in assembly. In addition to C and assembly, the C6x allows writing code in linear assembly. Figure 6-1 illustrates the code efficiency versus coding effort for three types of source files on the C6x: C, linear assembly, and hand-optimized assembly. As can be seen, linear assembly provides a good compromise between code efficiency and coding effort.

Figure 6-2 shows the steps involved for going from a source file (.c extension for C, .asm for assembly, and .sa for linear assembly) to an executable file (.out extension). Figure 6-3 lists the .c and .sa versions of the dot-product example to see what they look like. The assembler is used to convert an assembly file into an object file (.obj extension). The assembly optimizer and compiler are used to convert, respectively, a linear assembly file and a C file into an object file. The linker is used to combine object files, as instructed by the command file (.cmd extension), into an executable file. All the assembling, linking, compiling, and debugging steps have been incorporated into an integrated development environment (IDE) called Code Composer Studio (CCS). CCS provides an easy-to-use graphical user environment for building and debugging both C and assembly codes on various target DSPs.

FIGURE 6-1   Code efficiency vs. coding effort.[†]

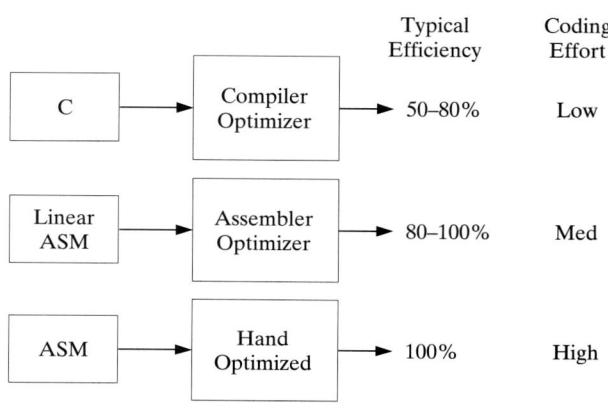

Introduction

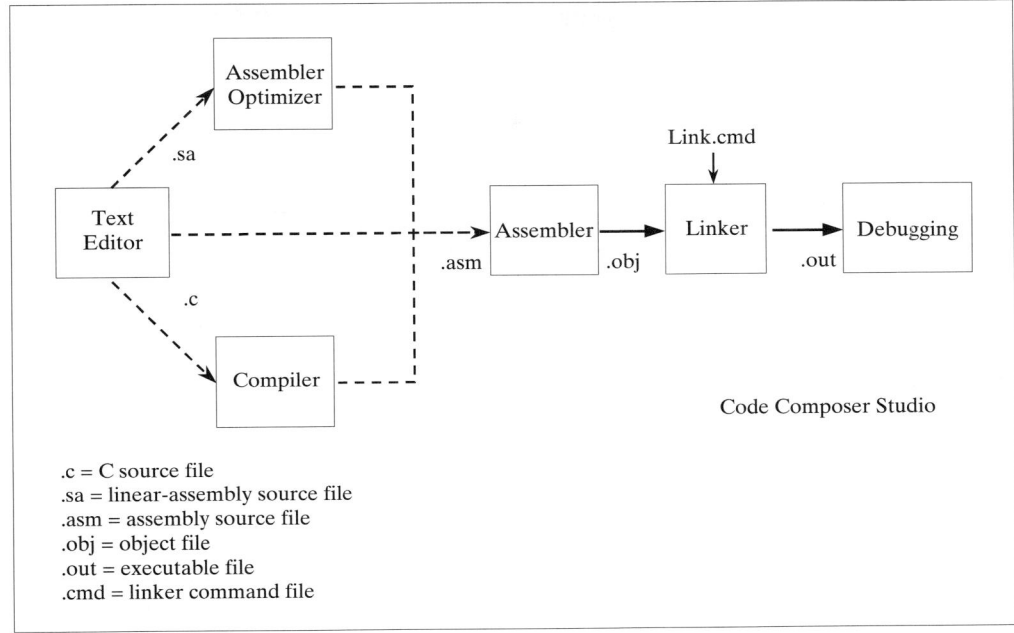

FIGURE 6-2 C6x software tools.

FIGURE 6-3 (a) .c and (b) .sa version of dot-product example.

```
void main()
{
        y=DotP( (int *) a, (int *) x, 40);
}
int DotP(int *m, int *n, short count)
{
        int sum, i;
        sum=0;

        for(i=0;i<count;i++)
                sum+=m[i] * n[i];

        return(sum);
}
```

(a)

```
            .title  "dot product"
            .def    dotp
            .sect   "code"
dotp:   .proc   A4,B4,A6,B6,A8,B3
        .reg    a, ai, b,bi,r,prod,sum,c,ci,i;
        MV      A4,c
        MV      B4,b
        MV      A6,a
        MV      B6,r
        MV      A8,i

loop:   .trip   40
        LDH     *a++, ai
        LDH     *b++,bi
        MPY     ai,bi,prod
        SHR     prod,15,sum
        ADD     ai,sum,ci
        STH     ci, *c++
[i]     SUB     i,1,i
[i]     B       loop
        .endproc B3
```

(b)

96 Chapter 6 Software Tools

6.1 EVM/DSK TARGET C6X BOARD

Upon the availability of either an EVM or a DSK board, an executable file can be run on an actual C6x processor. In the absence of such boards, CCS can be configured to simulate the execution process. As shown in Figure 6-4(a), the C6x EVM board is a complete DSP system, including a C6201 (or C6701) chip, some memory, A/D capabilities, and PC host interfacing components. The functional diagram of the EVM board appears in Figure 6-4(b). The board has a 16-bit codec that performs the analog-to-digital signal conversion with sampling frequencies from 5.5 kHz to 48 kHz.

The memory residing on the EVM board consists of two $1M \times 32$ SDRAM (synchronous dynamic RAM), running at 100MHz, and one $64K \times 32$ SBSRAM (synchronous burst static RAM), running at 133MHz. The latter is a faster, but more expensive, memory compared to SDRAM. A voltage regulator on the board is used to provide 1.8V or 2.5V for the C6x core and 3.3V for its memory and peripherals, and 5V for audio components. As shown in Figure 6-5, the DSK board includes a C6211 (or C6711) DSP chip operating at 150 MHz, 4 Mbytes of SDRAM and 128 Kbytes of flash memory.

6.2 ASSEMBLY FILE

Similar to other assembly languages, the C6x assembly consists of four fields: label, instruction, operands, and comment. (See Figure 5-4.) The first field is the label field. Labels must start in the first column and must begin with a letter. A label, if present, indicates an assigned name to a specific memory location that contains an instruction or

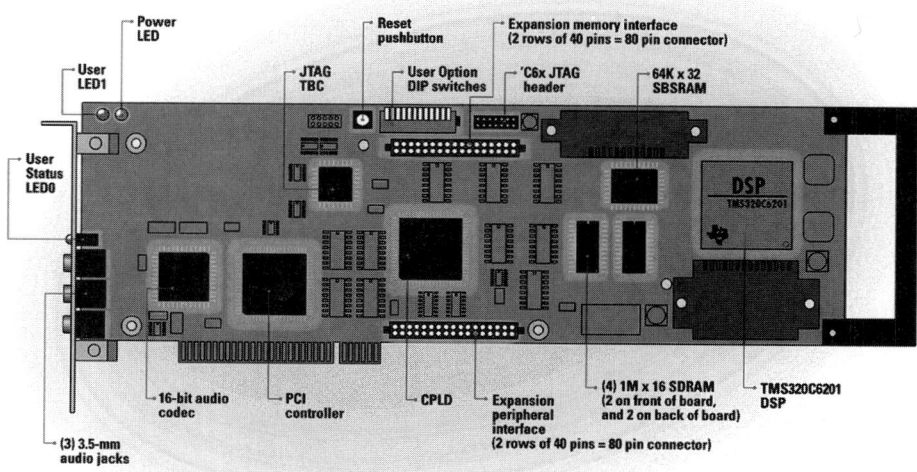

(a)

FIGURE 6-4 (a) An EVM board and (b) its functional diagram.[†]

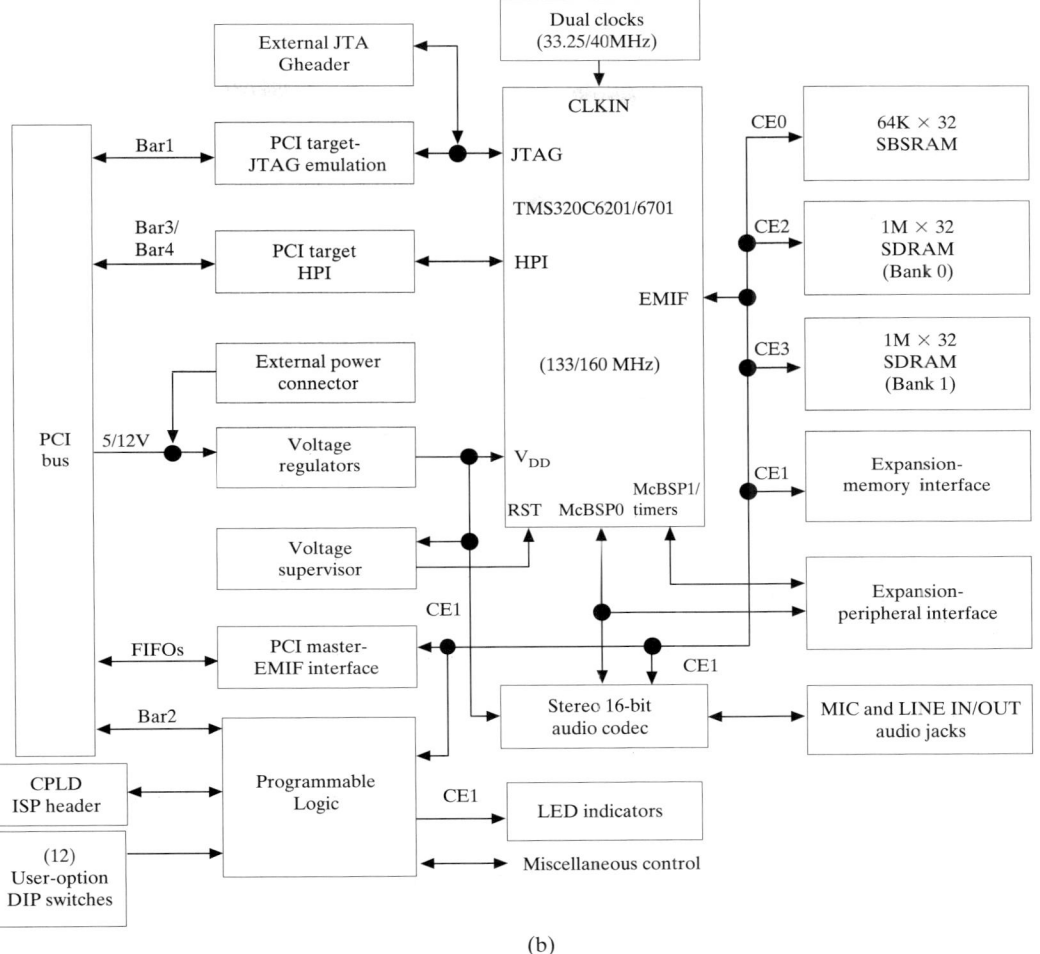

FIGURE 6-4 *(continued)*

data. Either a mnemonic or a directive constitutes the instruction field. It is optional for the instruction field to include the functional unit which performs that particular instruction. However, to make codes more understandable, the assignment of functional units is recommended. If a functional unit is specified, the data path must be indexed by 1 for the A side and 2 for the B side. A parallel instruction is indicated by a double-pipe symbol (||) and a conditional instruction by a register appearing in brackets in the instruction field. As the name operand implies, the operand field contains arguments of an instruction. Instructions require two or three operands. Except for store instructions, the destination operand must be a register. One of the source operands must be a register, the other either a register or a constant. After the operand field, there is an optional comment field that, if stated, should begin with a semicolon (;).

98 Chapter 6 Software Tools

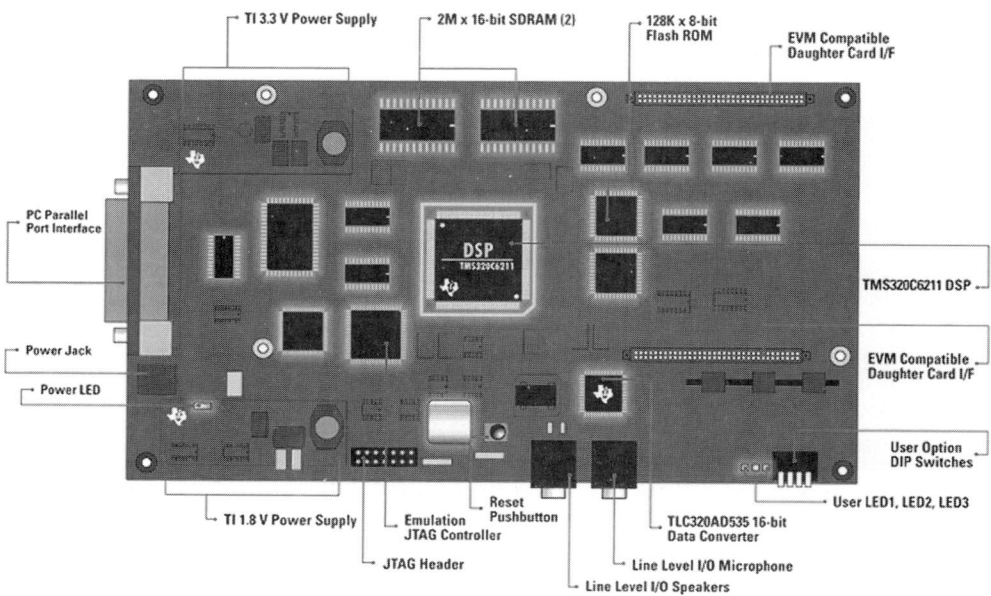

FIGURE 6-5 DSK board.[†]

6.2.1 Directives

Directives are used to indicate assembly code sections and to declare data structures. It should be noted that assembly statements appearing as directives do not produce any executable code. They merely control the assembling process by the assembler. Some of the widely used assembler directives are as follows:

.sect "name" directive, which defines a section of code or data named "name".
.int, .long, or .word directive which reserves 32 bits of memory initialized to a value.
.short or .half directive which reserves 16 bits of memory initialized to a value.
.byte directive which reserves 8 bits of memory initialized to a value.

Note that in the TI Common Object File Format (COFF), the directives .text, .data, and .bss are used to indicate code, initialized constant data, and uninitialized variables, respectively. Other directives often used include .set directive, for assigning a value to a symbol, .global or .def directive, to declare a symbol or module as global so that it may be recognized externally by other modules, and .end directive, to signal the termination of assembly code. The directive .global acts as a .def directive for defined symbols and as a .ref directive for undefined symbols.

FIGURE 6-6 Common compiler sections.

Section Name	Description
.text	Code
.switch	Tables for switch instructions
.const	Global and static string literals
.cinit	Initial values for global/static vars
.bss	Global and static variables
.far	Global and statics declared far
.stack	Stack (local variables)
.sysmem	Memory for malloc fcns (heap)
.cio	Buffers for stdio functions

At this point, it should be mentioned that the C compiler creates various sections indicated by the directives .text, .switch, .const, .cinit, .bss, .far, .stack, .sysmem, and .cio. Figure 6-6 lists some common compiler sections. For a complete listing of directives, refer to the *TI TMS320C6x Assembly Language Tools User's Guide* [11].

6.3 MEMORY MANAGEMENT

The external memory used by a DSP processor can be either static or dynamic. Static memory (SRAM) is faster than dynamic memory (DRAM), but it is more expensive, since it takes more space on silicon. DRAMs also need to be refreshed periodically. A good compromise between cost and performance is achieved by using SDRAM (Synchronous DRAM). Synchronous memory requires clocking, as compared to asynchronous memory, which does not.

Given that the address bus is 32 bits wide, the total memory space consists of $2^{32} = 4$ Gbytes. This space is divided, according to a memory map, into the internal program memory (PMEM), internal data memory (DMEM), internal peripherals, and four external memory spaces named CE0, CE1, CE2, and CE3. There are two memory map configurations: memory map 0 and memory map 1. Figures 6-7(a) and 6-7(b) illustrate these two memory maps. For the lab exercises in this book, the EVM board is configured on the basis of memory map 1 shown in Figure 6-7(c), and the DSK board is configured on the basis of memory map 1 shown in Figure 6-7(d).

As indicated in Figure 6-7, there are three 16M external memory ranges (CE0, CE2, CE3) and one 4M external-memory range (CE1). These external memory ranges support synchronous (SBSRAM, SDRAM) or asynchronous (SRAM, ROM, etc.) memory, accessible as bytes (8 bits), halfwords (16 bits), or words (32 bits). There exists 128 Kbytes of internal RAM, which is equally divided between program and data. The data portion

100 Chapter 6 Software Tools

Address	Memory Map 0	Block Size (Bytes)
0000 0000	External Memory Space CE0	16M
0100 0000	External Memory Space CE1	4M
0140 0000	Internal Program RAM	64K
0141 0000	Reserved	4M
0180 0000	Internal Peripheral Space	4M
01C0 0000	Reserved	4M
0200 0000	External Memory Space CE2	16M
0300 0000	External Memory Space CE3	16M
0400 0000	Reserved	1984M
8000 0000	Internal Data RAM	64K
8001 0000	Reserved	4M
8040 0000	Reserved	2044M
1 0000 000		

(a)

Address	Memory Map 0	Block Size (Bytes)
0000 0000	Internal Program RAM	64K
0001 0000	Reserved	4M
0040 0000	External Memory Space CE0	16M
0140 0000	External Memory Space CE1	4M
0180 0000	Same as Memory Map 0	
1 0000 000		

(b)

FIGURE 6-7 (a) C6x memory map 0, (b) map 1, (c) EVM map 1, and (d) DSK map 1.†

consists of 64 Kbytes, accessible as bytes, halfwords, and words. The program portion consists of either 2K fetch packets or 16K 32-bit instructions. The on-chip peripherals and control registers are memory mapped into the internal peripheral space. A listing of the memory-mapped registers is provided in Appendix A (Quick Reference Guide).

The internal data memory is organized into memory banks so that two loads (or stores) can be performed simultaneously. As long as data are accessed from different banks, no conflict occurs. However, if data are accessed from the same bank in one instruction, a memory conflict occurs and the CPU is stalled by one cycle.

If a program fits into the on-chip or internal program memory, it should be run from there to avoid delays associated with accessing off-chip or external memory. If a program is too big to fit into the internal memory, most of its time-consuming portions should be placed into the internal memory for efficient execution. For repetitive codes, it is recommended that the internal program memory be configured as cache memory. This al-

6.3 Memory Management

Address	Memory Map 1	Block Size (Bytes)	
0000 0000	Internal Program RAM	64K	4M
0001 0000	Reserved		
0040 0000	SBSRAM	256K	
0044 0000	Unused	Unused	16M
0140 0000	Asynchronous Expansion Memory	3M	
0170 0000	PCI add-on registers	64	64K
0170 0040	Unavailable		
0171 0000	PCI FIFO	4	64K
0171 0004	Unavailable		
0172 0000	Audio Codec Registers	16	64K
0172 0010	Unavailable		
0173 0000	Reserved	320K	
0178 0000	DSP control/status registers	32	64K
0178 0020	Unavailable		
0179 0000	Reserved	448K	
0180 0000	Internal Peripheral Space	4M	
01C0 0000	Reserved	4M	
0200 0000	SDRAM(Bank 0)	4M	
0240 0000	Reserved	12M	
0300 0000	SDRAM(Bank 1)	4M	
0340 0000	Reserved	12M	
0400 0000	Reserved	1984M	
8000 0000	Internal Data RAM	64K	4M
8001 0000	Reserved		
8040 0000	Reserved	2044M	
1000 0000			

(c)

Address	Memory Map 1	Block Size (Bytes)
0000 0000	Internal RAM (L2)	64K
0001 0000	Reserved	24M
0180 0000	EMIF control regs	32
0184 0000	Cache Configuration regs	4
0184 4000	L2 base addr & count regs	32
0184 4020	L1 base addr & count regs	32
0184 5000	L2 flush & clean regs	32
0184 8200	CE0 mem attribute regs	16
0184 8240	CE1 mem attribute regs	16
0184 8280	CE2 mem attribute regs	16
0184 82c0	CE3 mem attribute regs	16
0188 0000	HPI control regs	4
018C 0000	McBSP0 regs	40
0190 0000	McBSP1 regs	40
0194 0000	Timer0 regs	12
0198 0000	Timer1 regs	12
019C 0000	Interrupt selector regs	12
01A0 0000	EDMA parameter RAM	2M
01A0 FFE0	EDMA control regs	32
0200 0000	QDMA regs	20
0200 0020	QDMA psuedo-regs	20
3000 0000	McBSP0 data	64M
3400 0000	McBSP1 data	64M
8000 0000	CE0, SDRAM	16M
9000 0000	CE1 8-bit ROM	128K
9008 0000	CE1, 8-bit I/O port	4
A000 0000	CE2-Daughtercard	256M
B000 0000	CE3-Daughtercard	256M

(d)

FIGURE 6-7 *(continued)*

lows accessing the external memory as seldom as possible and hence avoiding delays associated with such accesses.

6.3.1 Linking

Linking places code, constant, and variable sections into appropriate locations in memory, as specified in the linker .cmd command file. Also, it combines several object files .obj into the final executable .out file. A typical command file corresponding to memory map 1 is shown in Figure 6-8.

The first part, MEMORY, provides a description of the type of physical memory, its origin, and its length. The second part, SECTIONS, specifies the assignment of various code sections to the available physical memory.

```
/* Memory Map 1 */
MEMORY
{
    PMEM        : origin = 0x00000000,    length = 0x00010000
    EXT2        : origin = 0x02000000,    length = 0x01000000
    EXT3        : origin = 0x03000000,    length = 0x01000000
    DMEM        : origin = 0x80000000,    length = 0x00010000

}

SECTIONS
{
        .vectors  > PMEM              ; PMEM = Program memory, DMEM = Data memory
        .text     > PMEM              ; EXT = External memory
        .bss      > DMEM
        .cinit    > DMEM
        .const    > DMEM
        .stack    > DMEM
        .cio      > DMEM
        .sysmem   > DMEM
        .far      > EXT2
        .mydata   > EXT3
}
```

FIGURE 6-8   A typical command file.

6.4 COMPILER UTILITY

The build feature of CCS can be used to perform the entire process of compiling, assembling, and linking in one step by activating the utility cl6x and stating the right options for it. The following command shows how this utility is used within CCS to build the source files *file1.c*, *file2.asm*, and *file3.sa*:

cl6x -gs *file1.c file2.asm file3.sa* -z -o *file.out* -m *file.map* -l *rts6201.lib*

The option -g adds debugger specific information to the object file for debugging purposes. The option -s provides an interlisting of C and assembly. For *file1.c*, the C compiler, for *file2.asm*, the assembler, and for *file3.sa*, the assembler optimizer (linear assembler) are invoked. The option -z invokes the linker, placing the executable code in *file.out* if the -o option is used. Otherwise, the default file *a.out* is created. The option -m provides a map file (*file.map*), which includes a listing of all the addresses of sections, symbols and labels. The option -l specifies the run-time support library *rts6201.lib* for linking files on the C6201 processor. Table 6-1 lists some frequently used options. Refer to the *TI Optimizing C Compiler* manual [17] for a complete list of available options.

The compiler allows four levels of optimizations to be invoked by using -o0, -o1, -o2, and -o3. Debugging and full-scale optimization cannot be performed together, since they are in conflict; that is, in debugging, information is added to enhance the debugging process, while in optimizing, information is minimized or removed to enhance code ef-

TABLE 6-1 Common compile options.[†]

Options	Description	Tool
-mv6701	Generate 'C67x code ('C62x is default)	Comp/Asm
-g	Enables src-level symbolic debugging	Comp/Asm
-mg	Enables minimum debug to allow profiling	Compiler
-s	Interlist C statements into assembly listing	Compiler
-o	Invoke optimizer (-o0, -o1, -o2/-o, -o3)	Compiler
-pm	Combine all C source files before compile	Compiler
-mt	No aliasing used	Compiler
-ms	Minimize code size (-ms0/-ms, -ms1, -ms2)	Compiler
-z	Invokes linker	Linker
-o	Output file name	Linker
-m	Map file name	Linker
-c	Auto-Init C variables (-cs turns off autoinit)	Linker
-l	Link-in libraries (small -L)	Linker

1. Compile without optimization.
   (Get the code functioning!)
   cl6x -g -s *file.c* -z

2. Compile with some optimization.
   (Verify code functionality, again)
   cl6x -g -o *file.c* -z

3. Compile with all optimizations.
   (Generate efficient code)
   cl6x -o3 -pm *file.c* -z

FIGURE 6-9 Programming approach.

ficiency. In essence, the optimizer changes the flow of C code, making program debugging very difficult.

As shown in Figure 6-9, a good programming approach would be first to verify that the code is properly functioning by using the compiler with no optimization (-gs option). Then, one should use full optimization to generate an efficient code (-o3 option). It is recommended that an intermediary step be taken in which some optimization is done without interfering with source level debugging (-go option). This intermediary step can reverify code functionality before performing full optimization. It should be pointed out that full optimization may change memory locations outside the scope of the C code. Such memory locations must be declared as "volatile" to prevent compiling errors.

As a step to further optimize C codes, it is recommended that intrinsics be used whenever possible. *Intrinsics* are functions similar to mathematics functions as part of the run-time support library. Intrinsics allow the C compiler to directly access the hardware

104 Chapter 6 Software Tools

```
short DotP(int *m, int *n, short count)
{
        short i, productl, producth, suml = 0, sumh = 0;

        for(i=0;i<count;i++)
        {
                productl = _mpy(m[i],n[i]);// _mpy intrinsic
                producth = _mpyh(m[i],n[i]);// _mpyh intrinsic
                suml += productl;
                sumh += producth;
        }
        suml += sumh;
        return(sum);
}
```

FIGURE 6-10 Intrinsic version of dot-product C code.

while preserving the C environment. As an example, instead of using the multiply operator * in C, the intrinsic _mpy() can be used to tell the compiler to use the C6x instruction MPY. Figure 6-10 shows the intrinsic version of the dot-product C code. A list of the C6x intrinsics is provided in Appendix A (Quick Reference Quide).

6.5 CODE INITIALIZATION

All programs start by going through a reset initialization code. Figure 6-11 illustrates both the C and assembly version of a typical reset initialization code. This initialization is for the purpose of starting at a previously defined initial location. Upon power-up, the system always goes to the reset location in memory, which normally includes a branch instruction to the beginning of the code to be executed. The reset code shown in Figure 6-11 takes the program counter to a globally defined location in memory named *init* or *_c_int00*.

```
              "ASM"                                    "C"

vectors.asm                              cvectors.asm

        .ref   init                              .global _c_int00
        .sect  "vectors"                         .sect   "vectors"
rst     MVK    .s2    init,B0          rst      B       _c_int00
        MVKH   .s2    init,B0                   NOP     ;additional NOP's
        B      .s2    B0                        NOP     ;to create a
        NOP                                     NOP     ;fetch packet
        NOP                                     NOP
        NOP                                     NOP
        NOP                                     NOP
        NOP                                     NOP
```

FIGURE 6-11 Reset code.

6.5 Code Initialization

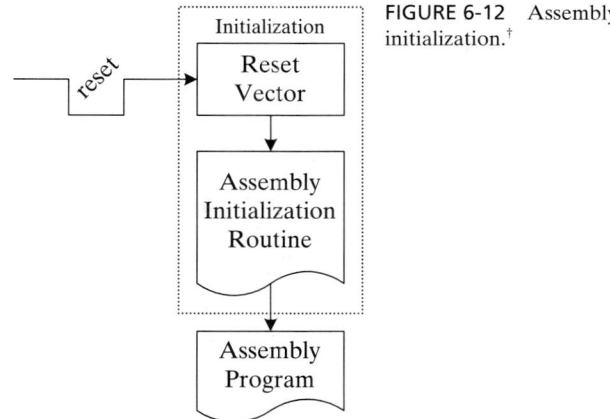

FIGURE 6-12 Assembly initialization.†

As indicated in Figure 6-12, when writing in assembly, an initialization code is needed to create initialized data and variables, and to copy initialized data into corresponding variables. Initialized values are specified by using .byte, .short, or .int directives. Uninitialized variables are specified by using .usect directive. The first, second, and third arguments of this directive denote section name, size in bytes, and data alignment in bytes, respectively. Before calling the main function or subroutine, another initialization code portion is usually needed to set up registers and pointers, and to move data to appropriate places in memory.

Figure 6-13 provides the initialization code for the dot-product example in which initialized data values appear for three initialized data arrays labeled table_a, table_x, and table_y. In addition, three variable sections called a, x, and y are declared. The second part of the initialization code copies the initialized data into the corresponding variables. The setup code for calling the dot-product routine is also shown in this figure.

As far as C coding is concerned, the C compiler uses *boot.c* in the run-time support library to perform the initialization before calling *main()*. The -c option activates *boot.c* to autoinitialize variables. This is illustrated in Figure 6-14.

6.5.1 Data Alignment

The C6x allows byte, half word, word addressing. Consider a word-format representation of memory as shown in Figure 6-15. There are four byte boundaries, two halfword (or short) boundaries, and one word boundary per word. The C6x always accesses data on these boundaries depending on the datatype specified; that is, it always accesses aligned data. When specifying an uninitialized variable section .usect, it is required to specify the alignment as well as the total number of bytes. The examples appearing in Figure 6-16 show the data alignment for both constants and variables.

```
        .def    init
        .ref    dotp

;Data initialization
;Initialize tables

        .sect   "init_tables"

table_a .short 40,39,38,37,36,35,34,33,32,31,30,29,28,27 ;Initialize table_a array with values
        .short 26,25,24,23,22,21,20,19,18,17
        .short 16,15,14,13,12,14,10,9,8,7,6,5,4,3,2,1
table_x .short 1,2,3,4,5,6,7,8,9,10,11,12,13,14,15    ;Initialize table_x array with values
        .short 16,17,18,19,20,21,22,23,24,25,26,27,28,29
        .short 30,31,32,33,34,35,36,37,38,39,40
table_y .short 0                                       ;table_y = 0

;Variable declaration
a       .usect "var", 80, 2                            ;define variables
x       .usect "var", 80, 2
y       .usect "var",  2, 2

;Initialization to copy data into variables
        .sect   "init_code"
init    mvk     .s1     table_a, A0    ;move address of table_a to register A0
        mvkh    .s1     table_a, A0
        mvk     .s2     a,B0           ;move address of a to register B0
        mvkh    .s2     a,B0
        mvk     .s2     40,B1          ;create a counter in register B1, B1=40
loop_a  ldh     .d1     *A0++,A1       ;load an element from the address pointed by A0 into A1
        sub     .l2     B1,1,B1        ;decrement counter
        nop             3
        sth     .d2     A1,*B0++       ;store the element to address pointed by B0
 [B1] b .s2     loop_a                 ;branch back to loop_a
        nop             5              ;required latency
init_x  mvk     .s1     table_x, A0    ;move address of table_x into register A0
        mvkh    .s1     table_x, A0
        mvk     .s2     x, B0          ;move address of x into register A0
        mvkh    .s2     x, B0
        mvk     .s2     40, B1         ;create a counter
loop_x  ldh     .d1     *A0++,A1       ;load the element from the address pointed
                                       ; by A0 into A1
        sub     .l2     B1,1,B1        ;decrement counter
        nop             3
        sth     .d2     A1,*B0++       ;store element to address pointed by B0
 [B1] b .s2     loop_x                 ;branch back to loop_x
        nop             5
init_y  mvk     .s1     table_y, A0    ;repeat above procedure for table_y
        mvkh    .s1     table_y, A0
        mvk     .s2     y, B0
        mvkh    .s2     y, B0
        ldh     .d1     *A0, A1
        nop             4
        sth     .d2     A1, *B0
```

(a)

FIGURE 6-13 (a) Initialization code for dot-product example, (b) setup code for calling dot-product routine, and (c) dot-prod uct routine.[†]

```
;Setup for calling dotp

start   mvk     .s1     a,A4            ;move a into register A4
        mvkh    .s1     a,A4
        mvk     .s2     x,B4            ;move x into register B4
        mvkh    .s2     x,B4
        mvk     .s1     40,A6           ;create a counter in A6, A6=40
        b       .s1     dotp            ;branch to routine dotp
        mvk     .s2     return,B3       ;store return address in B3
        mvkh    .s2     return,B3
        nop             3

;return from dotp here
return  mvk     .s1     y, A0           ;move y into register A0
        mvkh    .s1     y, A0
        sth     .d1     A4, *A0         ;store the result of dotp (returned in A4) to y

;infinite loop
end     b       .s1     end             ;infinite loop
        nop             5
```

(b)

```
;dotp

        .def    dotp

;A4 = &a, B4 = &x, A6 = 40 (iteration count) , B3 = return address

dotp    mv              A6,B0           ;move A6 to B0 (third argument passed from calling function)
        zero            A2              ;zero the sum register A2

loop    ldh     .d1     *A4++,A5        ;load an element from the location pointed by A4 into A5
        ldh     .d2     *B4++,B5        ;load an element from the location pointed by B4 into B5
        nop             4
        mpy     .m1x    A5,B5,A5        ;A5=B5*A5
        nop
        add     .l1     A2,A5,A2        ;A2 += A5
 [B0]   sub     .l2     B0,1,B0         ;decrement counter B0
 [B0]   b       .s1     loop            ;branch back to loop
        nop             5

        mv              A2,A4           ;move result in A2 to return register A4
        b       .s2     B3              ;branch back to calling address stored in B3
        nop             5
```

(c)

FIGURE 6-13 *(continued)*

108 Chapter 6 Software Tools

FIGURE 6-14 C initialization.[†]

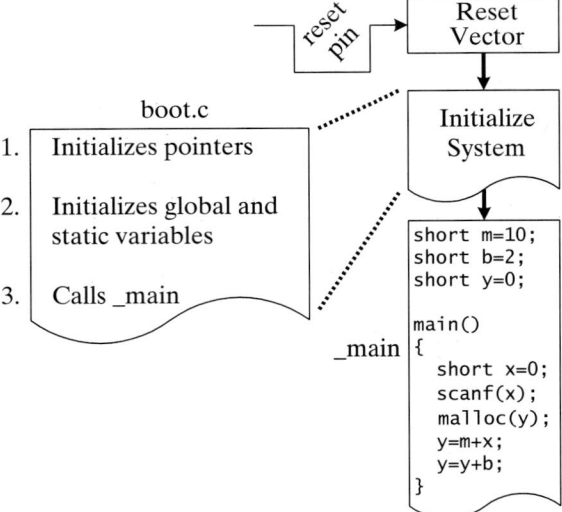

FIGURE 6-15 Data boundaries.

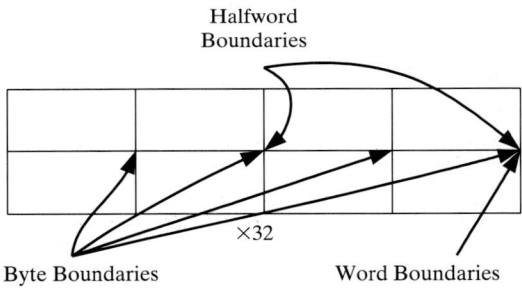

Constants are automatically aligned			
.sect "my_ const "			
A	.byte	11h	
B	.short	2222h	
C	.int	33333333h	

22	22	--	11
33	33	33	33
--	e	e	d
f	f	f	f
	g1		g0
	g3		g2

Variables need an alignment field

```
; label      usect      "section",   #bytes,    alignment

d           .usect     "vars ",     1          1
ee          .usect     "vars ",     2                     ;byte alignment by default
ffff        .usect     "vars ",     4          4
g_array     .usect     "vars ",     8          2
```

Note 1: vars and my_const sections are assumed contiguous.

Note 2: First declare words, then shorts and bytes to save memory space.

FIGURE 6-16 Constant and variable alignment examples.[†]

6.5 Code Initialization

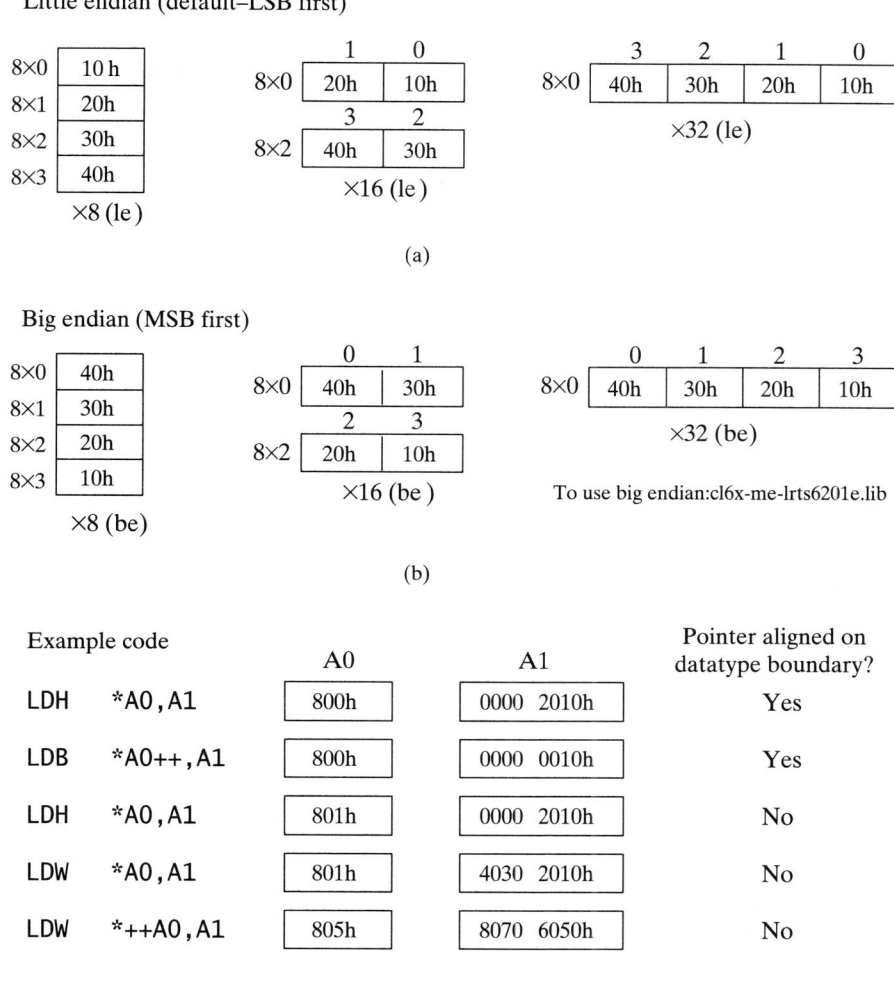

FIGURE 6-17 (a) Little endian, (b) big endian, and (c) more data alignment examples.[†]

Data in memory can be arranged either in little- or big-endian format. Little endian (le) means that the least significant byte is stored first. Figure 6-17(a) shows storing .int 40302010h in little-endian format for byte, halfword, and word access addressing. In big-endian (be) format, shown in Figure 6-17(b), the most significant byte is stored first. Little endian is the format used in most applications. Additional data alignment examples are shown in Figure 6-17(c), based on the little-endian data format appearing in Figure 6-17(a).

Lab 1: Getting Familiar with Code Composer Studio

Code Composer Studio (CCS) is a useful integrated development environment (IDE) that provides an easy-to-use software tool to build and debug programs. In addition, it allows real-time analysis of application programs. Figure L1-1 shows the phases associated with the CCS code development process. During its setup, CCS can be configured for different target DSP boards (e.g., C6211 DSK, C6xxx EVM, C6xxx Simulator).

CCS includes an integrated editor for editing both C and assembly files. For building application programs, it provides a file management environment. For debugging purposes, it provides breakpoints, data monitoring and graphing capabilities, profile points for benchmarking, and probe points to stream data to and from the target DSP. This tutorial lab introduces the basic features of CCS—those needed in order to build and debug an application program. To become familiar with all of its features, one should go through the *TI CCS Tutorial* manual [13]. The real-time analysis and scheduling features of CCS will be covered later, in Labs 7 and 8.

This lab demonstrates how a simple multifile algorithm can be compiled, assembled and linked by using CCS. First, several data values are consecutively written to memory. Then, a pointer is assigned to the beginning of the data so that they can be treated as an array. Finally, simple functions are added in both C and assembly to illustrate how function calling works. This method of placing data in memory is simple to use and can be used in applications in which constants need to be in memory, such as filter coefficients and FFT twiddle factors. Issues related to debugging and benchmarking are also covered in this lab.

L1.1 CREATING PROJECTS

Let us consider all of the files required to create an executable file; that is, .c, .asm, .sa source files, a .cmd command file, .h header, and .lib library files. The CCS code development process begins with the creation of a so-called Project to easily integrate and manage all these required files for generating and running an executable file. The Project View panel on the left-hand side of the CCS window provides an easy mechanism for doing so. In this panel, a project file (.mak extension) can be created or opened to contain not only the source and library files but also the compiler, assembler, and linker

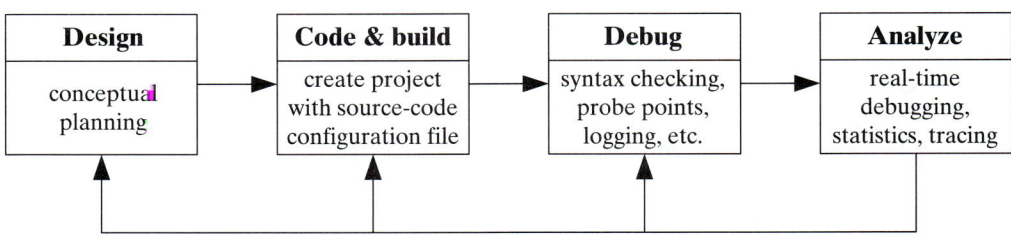

FIGURE L1-1 CCS code development process.[†]

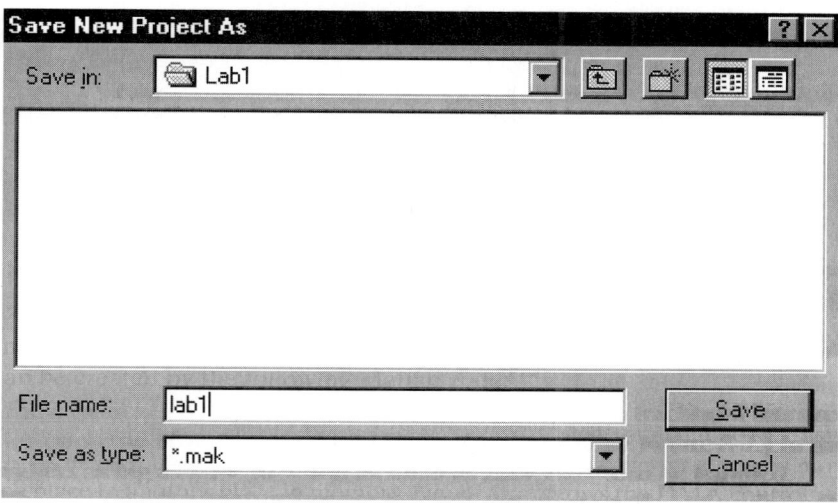

FIGURE L1-2 Creating a new project.

options for generating an executable file. As a result, one need not type command lines for compilation, assembling and linking, as was the case with the original software-development tools.

To create a project, choose the menu item Project–>New from the CCS menu bar. This brings up the dialog box Save New Project As, as shown in Figure L1-2. In the dialog box, navigate to the working folder and type a project name in the field Filename. Then, click the button Save for CCS to create a project file named *lab1.mak*. All of the files necessary to build an application should be added to the project.

CCS provides an integrated editor that allows the creation of source files. Some of the features of the editor are color syntax highlighting, C blocks marking in parentheses and braces, parenthesis/brace matching, control indentions, and find/replace/search capabilities. It is also possible to add files to the project from Windows Explorer, using the drag-and-drop approach. An editor window is brought up by choosing the menu item File–>New–>Source File. For this lab, let us type the following assembly code into the editor window:

```
        .sect ".mydata"
        .short  0
        .short  7
        .short  10
        .short  7
        .short  0
        .short  -7
        .short  -10
        .short  -7
        .short  0
        .short  7
```

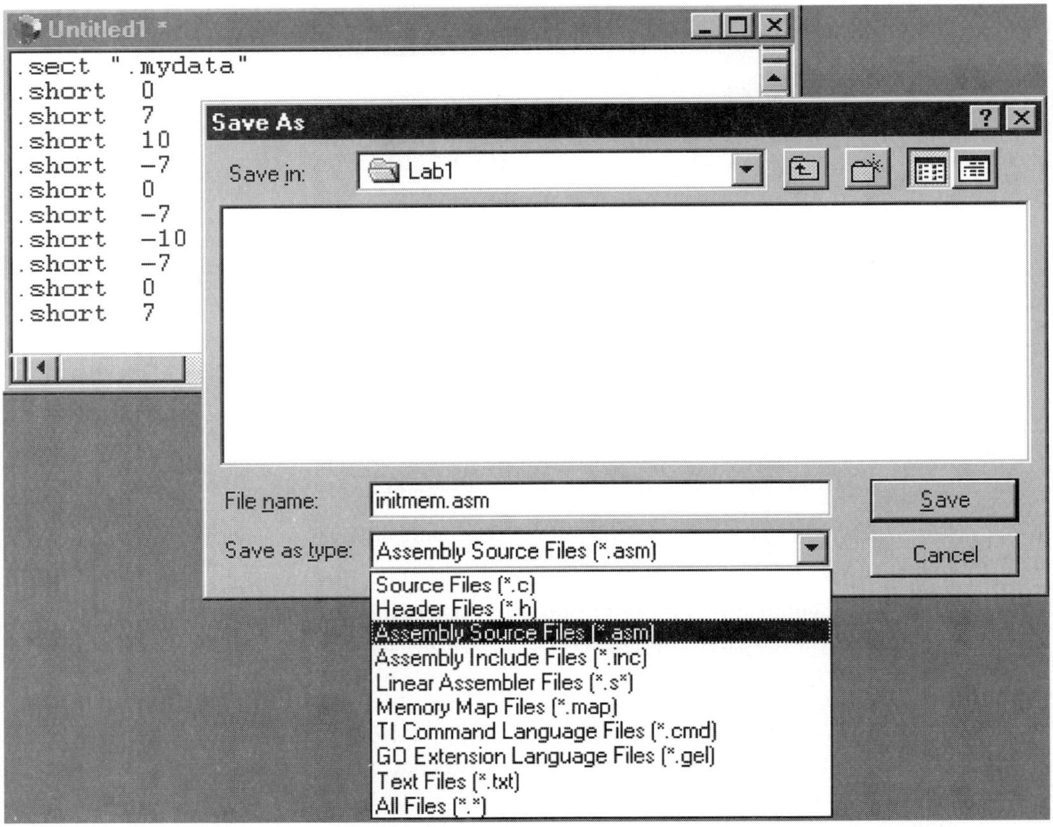

FIGURE L1-3 Creating a source file.

This code consists of the declaration of 10 values through the use of .short directives. Note that any of the data to memory allocation directives can be used to do the same. Assign a section named .mydata to the values by using a .sect directive. Save the created source file by choosing the menu item File->Save. This brings up the dialog box Save As, as shown in Figure L1-3. In the dialog box, go to the field Save as type and select Assembly Source Files(*.asm) from the pull-down list. Then, go to the field Filename, and type *initmem.asm*. Finally, click Save, so that the code can be saved into an assembly source file named *initmem.asm*.

In addition to source files, a linker command file must be specified both to create an executable file and to conform to memory specifics of the target DSP on which the executable file is going to run. Again, a linker command file can be created by choosing File->New->Source. For this lab, let us type the command file shown in Figure L1-4. This file can also be loaded from the attached CD. This command file is configured based on the mem-

```
/* Memory Map 1 */
MEMORY
{
    PMEM    : origin = 0x00000000,   length = 0x00010000
    EXT2    : origin = 0x02000000,   length = 0x01000000
    EXT3    : origin = 0x03000000,   length = 0x01000000
    DMEM    : origin = 0x80000000,   length = 0x00010000
}

SECTIONS
{
        .vectors > PMEM
        .text    > PMEM
        .bss     > DMEM
        .cinit   > DMEM
        .const   > DMEM
        .stack   > DMEM
        .cio     > DMEM
        .sysmem  > DMEM
        .far     > EXT2
        .mydata  > EXT3
}
```

FIGURE L1-4   Command file for Lab 1.

ory map 1. Since our intention is to place the array of values defined in *initmem.asm* into the memory, a space that will not be overwritten by the compiler should be selected. The external data space EXT2 or EXT3 can be used for this purpose. Let us assemble the data at the memory address 0x03000000 (0x denotes hex), located at the beginning of EXT3. To do this, assign the section named .mydata to EXT3 by adding .mydata > EXT3 to the SECTIONS part of the command file, as shown in Figure L1-4. The editor window should be saved into a linker command file by choosing File–>Save. This brings up the dialog box Save As. Go to the field Save as type and select TI Command Language Files (*.cmd) from the pull-down list. Then, type *lab1.cmd* in the field Filename and click Save.

Now that the source file *initmem.asm* and the linker command file *lab1.cmd* are created, they should be added to the project for assembling and linking. To do this, choose the menu item Project–>Add Files to Project. This brings up the dialog box Add Files to Project. In the dialog box, select *initmem.asm* and click the button Open. This adds *initmem.asm* to the project. In order to add *lab1.cmd*, choose Project->Add Files to Project. Then, in the dialog box Add Files to Project, set Files of type to Linker Command File (*.cmd), so that *lab1.cmd* appears in the dialog box. Now, select *lab1.cmd* and click the button Open. In addition to *initmem.asm* and *lab1.cmd* files, the run-time support library file should be added to the project. To do so, choose Project->Add Files to Project, go to the compiler library folder, select Library Files (*.lib) in the box Files of type, select rts6201.lib and click Open. If running on the floating-point DSP TMS320C6701, select rts6701.lib instead. For debugging purposes, let us use the following empty shell C program. Create a C source file *main.c*, enter the following lines and add *main.c* to the project in the same way as just described.

```
#include <stdio.h>

void main()
{
        printf("BEGIN\n");

        printf("END\n");
}
```

After adding all of the source files, the command file and the library file to the project, it is time to either build the project or to create an executable file for the target DSP. To do this, choose the Project->Build menu item. CCS compiles, assembles, and links all of the files in the project. Messages about this process are shown in a panel at the bottom of the CCS window. When the building process is completed without any errors, the executable *lab1.out* file is generated. It is also possible to perform incremental builds—that is, rebuilding only those files changed since the last build, by choosing the menu item Project->Rebuild. The CCS window provides shortcut buttons for frequently used menu options, such as build and rebuild .

Although CCS provides default build options, these options can be changed by choosing Project->Options. For instance, to change the executable filename to *test.out*, choose Project->Options, click Linker tab of the Build Options window, and type *test.out* in the field Output Filename, as shown in Figure L1-5a. Notice that the linker command file will include *test.out* as you click on this window.

All of the compiler, assembler, and linker options are set through the menu item Project->Options. Among many compiler options shown in Figure L1-5(b), pay particular attention to the optimization-level options. There are four levels of optimization (0, 1, 2, and 3), which control the type and degree of optimization. Level 0 optimization option performs control-flow-graph simplification, allocates variables to registers, eliminates unused code, and simplifies expressions and statements. Level 1 optimization performs all Level 0 optimizations, removes unused assignments, and eliminates local common expressions. Level 2 optimization performs all Level 1 optimizations, plus software pipelining, loop optimizations, and loop unrolling. It also eliminates global common subexpressions and unused assignments. Finally, Level 3 optimization performs all Level 2 optimizations, removes all functions that are never called, and simplifies functions with return values that are never used. It also inlines calls to small functions and reorders function declarations.

Note that in some cases, debugging is not possible due to optimization. Thus, it is recommended to first debug your program to make sure that it is logically correct before performing any optimization. Another important compiler option is the Target Version option. When the application is for the floating-point target DSP TMS320C6701, go to the Target Version field and select 670x from the pull-down list.

One common mistake in writing *initmem.asm* is to type directives `.sect` and `.short` in the first column. Because only labels can start in the first column, this will result in an assembler error. When a message stating a compilation error appears, click `Cancel` and scroll up in the `Build` area to see the syntax error message. Double click on the red text that describes the location of the syntax error. Notice that the *initmem.asm* file opens, and your cursor appears on the line that has caused the error, as shown in Figure L1-6. After correcting the syntax error, the file should be saved and the project rebuilt.

L1.2 DEBUGGING TOOLS

Once the build process is completed without any errors, the program can be loaded and executed on the target DSP. To load the program, choose `File->Load Program`, select the program *lab1.out* just rebuilt, and click `Open`. To run the program, choose the menu item `Debug->Run`. You should be able to see `BEGIN` and `END` appearing in the `Stdout` window.

FIGURE L1-5 (a) Build options and (b) compiler options. *(continues on next page)*

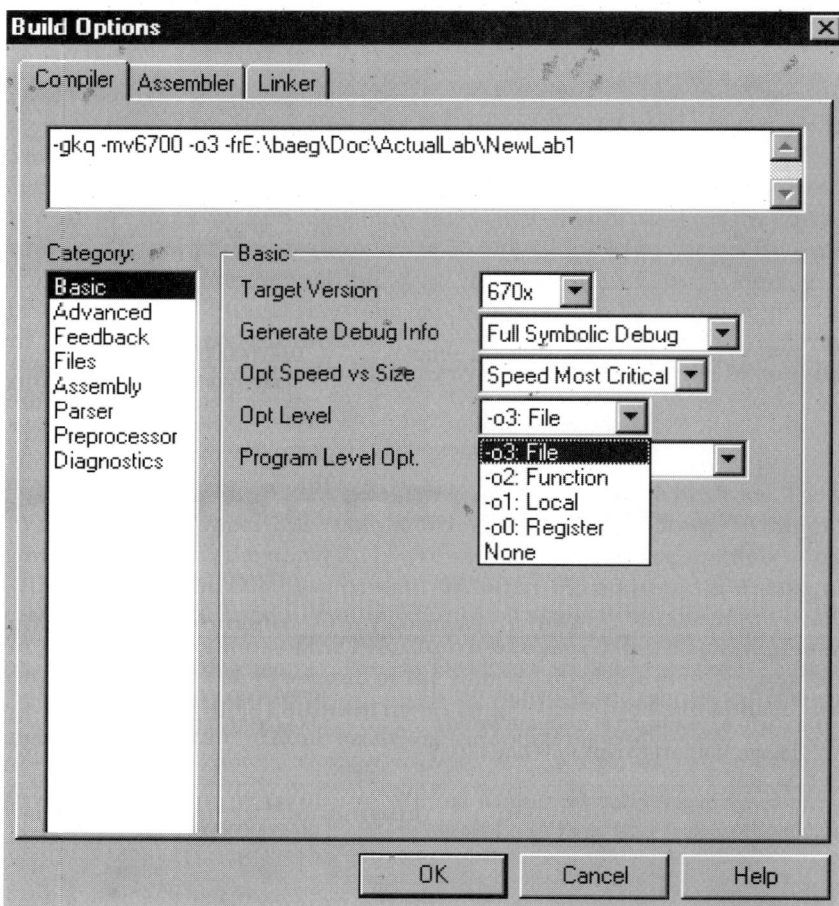

FIGURE L1-5   *(continued)*

Now, let us verify if the array of values is assembled into the specified memory location. CCS allows one to view the content of memory at a specific location. To view the content of memory at 0x03000000, select View->Memory from the menu. The dialog box Memory Window Options will appear. This dialog box allows one to specify various attributes of the Memory window. Go to the Address field and enter 0x03000000. Then, select 16bit Signed Int from the pull-down list in the Format field and click OK. A Memory window appears, as shown in Figure L1-7. The contents of CPU, peripheral, DMA, and serial-port registers can also be viewed by selecting View->CPU Registers, for example.

There is another way to load data from a PC file to the DSP memory. CCS provides a probe point capability, so that a stream of data can be moved from the host PC file to the DSP or vice versa. In order to use this capability, a probe point should be set within the program by placing the mouse cursor at the line where a stream of data needs to be transferred and clicking the button Toggle Probe Point. Then,

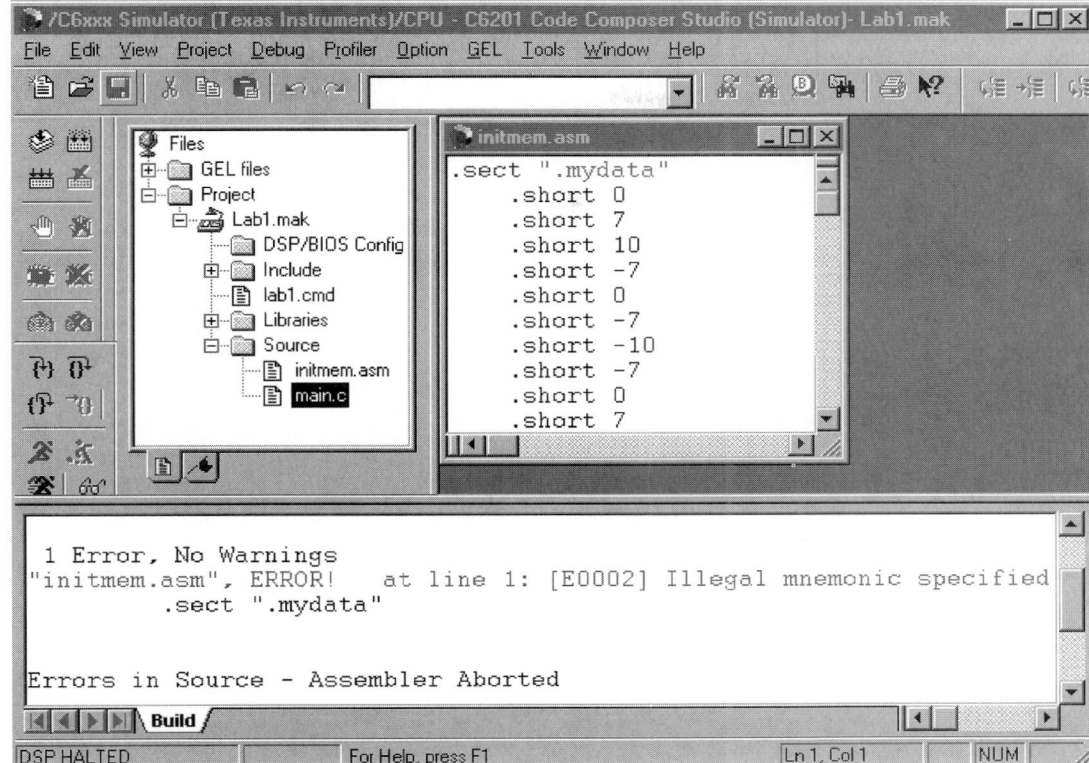

FIGURE L1-6 Build Error.

choose `File->File I/O` to invoke the dialog box `File I/O`. Click the button `Add File` and select the datafile to load. Now the file should be connected to the probe point by clicking the button `Add Probe Point`. From the `Probe Points` tab, select the probe point to be connected to a PC file through `File In:...` in the field `Connect To`. Click the button `Replace` and then the button `OK`. Finally, enter the memory location in the `Address` field and the number of data in the `Length` field. Note that a variable name can be used in the `Address` field. The probe point capability is frequently used to simulate the execution of an application program in the absence of any live signals. A valid PC file should have the correct file header and extension. The file header should conform to the following format:

`MagicNumber Format StartingAddress PageNum Length`

`MagicNumber` is fixed at 1651. `Format` indicates the format of samples in the file: 1 for hexadecimal, 2 for integer, 3 for long, and 4 for float. `StartingAddress` and `PageNum` are determined by CCS when a stream of data is saved into a PC file. `Length` indicates the number of samples in the memory. A valid data file should have the extension `.dat`.

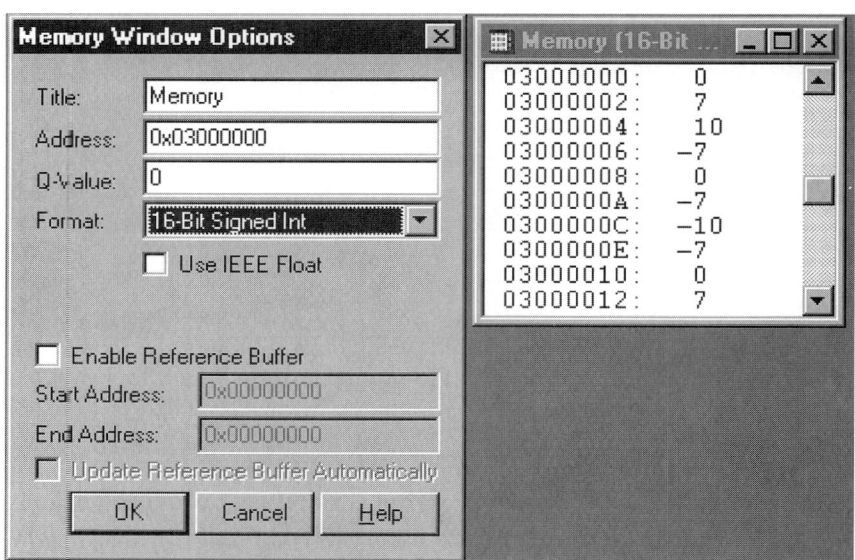

FIGURE L1-7 Memory Window Options dialog box and Memory window.

A graphical display of data often provides better feedback about the behavior of a program. CCS provides a signal analysis interface to monitor a signal or data. Let us display the array of values at 0x03000000 as either a signal or a time graph. To do so, select View->Graph->Time/Frequency to view the Graph Property Dialog box. Field names appear in the left column. Go to the Start Address field, click it and type 0x03000000. Then, go to the Acquisition Buffer Size field, click it and enter 10. Finally, click on DSP Data Type, select 16-bit Signed Integer from the pull-down list, and click OK. A graph window appears with the properties selected. This is illustrated in Figure L1-8. You can change any of these parameters from the graph window by right clicking the mouse, selecting Properties, and adjusting the properties as needed. The properties can be updated at any time during the debugging process.

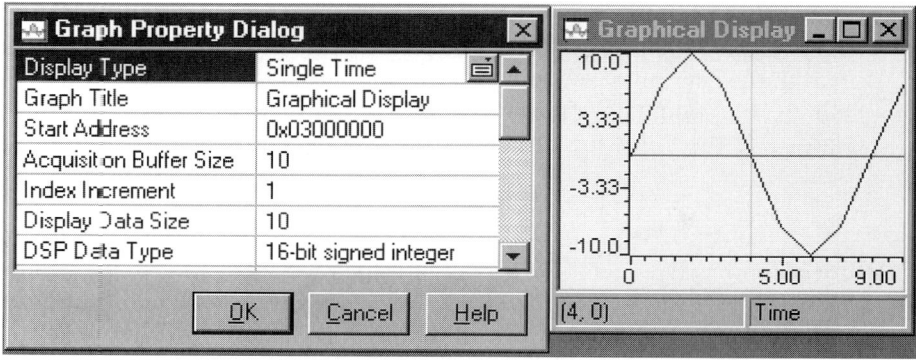

FIGURE L1-8 Graph Property Dialog box and Graphical Display window.

To assign a pointer to the beginning of the assembled memory space, the memory address can be typed in directly to a pointer. It is necessary to typecast the pointer to short, since the values are of that type. The following code can be used to assign a pointer to the beginning of the values and loop through them to print each onto the Stdout window:

```c
#include <stdio.h>

void main()
{
    int i;

    short *point;

    point= (short *) 0x03000000;

    printf("BEGIN\n");

    for(i=0;i<10;i++)
    {
        printf("[%d] %d\n",i, point[i]);
    }

    printf("END\n");
}
```

Instead of creating a new source file, we can modify the existing *main.c* by double clicking on the *main.c* file in the Project View panel, as shown in Figure L1-9. This action will bring up the *main.c* source file in the right-half part of the CCS window. Then, enter the code and rebuild it. Before running the executable file, make sure you reload the file *lab1.out*. By running this file, you should be able to see the values in the Stdout window. Double clicking in the Project View panel provides an easy way to bring up any source or command file for reviewing or modifying purposes.

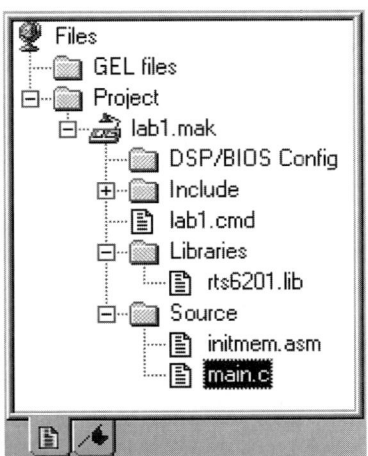

FIGURE L1-9 Project View panel.

When developing and testing programs, one often needs to check the value of a variable during program execution. This can be achieved by using breakpoints and watch windows. To view the values of the pointer in *main.c* before and after the pointer assignment, choose `File->Reload Program` to reload the program. Then, double click on *main.c* in the `Project View` panel. You may want to make the window larger so that you can see more of the file in one place. Next, put your cursor on the line that says `point = (short *) 0x03000000` and press F9 to set a breakpoint. To open a watch window, choose `View->Watch Window` from the menu bar. This will bring up a `Watch Window`. To add a new expression to the `Watch Window`, press the right mouse button in the `Watch Window` and select `Insert New Expression`. A `Watch Add Expression` dialog box appears. In the `Expression` field, type `point` (or any expression you want to examine) and click `OK`. Then, choose `Debug->Run` or press F5. The program stops at the breakpoint and the `Watch Window` displays the value of the pointer. This is the value before the pointer is set to 0x03000000. By pressing F10 to step over the line, or the shortcut button, you should be able to see the value 0x03000000 in the `Watch Window`.

To add a simple C function that sums the values, we can simply pass the pointer to the array and have a return type of integer. For the time being, considering that we are not using assembly, what is of concern is not how the variables are passed, but rather how much time it takes to perform the operation.

The following simple function can be used to sum the values and return the result:

```
#include <stdio.h>

void main()
{
        int i,ret;

        short *point;

        point = (short *) 0x03000000;

        printf("BEGIN\n");

        for(i=0;i<10;i++)
        {
                printf("[%d] %d\n",i, point[i]);
        }

        ret = ret_sum(point,10);

        printf("Sum = %d\n",ret);

        printf("END\n");
}

int ret_sum(const short* array,int N)
```

```c
{
    int count,sum;

    sum=0;

    for(count=0 ; count < N ; count++)
            sum += array[count];

    return(sum);
}
```

As part of the debugging process, it is normally required to benchmark or time the application program. In this lab, let us determine how much time it takes for the function *ret_sum()* to run. To achieve the benchmarking, reload the program and choose Profiler->Enable Clock to count instruction cycles. Then, double click on the *main.c* file in the Project View panel and choose View->Mixed Source/ASM to list assembled instructions following each C code line. A profile point is set at the calling line by placing the cursor on the line that reads ret = ret_sum(point,10), right clicking and selecting Toggle Profile Pt from the pop-up menu. Another profile point should be set at the next line, as shown in Figure L1-10. Once the profile points are set, choose Profiler->View Statistics to bring up a window that displays statistics about the number of instruction cycles. Resize this window by dragging its edges so that you can see all of the

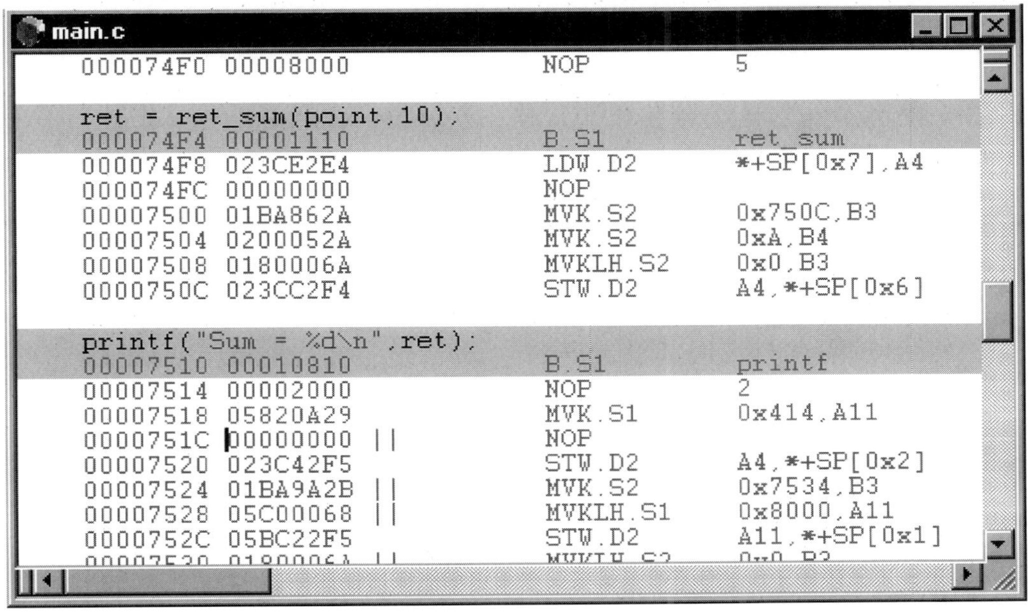

```
main.c
000074F0  00008000             NOP         5
ret = ret_sum(point,10);
000074F4  00001110             B.S1        ret_sum
000074F8  023CE2E4             LDW.D2      *+SP[0x7],A4
000074FC  00000000             NOP
00007500  01BA862A             MVK.S2      0x750C,B3
00007504  0200052A             MVK.S2      0xA,B4
00007508  0180006A             MVKLH.S2    0x0,B3
0000750C  023CC2F4             STW.D2      A4,*+SP[0x6]
printf("Sum = %d\n",ret);
00007510  00010810             B.S1        printf
00007514  00002000             NOP         2
00007518  05820A29             MVK.S1      0x414,A11
0000751C  00000000       ||    NOP
00007520  023C42F5             STW.D2      A4,*+SP[0x2]
00007524  01BA9A2B       ||    MVK.S2      0x7534,B3
00007528  05C00068       ||    MVKLH.S1    0x8000,A11
0000752C  05BC22F5             STW.D2      A11,*+SP[0x1]
00007530  0180006A       ||    MVKLH.S2    0x0,B3
```

FIGURE L1-10 Profiling code execution time.

Chapter 6 Lab 1: Getting Familiar with Code Composer Studio

Profile Statistics					
Location	Count	Average	Total	Maximum	Minimum
main.c line 19	1	35871.0	35871	35871	35871
main.c line 21	1	473.0	473	473	473

FIGURE L1-11 Profile Statistics window.

columns. Finally, press F5 to run the program. Examine the number of cycles shown in Figure L1-11 for the second profile point. It should be about 473 cycles (the exact number may slightly vary). This is the number of cycles it takes to execute the function *ret_sum()* and return.

The file shown next is an assembly program for calculating the sum of the values. Here, the two arguments of the sum function are passed in registers A4 and B4. The return value is stored in A4 and the return address in B3. The order in which the registers are chosen is governed by the passing argument convention, discussed later. The name of the function should be preceded by an underscore as .global _sum. Create a new source file *sum.asm*, as shown, and add it to the project so that *main()* can call the function *sum()*.

sum.asm

```
        .global_sum

_sum:
        ZERO    .L1     A9              ;Sum register
        MV      .L1     B4,A2           ;initialize counter with passed argument

loop:   LDH     .D1     *A4++, A7       ;load value pointed by A4 into register A7
        NOP     4
        ADD     .L1     A7,A9,A9        ;A9 += A7
[A2]    SUB     .L1     A2,1,A2         ;decrement counter
[A2]    B       .S1     loop            ;branch back to loop
        NOP     5

        MV      .L1     A9,A4           ;move result into return register A4
        B       .S2     B3              ;branch back to address stored in B3
        NOP     5
```

To save the file, go to the Save as type field and select Assembly Source Files (*.asm) from the pull-down list.

main.c

```
#include <stdio.h>

void main()
{
    int i,ret;

    short *point;

    point = (short *) 0x03000000;

    printf("BEGIN\n");

    for(i=0;i<10;i++)
    {
        printf("[%d] %d\n",i, point[i]);
    }

    ret = ret_sum(point,10);

    printf("C program Sum = %d\n",ret);

    ret = sum(point,10);

    printf("Assembly program Sum = %d\n", ret);

    printf("END\n");
}
int ret_sum(const short* array,int N)
{
    int count,sum;

    sum=0;

    for(count=0 ; count < N ; count++)
        sum += array[count];

    return(sum);
}
```

FIGURE L1-12 Lab1 complete program.

The program *main()* must also be modified by adding a function call to the assembly function *sum()*. This program is shown in Figure L1-12. Build the program and run it. You should be able to see the same return value.

Table L1-1 provides the number of cycles it takes to run the sum function using several different builds. When a program is too big to fit into the internal program memory, it has to be placed into the external memory. Although the program in this lab is small enough to fit into the internal program memory, it is placed in the external memory to study the change in the number of cycles. To move the program into the external memory, open the *lab1.cmd* file and replace the line `.text > PMEM` with `.text > EXT2`. As seen in Table L1-1, this build slows down the execution to 1433 cycles. In the second build, the program resides in the internal memory and the number of cycles is hence reduced to 473. By increasing the optimization level, the number of cycles can be further decreased to 204. The assembly version of the program gives 323 cycles. This is slower than the fully optimized C program because the former is not yet optimized. Optimization of assembly codes will be discussed later.

As a final note, it is worthwhile to mention EVM board resetting. Sometimes, you may notice that your program cannot be loaded into the DSP, even though there is nothing wrong with it. Under such circumstances, you need to reset the EVM board to fix the problem. However, you have to close Code Composer Studio before you reset the board. Otherwise, the problem will not be resolved.

TABLE L1-1 Number of cycles for different builds.

Type of Build	Number of Cycles
C program in external memory	1433
C program in internal memory	473
-o0/1	348
-o2	227
-o3	204
Assembly program	323

CHAPTER 7

Interrupt Data Processing

On a DSP processor, the processing of samples can be done within an ISR (interrupt service routine). Let us first discuss interrupts. As the name implies, an interrupt causes the processor to halt whatever it is processing in order to execute an ISR. An interrupt can be issued externally or internally. There are 12 CPU interrupts available on the C6x. The priorities of these interrupts are shown in Table 7-1. RESET is the highest priority interrupt. It halts the CPU and initializes all the registers to their default values. Nonmaskable interrupt (NMI) is used for nonmaskable or uninterruptible processing provided that the NMIE bit of the control register CSR (Control Status Register) is set to zero. As indicated in Figure 7-1, there are a total of 16 interrupt sources, while there are only 12 CPU interrupts. As a result, an interrupt source must be mapped to a CPU

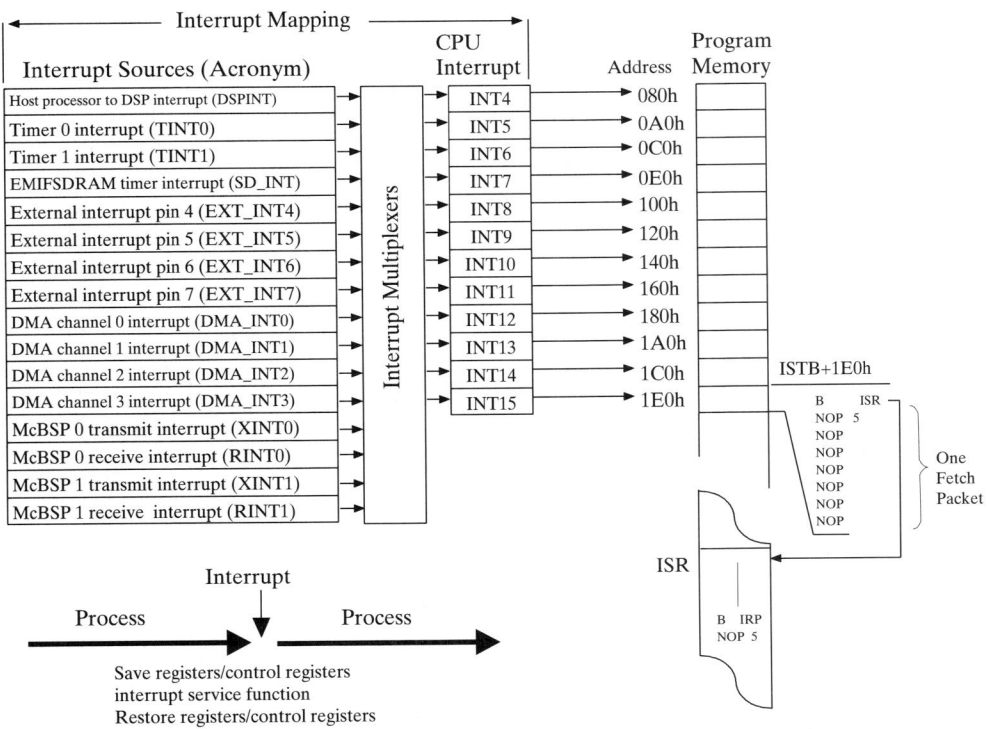

FIGURE 7-1 Interrupt mapping and operation.

interrupt. This is done by setting appropriate bits of the two memory mapped Interrupt Multiplex registers.

Interrupt name	Priority
RESET	Highest
NMI	
INT4	
INT5	
INT6	
INT7	
INT8	
INT9	
INT10	
INT11	
INT12	
INT13	
INT14	
INT15	Lowest

Interrupts can be enabled or disabled by setting or clearing appropriate bits in Interrupt Enable Register (IER). There is a master switch, the Global Interrupt Enable bit (GIE) as part of CSR, which can be used to turn all interrupts on or off. For example, the assembly code shown in Figure 7-2 indicates how to enable INT4 and the GIE bit. Here the instruction MVC (move to and from a control register) is used to transfer the contents of a control register to a CPU register for bit manipulation. Another register called Interrupt Flag Register (IFR) allows one to check if or what interrupt has occurred. (Refer to the *TI TMS320C6x CPU* manual [14] for more details on the interrupt registers.)

The location where the processor will go to after an interrupt occurs is specified by a predefined offset for that interrupt added to the Interrupt Service Table Base (ISTB) bits as part of the Interrupt Service Table Pointer (ISTP) register. As an example, for the CPU INT15, the processor will go to the location ISTB+1E0h. At this location, there is normally a branch instruction that would take the processor to a receive ISR somewhere in memory, as shown in Figure 7-1.

In general, an ISR includes three parts. The first and last part incorporate saving and restoring registers, respectively. The actual interrupt routine makes up the second part. If needed, saving and restoring are done to restore the status of the processor to the time when the interrupt was issued.

```
MVK    .S2    0010h, B3        ;bit4="1"
MVC    .S2    IER, B4          ;get IER
OR     .L2    B3, B4, B4       ;set bit4
MVC    .S2    B4, IER          ;write IER
MVC    .S2    CSR, B5          ;get CSR_GIE
OR     .L2    1, B5, B5        ;bit0="1"
MVC    .S2    B5, CSR          ;set GIE
```

FIGURE 7-2 Setup code to turn on INT4 and GIE.

Lab 2: Audio Signal Sampling

The purpose of this lab is to use a C62 or C67 board to sample an analog audio signal in real-time. A common approach to handling real-time signals, which is the use of an interrupt service routine, is utilized here to obtain and process signal samples.

The C62 EVM has a 16-bit stereo audio codec CS4231A, which can handle sampling rates from 5.5 kHz to 48 kHz. There are three 3.5 mm audio jacks on the back of the EVM for a microphone-in, a line-in, and a line-out. The codec is connected to the C6201 DSP through its multichannel buffered serial port (McBSP). Each of the audio jacks inputs has its own amplifying and filtering capabilities. A block diagram of the EVM stereo interface is shown in Figure L2-1.

All the adjustments to the codec can be done through the DSP support software provided by Texas Instruments. The DSP support software contains C functions for accessing and setting up the EVM board, McBSP, and codec. The codec library is supplied in the archived object library file *drv6x.lib* (*drv6xe.lib* is the big-endian version of this library). The corresponding source file is *drv6x.src*. The codec library contains API (application programmer interface) functions that can be used to configure and control the operation of the codec. The functional descriptions of these functions can be found in the *TI EVM Reference Guide* [16] under TMS320C6x EVM DSP Support Software. The functions are utilized here to write a sampling program for the C62 or C67 EVM.

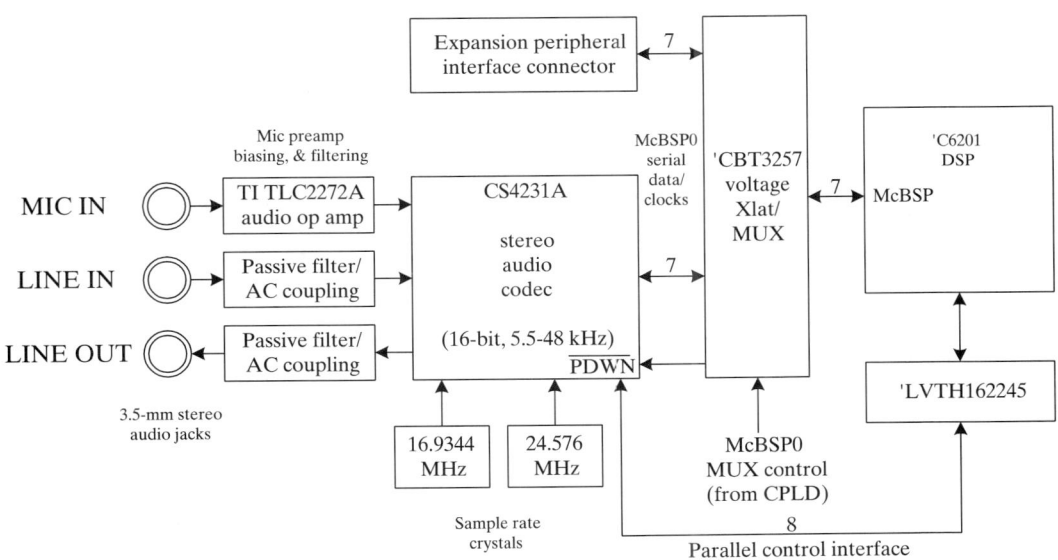

FIGURE L2-1   EVM stereo audio interface[†]

128 Chapter 7 Lab 2: Audio Signal Sampling

L2.1 INITIALIZATION OF EVM AND CODEC

In writing a program that uses the codec to sample an incoming analog signal, several initializations have to be performed. Among these are the initialization of the EVM, McBSP, and codec. The sampling rate for the codec has to be set as well as any gain adjustments of its 16-bit data converters. The API functions can be used to achieve all of these mentioned adjustments. Once the required initializations are made, an interrupt needs to be assigned to the receive register of the codec to halt the processor and jump to a defined interrupt service routine. The final program will output the same input sample back to the codec. The following program includes an order of API functions that achieves all of the foregoing mentioned initializations (Figure L2-2 shows the flowchart of the steps involved):

```
#include <stdlib.h>
#include <stdio.h>
#include <string.h>
#include <common.h>
#include <mcbspdrv.h>
#include <intr.h>
#include <board.h>
#include <codec.h>
#include <mcbsp.h>
#include <mathf.h>
void hookint(void);
interrupt void serialPortRcvISR(void);

int main()
{
        Mcbsp_dev dev;
        Mcbsp_config mcbspConfig;
        int sampleRate, status;
        /*************************************************/
        /* Initialize EVM                                */
        /*************************************************/
        status = evm_init();
        if(status == ERROR)
                return (ERROR);
        /*************************************************/
        /* Open MCBSP for subsequent Examples            */
        /*************************************************/
        mcbsp_drv_init();
        dev= mcbsp_open(0);
        if (dev == NULL)
```

```
        {
                printf("Error opening MCBSP 0    /n ");
                return(ERROR);
        }
        /**************************************************/
        /* configure McBSP                              */
        /**************************************************/
        memset(&mcbspConfig,0,sizeof(mcbspConfig));
        mcbspConfig.loopback         = FALSE;
        mcbspConfig.tx.update        = TRUE;
        mcbspConfig.tx.clock_mode    = CLK_MODE_EXT;
        mcbspConfig.tx.frame_length1 = 0;
        mcbspConfig.tx.word_length1  = WORD_LENGTH_32;
        mcbspConfig.rx.update        = TRUE;
        mcbspConfig.rx.clock_mode    = CLK_MODE_EXT;
        mcbspConfig.rx.frame_length1 = 0;
        mcbspConfig.rx.word_length1  = WORD_LENGTH_32;
        mcbsp_config(dev,&mcbspConfig);
        MCBSP_ENABLE(0, MCBSP_BOTH);
        /**************************************************/
        /* configure CODEC                              */
        /**************************************************/
        codec_init();
        /* A/D 0.0 dB gain, turn off 20dB mic gain, sel (L/R)LINE input
*/
        codec_adc_control(LEFT,0.0,FALSE,LINE_SEL);
        codec_adc_control(RIGHT,0.0,FALSE,LINE_SEL);
        /* mute (L/R)LINE input to mixer      */
        codec_line_in_control(LEFT,MIN_AUX_LINE_GAIN,TRUE);
        codec_line_in_control(RIGHT,MIN_AUX_LINE_GAIN,TRUE);
        /* D/A 0.0 dB atten, do not mute DAC outputs    */
        codec_dac_control(LEFT, 0.0, FALSE);
        codec_dac_control(RIGHT, 0.0, FALSE);
        sampleRate = 44100;
        codec_change_sample_rate(sampleRate, TRUE);
        codec_interrupt_enable();
        hookint();
        /**************************************************/
        /* Main Loop, wait for Interrupt       */
        /**************************************************/
        while (1)
        {
        }
}
```

```
void hockint()
{
intr_init();
intr_map(CPU_INT15, ISN_RINT0);
intr_hook(serialPortRcvISR, CPU_INT15);
INTR_ENABLE(15);
INTR_GLOBAL_ENABLE();
return;
}
interrupt void serialPortRcvISR(void)
{
int temp
temp = MCBSP_READ(0);
MCBSP_WRITE(0, temp);
}
```

Let us explain this program in a step-by-step fashion. We must first initialize the EVM, which is done by using the function *evm_init()* before calling any other support functions. This function configures the EVM base address variables and initializes the external memory interface (EMIF). The return value of the function indicates success or failure of the EVM initialization.

Once the EVM has been successfully initialized, the next step is to open a handle to the McBSP in order to send and receive data. The McBSP API functions are used for this purpose. The API function *mcbsp_drv_init()* initializes the McBSP driver and allocates memory for the device handle. The return value of this function also indicates success or failure. As a result of the initialization of the McBSP driver, the data structure elements that control the behavior of the McBSP are set to their default values (for more details, refer to the *EVM Reference Guide* [16]). Then, the McBSP needs to be actually opened to get a handle to it. The API function *mcbsp_open()* is used to return the handle dev for controlling the McBSP.

The next step is to adjust the parameters of the McBSP. The loop-back property of the McBSP is turned off or set to FALSE in order to disable the serial port test mode, in which the receive pins get connected internally to the transmit pins. The update property is set to TRUE for setting properties. The source signal for clocking the serial port transfers is made external by setting the clock mode to CLK_MODE_EXT. The frame and word lengths are set to 0 and WORD_LENGTH_32, respectively. The adjustments to the McBSP are made by allocating memory to the structure mcbsp_Config using the function *memset()*. The address of this structure is passed as an argument to the function *mcbsp_config()*, which performs the required adjustments.

Finally, the McBSP needs to be activated. This is done by using the macro MCBSP_ENABLE(), which is defined in the header file *mcbsp.h*. A macro is a collection of instructions that gets substituted for the macro in the program by the assembler. In this lab, the macro MCBSP_ENABLE(0) places the selected port 0 in the general-purpose I/O mode. The following lines of code are used to make these adjustments:

L2.1 Initialization of EVM and Codec 131

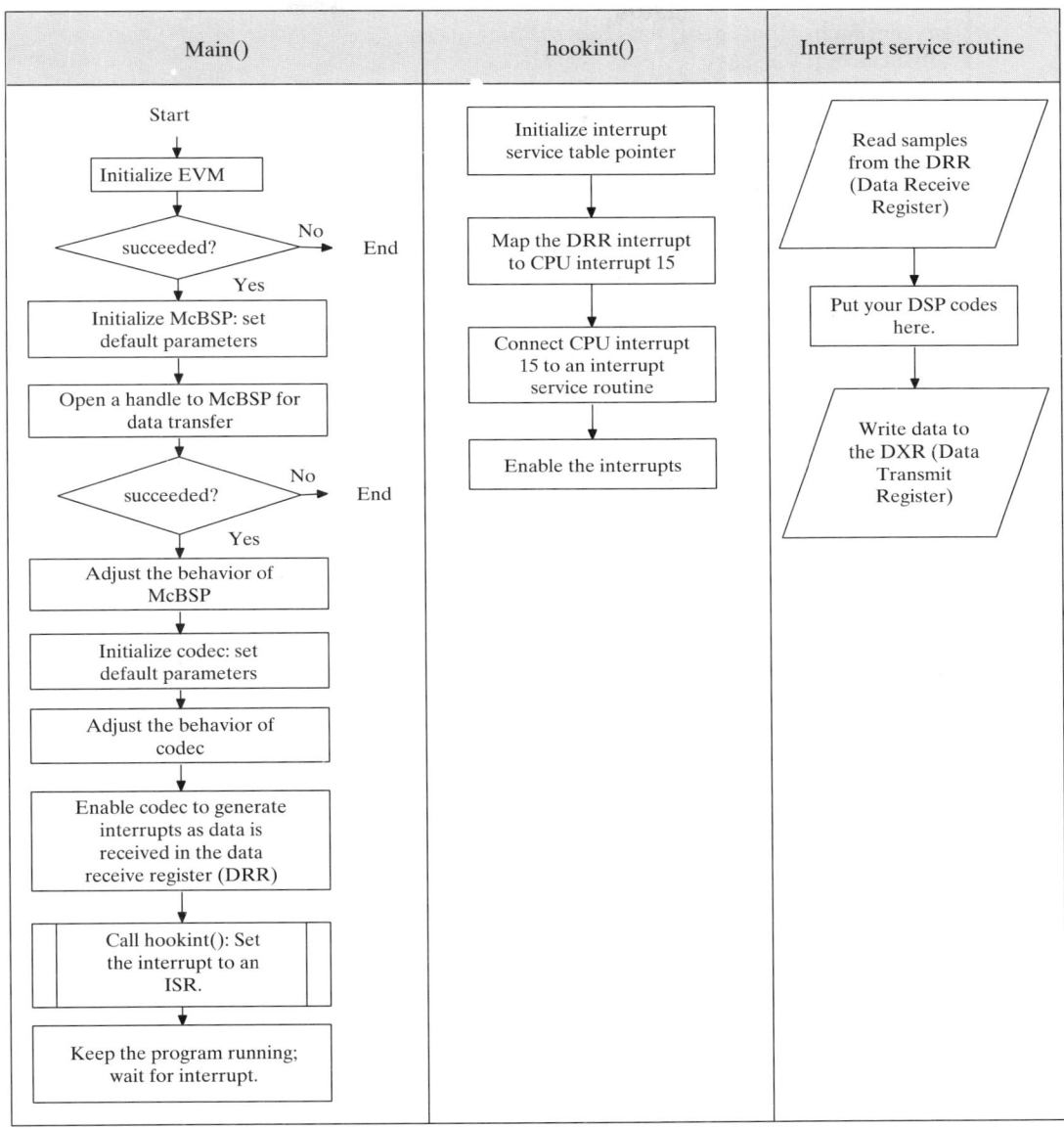

FIGURE L2-2 Flowchart of sampling program.

Chapter 7 Lab 2: Audio Signal Sampling

```
/************************************************************/
/* configure McBSP                                          */
/************************************************************/
memset(&mcbspConfig,0,sizeof(mcbspConfig));
mcbspConfig.loopback           = FALSE;
mcbspConfig.tx.update          = TRUE;
mcbspConfig.tx.clock_mode      = CLK_MODE_EXT;
mcbspConfig.tx.frame_length1   = 0;
mcbspConfig.tx.word_length1    = WORD_LENGTH_32;
mcbspConfig.rx.update          = TRUE;
mcbspConfig.rx.clock_mode      = CLK_MODE_EXT;
mcbspConfig.rx.frame_length1   = 0;
mcbspConfig.rx.word_length1    = WORD_LENGTH_32;
mcbsp_config(dev,&mcbspConfig);
MCBSP_ENABLE(0, MCBSP_BOTH);
```

The next step is to adjust the parameters of the codec. The codec is initialized by using the codec API function *codec_init()*. This function sets the codec to its default parameters. The main item to adjust here is sampling rate. It is adjusted by using the API function *codec_change_sample_rate()*. This function sets the sampling rate of the codec to the closest allowed sampling rate of the passed argument. The return value from this function will be the actual sampling rate. Table L2-1 lists the sampling rates supported by the codec. The other required adjustments are the selection of line-in or mic-in and the adjustment of their gain settings. To have stereo input, both channels should be selected and their gains adjusted to 0 dB settings. The API functions that accomplish these tasks are *codec_adc_control()*, *codec_line_in_control()*, and *codec_dac_control()*. The codec is also required to generate interrupts as data is received in the DRR (Data Receive Register). Hence, the interrupt processing capability of the codec must be enabled. This is accomplished by using the API function *codec_interrupt_enable()*. The following lines of code are used for the purpose of initializing the codec as just described:

TABLE L2-1 Sample rates allowed by CS4231A stereo audio codec.

Sample rate (KHz)	
5.5125	22.0500
6.6150	27.4286
8.0000	32.0000
9.6000	33.0750
11.0250	37.8000
16.0000	44.1000
18.9000	48.0000

```
/************************************************************/
/* configure CODEC                                          */
/************************************************************/
codec_init();
// ADC 0.0 dB gain, turn off 20dB mic gain, sel (L/R)LINE input
codec_adc_control(LEFT,0.0,FALSE,LINE_SEL);
codec_adc_control(RIGHT,0.0,FALSE,LINE_SEL);
// (L/R) LINE input to mixer
codec_line_in_control(LEFT,MIN_AUX_LINE_GAIN,FALSE);
codec_line_in_control(RIGHT,MIN_AUX_LINE_GAIN,FALSE);
// DAC 0.0 dB atten, do not mute DAC outputs
codec_dac_control(LEFT, 0.0, FALSE);
codec_dac_control(RIGHT, 0.0, FALSE);
sampleRate = 44100;
actualrate = codec_change_sample_rate(sampleRate, TRUE);
codec_interrupt_enable();
```

The initialization of the EVM, McBSP, and codec is now complete. Next, let us focus our attention on setting up an interrupt that will branch to a simple ISR to handle an incoming sampled signal.

L2.2 INTERRUPT SERVICE ROUTINE

The idea of using interrupts is commonly used for real-time data processing. This approach is widely used, since it eliminates the need for complicated synchronization schemes. In our case, the interrupt occurs when a new data sample arrives in the DRR of the serial port. The generated interrupt causes a branch to an ISR, which is then used to process the sample and send it back out. To do this, the interrupt capabilities of the EVM must be enabled and adjusted so that an unused interrupt is assigned to the DRR event of the serial port.

The first task at hand is to initialize the interrupt service table pointer (ISTP) register with the address of the global vec_table to be resolved at the link time. This is done by placing the base address of the vector table in the ISTP register. The function *intr_init()* is used for this purpose. Next, we need to select an interrupt number and map it to a CPU interrupt, in our case the DRR receive interrupt. Here, the CPU interrupt 15 is used and mapped to the DRR interrupt by using the function *intr_map()*. To connect an ISR to this interrupt, the function *intr_hook()* is used, where the name of the function that we wish to use is passed to it. The last thing is to enable the interrupts by using the macros INTR_ENABLE and INTR_GLOBAL_ENABLE. The following lines of code map the CPU interrupt 15 to the DRR interrupt and then hook it to an ISR named *serialPortRcvISR*:

```
intr_init();
intr_map(CPU_INT15, ISN_RINT0);
intr_hook(serialPortRcvISR, CPU_INT15);
INTR_ENABLE(15);
INTR_GLOBAL_ENABLE();
```

A simple ISR can now be written to receive samples from the McBSP and send them back out, unprocessed for the time being. To write an ISR, we need to state an interrupt declaration with no arguments. The macros MCBSP_READ and MCBSP_WRITE are then used to read samples from the DRR and write them to the DXR (Data Transmit Register) of the McBSP. Since port 0 was selected during the configuration of the McBSP, this port number should be specified as an argument in both MCBSP_READ and MCBSP_WRITE. Our ISR is as follows:

```
interrupt void serialPortRcvISR (void)
{
int temp;
temp = MCBSP_READ(0);
MCBSP_WRITE(0, temp);
}
```

Considering that the CPU is not actually doing anything as it waits for a new data sample, an infinite loop is set up inside the main program to keep it running. As an interrupt occurs, the program branches to the ISR, performs it, and then returns to its wait state. This is accomplished by using a simple while(1){} statement.

Now the complete program for sampling an analog signal is ready for use. Basically, the program services interrupts to read in samples of an analog signal, such as the output from a CD player connected to the line-in of the EVM.

To build this program in Code Composer Studio, the project should include three libraries: *rts6201.lib* (or *rts6701.lib* if the target board is C67x and floating-point values are used), *drv6x.lib*, and *dev6x.lib*. The library *rts6201.lib* is the run-time support library containing the run-time support functions such as math functions. The library *dev6x.lib* is a collection of macros and functions for programming the C6x registers and peripherals. This library allows the programmer to control interrupt functionality, CPU operational modes, and internal peripherals, including McBSPs. In addition to these libraries files, a linker command file needs to be added into the project. The following command file is used in this lab:

link.cmd

```
MEMORY
{
INT_PROG_MEM     (RX)       : origin = 0x00000000 length = 0x00010000
SBSRAM_PROG_MEM  (RX)       : origin = 0x00400000 length = 0x00014000
SBSRAM_DATA_MEM  (RW)       : origin = 0x00414000 length = 0x0002C000
SDRAM0_DATA_MEM  (RW)       : origin = 0x02000000 length = 0x00400000
SDRAM1_DATA_MEM  (RW)       : origin = 0x03000000 length = 0x00400000
INT_DATA_MEM     (RW)       : origin = 0x80000000 length = 0x00010000
```

```
}
SECTIONS
{
 .vec:     load = 0x00000000
 .text:    load = SBSRAM_PROG_MEM
 .const:   load = INT_DATA_MEM
 .bss:     load = INT_DATA_MEM
 .data:    load = INT_DATA_MEM
 .cinit:   load = INT_DATA_MEM
 .pinit:   load = INT_DATA_MEM
 .stack:   load = INT_DATA_MEM
 .far      load = INT_DATA_MEM
 .sysmem   load = SDRAM0_DATA_MEM
 .cio      load = INT_DATA_MEM
 sbsbuf    load = SBSRAM_DATA_MEM
        {_SbsramDataAddr=.;_SbsramDataSize = 0x0002C000; }
}
```

Figure L2-3 shows the Project view after the necessary files are added into the project. To build an executable file from these files, the button Rebuild All needs to be clicked. The executable file can then get loaded by choosing the menu item File–>Load Program. By running the executable and connecting the output of a CD player to the line-in and a pair of powered speakers to the line-out, CD quality sound should be heard. Figure L2-4 shows the block diagram of this setup.

The effect of the sampling rate on the sound quality can be studied by adding the following lines in the main program *codec.c*:

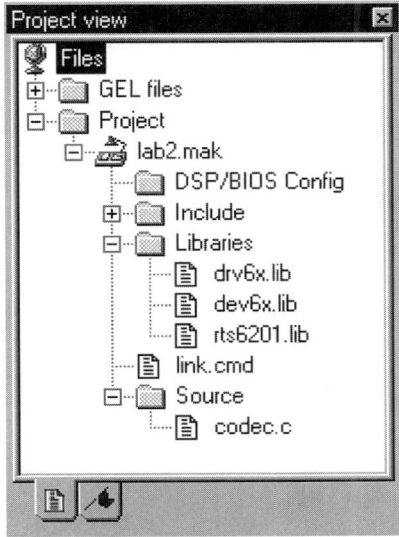

FIGURE L2-3 Project view for Lab2.

136 Chapter 7 Lab 2: Audio Signal Sampling

FIGURE L2-4 Block diagram of Lab 2 setup.

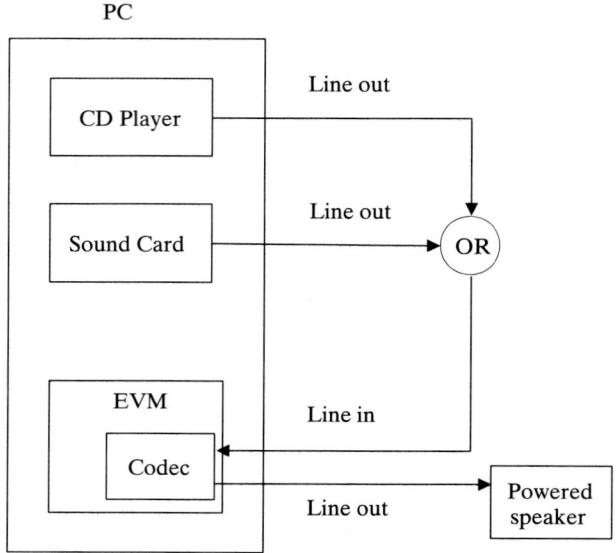

```
int main()
{
        ...
        ...
        sampleRate = 5500; /* Change the sampling rate */
        codec_change_sample_rate(sampleRate, TRUE);
}
```

This will change the sampling rate to 5.5 kHz. By rebuilding, reloading, and running the executable, a degradation in the sound quality can be heard due to the deviation from the Nyquist rate.

It is possible to manipulate the audio signal. For example, the sound volume can be controlled by multiplying a volume gain factor with the sound samples. The code for doing so is as follows:

```
int volumeGain;
int main()
{
        volumeGain = 1; /* Initialize */
        ...
        ...
}
interrupt void serialPortRcvISR (void)
{
        int temp;
```

L2.3 DSK 137

```
            temp = MCBSP_READ(0);
            temp = temp * volumeGain;
            MCBSP_WRITE(0, temp);
}
```

The variable volumeGain is declared as a global variable in order to be accessed at run time. In order to change the volume at run time, the option Edit->Edit Variable should be chosen, which brings up an Edit Variable dialog box. As shown in Figure L2-5, by entering volumeGain in the Variable field and a desired gain value in the Value field of this dialog box, the sound volume can be altered.

L2.3 DSK

Any program using C6x instructions can be run on a DSK board. However, it should be realized that the codec on the DSK board is different from the one on the EVM board. At the time of this writing, there are no available DSK I/O APIs similar to the ones used in the aforementioned sampling shell program. The sample rate on the DSK board is fixed at 11.025 kHz and cannot be changed by software. Therefore, to run the lab programs on a DSK board, it is advised that the probe point feature of CCS is used to read in samples from a file. The files for running the labs on a DSK board are provided on the attached CD under the subdirectory DSK.

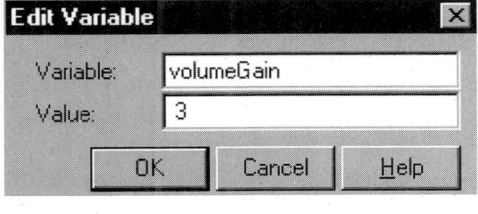

FIGURE L2-5 Editing value of a variable.

CHAPTER 8

Fixed Point vs. Floating Point

One important feature that distinguishes different DSP processors is whether their CPUs perform fixed-point or floating-point arithmetic. In a fixed-point processor, numbers are represented and manipulated in integer format. In a floating-point processor, in addition to integer arithmetic, floating-point arithmetic can be handled. This means that numbers are represented by the combination of a mantissa (or a fractional part) and an exponent part, and the CPU possesses the necessary hardware for manipulating both of these parts. As a result, in general, floating-point processors are more expensive and slower than fixed-point ones.

In a fixed-point processor, one needs to be concerned with the dynamic range of numbers, since a much narrower range of numbers can be represented in integer format as compared to floating-point format. For most applications, such a concern can be virtually ignored when using a floating-point processor. Consequently, fixed-point processors usually demand more coding effort than do their floating-point counterparts.

8.1 Q-FORMAT NUMBER REPRESENTATION ON FIXED-POINT DSPS

The decimal value of a 2's-complement number $B = b_{N-1}b_{N-2}\ldots b_1b_0, b_i \in \{0,1\}$, is given by

$$D(B) = -b_{N-1}2^{N-1} + b_{N-2}2^{N-2} + \ldots + b_1 2^1 + b_0 2^0. \quad (8.1)$$

2's-complement representation allows a processor to perform integer addition and subtraction by using the same hardware. When using unsigned integer representation, the sign bit is treated as an extra bit. This way, only positive numbers can be represented.

There is a limitation to the dynamic range of the foregoing integer representation scheme. For example, in a 16-bit system, it is not possible to represent numbers larger than $+2^{15} - 1 = 32767$ and smaller than $-2^{15} = -32768$. To cope with this limitation, numbers are normalized between -1 and 1. In other words, they are represented as fractions. This normalization is achieved by the programmer moving the implied or imaginary binary point (note that there is no physical memory allocated to this point) as indicated in Figure 8-1. This way, the fractional value is given by

$$F(B) = -b_{N-1}2^0 + b_{N-2}2^{-1} + \ldots + b_1 2^{-(N-2)} + b_0 2^{-(N-1)}. \quad (8.2)$$

This representation scheme is referred to as Q-format or fractional representation. The programmer needs to keep track of the implied binary point when manipulating Q-format numbers. For instance, let us consider two Q-15 format numbers, given that we have a 16-bit wide memory. Each number consists of 1 sign bit plus 15 fractional bits. When these numbers are multiplied, a Q-30 format number is obtained (the

8.1 Q-format Number Representation on Fixed-point DSPs 139

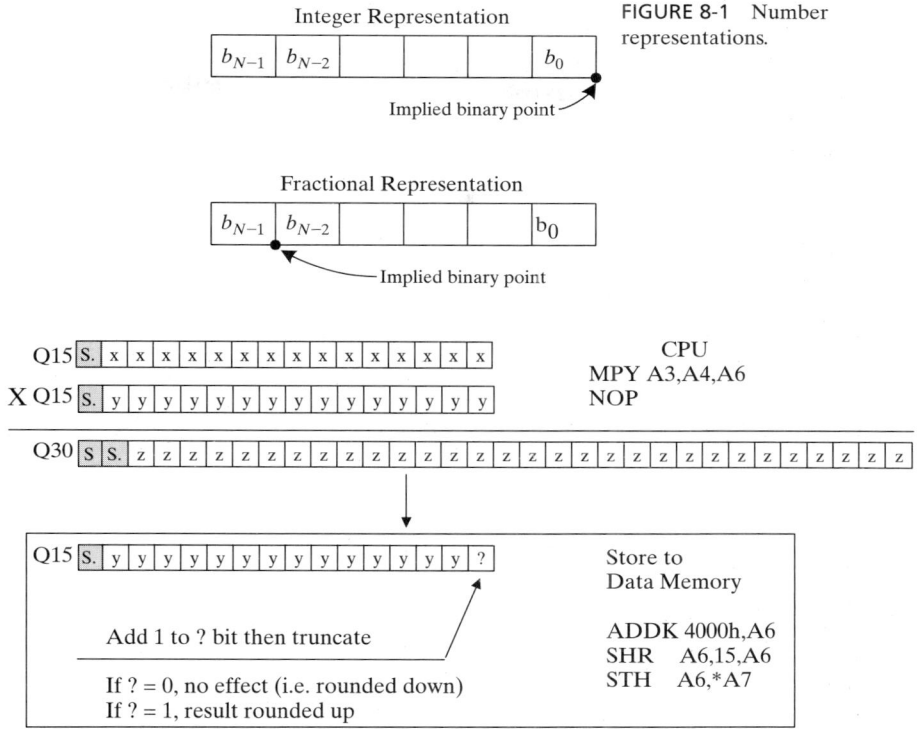

FIGURE 8-1 Number representations.

FIGURE 8-2 Multiplying and storing Q-15 numbers.†

product of two fractions is still a fraction), with bit 31 being the sign bit and bit 32 another sign bit (called an *extended* sign bit). If not enough bits are available to store all 32 bits, and only 16 bits can be stored, it makes sense to store the most significant bits. This translates into storing the upper portion of the 32-bit product register by doing a 15-bit right shift (SHR). In this manner, the product would be stored in Q-15 format. (See Figure 8-2.)

Based on 2's-complement representation, a dynamic range of $-(2^{N-1}) \leq D(B) \leq 2^{N-1} - 1$ can be achieved, where N denotes the number of bits. For illustration purposes, let us consider a 4-bit system in which the most negative number is -8 and the most positive number is 7. The decimal representations of the numbers are shown in Figure 8-3. Notice how the numbers change from most positive to most negative with the sign bit. Since only the integer numbers falling within the limits -8 and 7 can be represented, it is easy to see that any multiplication or addition resulting in a number larger than 7 or smaller than -8 will cause overflow. For example, when 6 is multiplied by 2, we get 12. Hence, the result is greater than the representation limits and will be wrapped around the circle to 1100, which is -4.

Q-format representation solves this problem by normalizing the dynamic range between -1 and 1. Any resulting multiplication will be within the limits of this range. In other words, the dynamic range gets divided into 2^N sections, where $2^{-(N-1)}$ is the

140 Chapter 8 Fixed Point vs. Floating Point

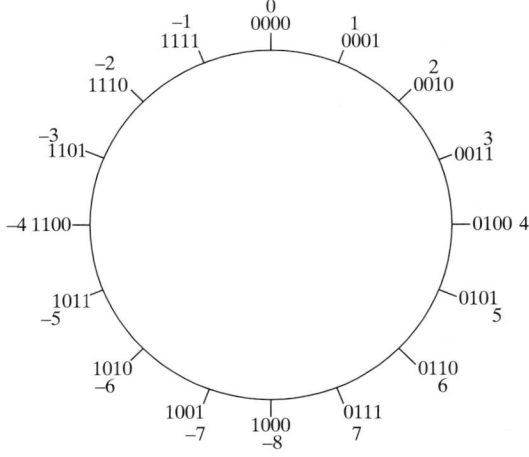

FIGURE 8-3 Four-bit binary representation.

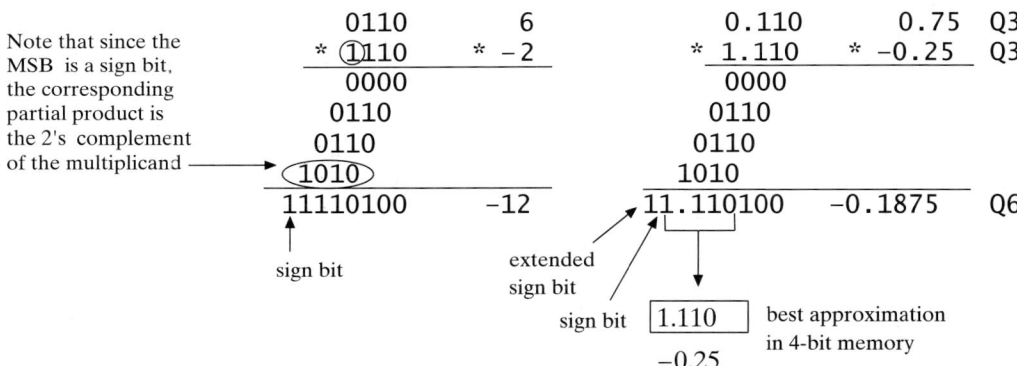

FIGURE 8-4 Binary and fractional multiplication.

size of a section. The most negative number is always -1, and the most positive number is $1 - 2^{-(N-1)}$.

The following example helps one to see the difference in the two representation schemes. As shown in Figure 8-4, the multiplication of 0110 by 1110 in binary is the equivalent of multiplying 6 by −2 in decimal, giving an outcome of −12, a number exceeding the dynamic range of the 4-bit system. Based on the Q-3 representation, these numbers correspond to 0.75 and −0.25, respectively. The result is −0.1875, which falls within the fractional range. Notice that the hardware generates the same 1's and 0's, what is different is the interpretation of the bits.

When multiplying Q-N numbers, it should be remembered that the result will consist of 2N fractional bits, one sign bit, and one or more extended sign bits. Based on the datatype used, the result has to be shifted accordingly. If two Q-15 numbers are multiplied, the result will be 32-bits wide, with the MSB being the extended sign bit followed by the sign bit. The imaginary decimal point will be after the 30th bit. So a right shift of 15 is required to store the result in a 16-bit memory location as a Q-15 number. It should be realized that some precision is lost, of course, as a result of discarding the smaller fractional bits. Since only 16 bits can be stored, the shifting allows one to retain the higher precision fractional bits. If a 32-bit storage capability is available, a left shift of 1 can be performed to remove the extended sign bit to store the result as a Q-31 number.

To further understand a possible precision loss when manipulating Q-format numbers, let us consider another example in which two Q12 numbers corresponding to 7.5 and 7.25 are multiplied. As can be seen from Figure 8-5, the resulting product must be left shifted by 4 bits to store all the fractional bits corresponding to Q12 format. However, doing so would result in a product value of 6.375, which is different than the correct value of 54.375. If the product is stored in a lower precision Q format—say, in Q8 format—then the correct product value can be stored.

Although Q-format solves the problem of overflow in multiplication, addition and subtraction still pose a problem. When two Q15 numbers are added, the sum could exceed the range of Q15 representation. To solve this problem, the scaling approach, discussed later in this chapter, needs to be employed.

8.2 FINITE WORD LENGTH EFFECTS ON FIXED-POINT DSPS

Because memory or registers have a finite number of bits, there could be a noticeable error between desired and actual outcomes on a fixed-point processor. The finite word length quantization effect is similar to the input data quantization effect introduced by an A/D converter.

Consider fractional numbers quantized by a $b + 1$ bit converter. When these numbers are manipulated and stored in an $M + 1$ bit memory, with $M < b$, there is going to be an error (simply because $b - M$ of the least significant fractional bits are discarded or truncated). This finite word length error could alter the behavior of a system to an unacceptable degree. The range of the magnitude of truncation error ε_t is given by $0 \leq |\varepsilon_t| \leq 2^M - 2^b$. The lowest level of truncation error corresponds to the situation when all the thrown-away bits are zeros, and the highest level to the situation when all the thrown-away bits are ones.

This effect has been extensively studied for FIR and IIR filters. (For example see [5].) Since the coefficients of such filters are represented by a finite number of bits, the roots of their transfer function polynomials or the positions of their zeros and poles

```
Q12  →  7.5           0111. 1000 0000 0000
Q12  →  7.25       *  0111. 0100 0000 0000
Q24  →  54.375     0011 0110. 1100 0000 0000 0000
                              |_____|
                                  Q12  →  6.375
                   |_____|
                          Q8  →  54.375
```

FIGURE 8-5   Q-format precision loss example.

shift in the complex plane. The amount of shift in the positions of poles and zeros can be related to the amount of quantization error in the coefficients. For example, for an Nth-order IIR filter, the sensitivity of the ith pole p_i with respect to the kth coefficient A_k can be derived to be (see [5]),

$$\frac{\partial p_i}{\partial A_k} = \frac{-p_i^{N-k}}{\prod_{\substack{l=1 \\ l \neq i}}^{N}(p_i - p_l)}. \tag{8.3}$$

This means that the change in the position of a pole is influenced by the positions of all the other poles. That is the reason the implementation of an Nth order IIR filter is normally achieved by having a number of second-order IIR filters in series in order to decouple this dependency of poles.

Also, note that as a result of coefficient quantization, the actual frequency response $\hat{H}(e^{j\theta})$ would become different than the desired frequency response $H(e^{j\theta})$. For example, for a FIR filter having N coefficients, it can be easily shown that the amount of error in the magnitude of the frequency response, $|\Delta H(e^{j\theta})|$, is bounded by

$$|\Delta H(e^{j\theta})| = |H(e^{j\theta}) - \hat{H}(e^{j\theta})| \leq N2^{-b}. \tag{8.4}$$

In addition to the preceding effects, coefficient quantization can lead to limit cycles. This means that in the absence of an input, the response of a supposedly stable system (poles inside the unit circle) to a unit sample is oscillatory instead of diminishing in magnitude.

8.3 FLOATING-POINT NUMBER REPRESENTATION

Due to the relatively limited dynamic ranges of fixed-point processors, when using such processors, one should be concerned with the scaling issue, or how big the numbers get, in the manipulation of a signal. Scaling is not an issue when using floating-point processors, since the floating-point hardware provides a much wider dynamic range. The C67x processor is the floating-point version of the C6x family with many additional floating-point instructions. [Appendix A (Quick Reference Guide) includes a listing of the C67x floating-point instructions.]

There are two floating-point data representations on the C67x processor: single precision (SP) and double precision (DP). In the single-precision format, a value is expressed as

$$-1^s * 2^{exp-127} * 1.frac, \tag{8.5}$$

where s denotes the sign bit (bit 31), exp the exponent bits (bits 23 through 30), and $frac$ fractional or mantissa bits (bits 0 through 22). (See Figure 8-6.)

31	30	23 22	0
s	exp		frac

FIGURE 8-6 Floating point data representation.

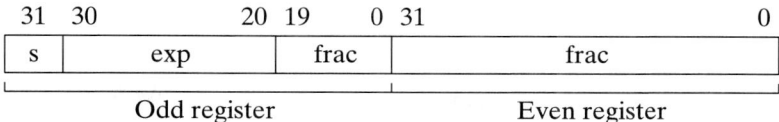

FIGURE 8-7 Double precision floating point representation.

Consequently, numbers as big as $3.4 * 10^{38}$ and as small as $1.175 * 10^{-38}$ can be processed.

In the double-precision format, more fractional and exponent bits are made available by using two words as

$$-1^s * 2^{exp-1023} * 1.frac, \qquad (8.6)$$

where the exponent bits are from bits 20 through 30 and the fractional bits are all the bits of one word and bits 0 through 19 of the other word. (See Figure 8-7.) In this manner, numbers as big as $1.7 * 10^{308}$ and as small as $2.2 * 10^{-308}$ can be handled.

When using a floating-point processor, all the steps needed to perform floating-point arithmetic are done by the CPU floating-point hardware. For example, consider adding two floating-point numbers represented by

$$a = a_{frac} * 2^{a_{exp}}$$

and

$$b = b_{frac} * 2^{b_{exp}}. \qquad (8.7)$$

The floating-point sum c has the following exponent and fractional parts:

$$\begin{aligned} c &= a + b \\ &= \left(a_{frac} + \left(b_{frac} * 2^{-(a_{exp}-b_{exp})}\right)\right) * 2^{a_{exp}} \quad \text{if } a_{exp} \geq b_{exp} \qquad (8.8) \\ &= \left(\left(a_{frac} * 2^{-(b_{exp}-a_{exp})}\right) + b_{frac}\right) * 2^{b_{exp}} \quad \text{if } a_{exp} < b_{exp}. \end{aligned}$$

These parts are computed by the floating-point hardware. This shows that, though possible, it is inefficient to perform floating-point arithmetic on fixed-point processors, since all the operations involved, such as those in Eq. (8.8), must be done in software.

The instructions ending in SP denote single-precision data format and in DP double-precision data format (e.g., MPYSP and MPYDP). It should be noted that some of these instructions require additional execute (E) cycles or latencies compared with fixed-point instructions. (See Figure 5-8.) For example, MPYSP requires three delays or NOPs and MPYDP nine delays or NOPs compared with one delay or NOP for fixed-point multiplication MPY.

As illustrated in Figure 8-8, the C62x can support 40-bit and the C67x 64-bit operations by concatenating two registers. Table 8-1 shows a listing of all the C6x datatypes.

8.4 OVERFLOW AND SCALING

As stated before, fixed-point processors have a much smaller dynamic range than their floating-point counterparts. Even though the C62 is considered to be a 32-bit device, its multiplier can multiply only 16-bit numbers. It is due to this limitation that the Q-15

144 Chapter 8   Fixed Point vs. Floating Point

TABLE 8-1   C6x datatypes.†

Type	Size	Representation
Char, signed char	8 bits	ASCII
unsigned char	8 bits	ASCII
short	16 bits	2's complement
unsigned short	16 bits	binary
int, signed int	32 bits	2's complement
unsigned int	32 bits	binary
long, signed long	40 bits	2's complement
unsigned long	40 bits	binary
enum	32 bits	2's complement
float	32 bits	IEEE 32-bit
double	64 bits	IEEE 64-bit
long double	64 bits	IEEE 64-bit
pointers	32 bits	binary

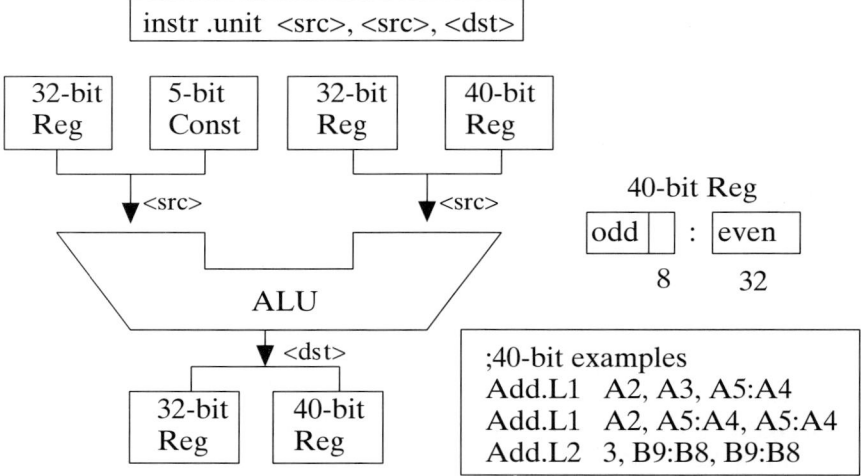

FIGURE 8-8   40-bit operations.†

representation of numbers is normally considered. The 16-bit multiplier can multiply two Q-15 numbers and produce a 32-bit product. Then the product can be stored in 32 bits or shifted back to 16 bits for storage or further processing.

When multiplying two Q-15 numbers, which are in the range of −1 to 1, it is clear that the resulting number will always be in the same range. However, when two Q-15 numbers are added, the sum may fall outside this range, leading to overflow. Overflow can cause major problems by generating erroneous results. In using a fixed-point processor, the range of numbers must be closely examined and adjusted to compensate for overflow. The simplest correction method for overflow is scaling.

8.4 Overflow and Scaling

The idea of scaling can be applied to most filtering and transform operations, where the input is scaled down for processing and the output is then scaled back up to the original size. An easy way to do scaling on the C62 is by shifting. Since a right shift of 1 is equivalent to a division by 2, we can scale the input repeatedly by 0.5 until all overflows disappear. The output can then be rescaled back to the total scaling amount.

As far as FIR and IIR filters are concerned, it is possible to scale coefficients to avoid overflow. Let us consider the output of a filter $y[n] = \sum_{k=0}^{N-1} h[k] * x[n-k]$, where the h's denote coefficients or unit sample response terms and the x's designate input samples. In case of IIR filters, for a large enough N, the terms of the unit sample response become so small that they can be ignored. Let us suppose that x's are in Q-15 format (i.e., $|x[n-k]| \leq 1$). Therefore, we can write $|y[n]| \leq \sum_{k=0}^{N-1} |h[k]|$. This means that, to ensure no output overflow (i.e., $|y[n]| \leq 1$), the condition $\sum_{k=0}^{N-1} |h[k]| \leq 1$ must be satisfied. This condition can be satisfied by repeatedly scaling (dividing by 2) coefficients or unit sample response terms.

The C62 provides a saturation flag bit, which is bit 9 of the CSR register. To cope with addition overflows, the saturated add instruction SADD can be used to see whether the saturation bit SAT is set to unity, indicating an overflow. Assuming Q-15 format values, the following function can be used to check the status of the SAT bit after using the _sadd() intrinsic:

```
short safe_add(short A, short B,int *status)
{
        int X,Y,result, SAT_BIT;
        X = A << 16;
        Y = B << 16;
        result = _sadd(X,Y);
        SAT_BIT=GET_REG_BIT(CSR,9);
        if(SAT_BIT==1){
                //Overflow Occured
                RESET_REG_BIT(CSR,9);       //Reset Sat Bit
                *status = 1;
        }
        else
                *status = 0;

        return (result >> 16);
}
```

This function adds two 16-bit numbers and reports any occurring overflow. If an overflow occurs, it also clears the SAT bit in the CSR.

148 Chapter 8 Fixed Point vs. Floating Point

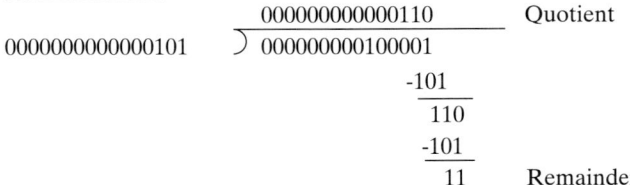

```
              MV       y, quot
  [sign]      NEG      quot, quot        ;incorporate quotient

              MV       quot, A4

              .endproc A4, B3

              B        B3
              NOP      5
```

Long Division:

```
                    000000000000110        Quotient
  0000000000000101 ) 000000000100001
                           -101
                           ────
                            110
                           -101
                           ────
                             11           Remainder
```

SUBC Method:

31 HIGH register	LOW register 0	comment

0000000000000000
       -10
──────────────
      -10
	0000000000100001
	1000000000000000
	──────────────
	0111111110011111
		(1) Dividend is loaded into register. The divisor is left-shifted 15 and subtracted from register. The subtraction is negative, so discard the result and shift the register left one bit.

0000000000000000
       -10
──────────────
      -10
	0111111101111110
	1000000000000000
	──────────────
	0111111110111110
		(2) 2nd subtract produces negative answer, so discard result and shift register (dividend) left.

⋮	⋮	⋮
0000000000000100 -10 ────────────── 0000000000000001	0010000000000000 1000000000000000 ────────────── 1010000000000000	(14) 14th SUBC command. The result is positive. Shift result left and replace LSB with 1.
0000000000000011 -10 ────────────── 0000000000000000	0100000000000001 1000000000000000 ────────────── 1100000000000001	(15) Result is again positive. Shift result left and replace LSB with 1.
0000000000000001 -10	1000000000000011 1000000000000000 ────────────── 1111111111111101	(16) Last subtract. negative answer, so discard result and shift register left.
0000000000000011	0000000000000110	Answer reached after 16 SUBC instructions.
Remainder	Quotient	

FIGURE 8-9   SUBC division example 33 by 5.†

8.5 Some Useful Arithmetic Operations

8.5.2 Sine and Cosine

Trigonometric functions such as sine and cosine can be approximated by using the Taylor series expansion. For sine, we can write the expansion as

$$\sin(x) \cong x - \frac{x^3}{3!} + \frac{x^5}{5!} - \frac{x^7}{7!} + \frac{x^9}{9!} \tag{8.10}$$

for the first five terms. For more precision, higher order terms need to be considered. For implementation purposes, this expansion can be rewritten as follows:

$$\sin(x) \cong x * \left(1 - \frac{x^2}{2*3}\left(1 - \frac{x^2}{4*5}\left(1 - \frac{x^2}{6*7}\left(1 - \frac{x^2}{8*9}\right)\right)\right)\right). \tag{8.11}$$

Similarly, for cosine, we can write

$$\cos(x) \cong 1 - \frac{x^2}{2} + \frac{x^4}{4!} - \frac{x^6}{6!} + \frac{x^8}{8!}$$

$$= 1 - \frac{x^2}{2}\left(1 - \frac{x^2}{3*4}\left(1 - \frac{x^2}{5*6}\left(1 - \frac{x^2}{7*8}\right)\right)\right). \tag{8.12}$$

Furthermore, to generate sine and cosine waves, the following recursive formulas can be used:

$$\sin nx = 2\cos x * \sin(n-1)x - \sin(n-2)x;$$
$$\cos nx = 2\cos x * \cos(n-1)x - \cos(n-2)x. \tag{8.13}$$

8.5.3 Square Root

Square root $sqrt(y)$ can be approximated by the following Taylor series expansion, assuming that $y^{0.5} = (x+1)^{0.5}$:

$$sqrt(y) \cong 1 + \frac{x}{2} - \frac{x^2}{8} + \frac{x^3}{16} - \frac{5x^4}{128} + \frac{7x^5}{256}$$

$$= 1 + \frac{x}{2} - 0.5\left(\frac{x}{2}\right)^2 + 0.5\left(\frac{x}{2}\right)^3 - 0.625\left(\frac{x}{2}\right)^4 + 0.875\left(\frac{x}{2}\right)^5. \tag{8.14}$$

Here, it is assumed that x is in Q15 format. For more precision, higher order terms need to be included. In this equation, the estimation error would be small for x values near unity. Hence, to improve accuracy in applications where the range of x is known, x can be scaled by a^2 to bring it close to unity (i.e., $sqrt(a^2 x)$, where $a^2 x \cong 1$). The result should then be scaled back by $1/a$.

It is also possible to compute the square root by using the recursive equation

$$v[n+1] = v[n] * (1.5 - (x/2) * v[n] * v[n]). \tag{8.15}$$

8.5.4 Lookup Table

An example of a lookup table is given next to show how this approach works. In this example, the arctangent function $arctan(x)$ is computed on the basis of a previously stored table of length 1024. Since $arctan(-x) = \pi/2 - arctan(x)$, the table needs to include only the entries for positive x values. Arctangent values vary from $-\pi/2$ to $\pi/2$ for x values from -1 to 1.

```
shr    x,5,index
ldh    *+p_arctan[index],arctan
```

Lab 3: Integer Arithmetic

Implementing algorithms on a fixed-point DSP requires that the range of numbers be closely examined in order to make necessary adjustments to avoid overflows. The simplest approach to correct for overflow is by scaling the input. This lab demonstrates the scaling approach to correct for overflow.

L3.1 OVERFLOW HANDLING

Overflow occurs when the result of an operation is too large or too small for the CPU to handle. In a 16-bit system, when manipulating integer numbers, they must remain in the range of −32768 to 32767. Otherwise, any operation resulting in a number smaller than −32768 or larger than 32767 will cause overflow. For example, when 32767 is multiplied by 2, we get 65534, which is beyond the representation limit of a 16-bit system.

Consider the following program:

```
#include <stdio.h>
#define SIZE 16
short SIGNAL[SIZE] = {
        11474,
        21204,
        27709,
        29999,
        27727,
        21238,
        11519,
           47,
        -11430,
        -21170,
        -27691,
        -29999,
        -27746,
        -21272,
        -11563,
          -95} ; /* Original data */
short NEWSIGNAL[SIZE]; /* Data after multiplication */
main()
{
int i;
/* Multiplication */
for(i=0;i<SIZE;i++)
{
        NEWSIGNAL[i] = SIGNAL[i] * 2; // multiply by 2
 }
}
```

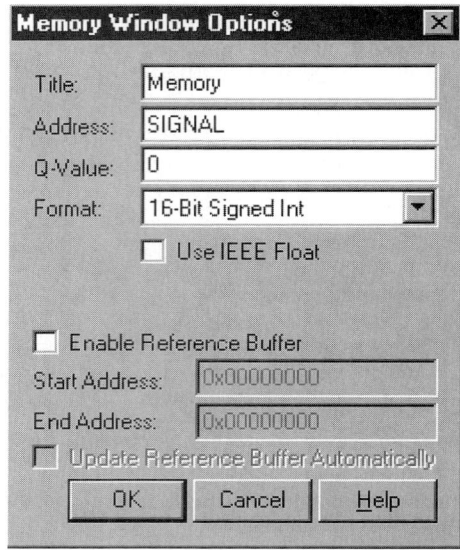

FIGURE L3-1 Memory Window Options dialog box.

In this program, the array SIGNAL contains samples of a sinusoidal signal. These sample values are multiplied by 2, and the results are placed into the array NEWSIGNAL. Let us examine whether any overflow is caused by these multiplications. In order to monitor the values, it is possible to use either the Watch Window or View Memory feature of CCS. Let us use the View Memory feature by choosing View->Memory from the menu bar. As a result, the dialog box as shown in Figure L3-1 will appear. In the dialog box, enter SIGNAL in the Address field, select 16-Bit Signed Int from the pop-down list of the Format field, and then click OK. A memory window displaying the values of SIGNAL will appear, as shown in Figure L3-2(a). Repeat these steps to see the values of NEWSIGNAL, as shown in Figure L3-2(b). From Figure L3-2, it can be seen that the array NEWSIGNAL includes wrong values due to overflows. For example, −23128 is indicated to be the result of multiplying 21204 by 2, which is incorrect.

As shown in Figure Figure L3-3, the CCS View Graph feature can be used to display SIGNAL and NEWSIGNAL. The multiplication of SIGNAL by 2 is expected to generate another sinusoidal signal with twice the amplitude. However, as seen from Figure L3-3(b), NEWSIGNAL is distorted and clipped when the multiplication results are beyond the 16-bit (short datatype) range.

L3.2 SCALING APPROACH

Scaling samples is the most widely used way to overcome the overflow problem. In order to see how scaling works, let us consider a simple multiply/accumulate operation. Suppose there are four constants or coefficients that need to be multiplied by samples of an input analog signal. The worst possible overflow case would be the one where all the multiplicants (C_k's and $x[n]$'s) are 1. For this case, the result $y[n]$ will be 4, given that

L3.2 Scaling Approach

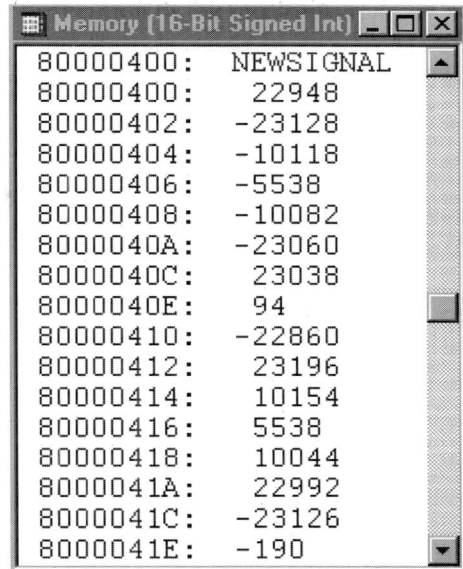

80000420:	SIGNAL
80000420:	11474
80000422:	21204
80000424:	27709
80000426:	29999
80000428:	27727
8000042A:	21238
8000042C:	11519
8000042E:	47
80000430:	-11430
80000432:	-21170
80000434:	-27691
80000436:	-29999
80000438:	-27746
8000043A:	-21272
8000043C:	-11563
8000043E:	-95

80000400:	NEWSIGNAL
80000400:	22948
80000402:	-23128
80000404:	-10118
80000406:	-5538
80000408:	-10082
8000040A:	-23060
8000040C:	23038
8000040E:	94
80000410:	-22860
80000412:	23196
80000414:	10154
80000416:	5538
80000418:	10044
8000041A:	22992
8000041C:	-23126
8000041E:	-190

FIGURE L3-2 Memory windows showing array values.

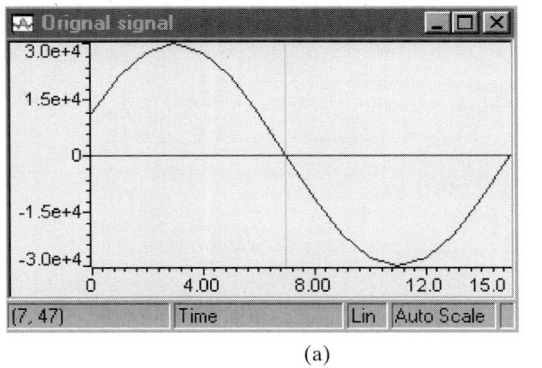
(a)

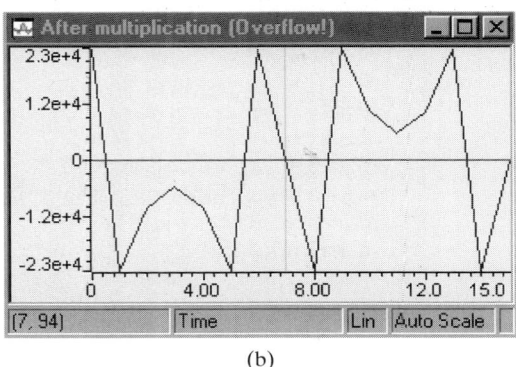
(b)

FIGURE L3-3 Signal distorted by overflow: (a) original, and (b) distorted.

$y[n] = \sum_{k=1}^{4} C_k * x[n - k]$. Assuming that we have control only over the input, the input samples should be scaled so that the result $y[n]$ will fall into the allowed range. A single right shift reduces the input by one-half, and a double shift reduces it further by one-quarter. Of course, this leads to less precision, but it is better than getting erroneous results.

154 Chapter 8 Lab 3: Integer Arithmetic

A simple method for implementing the scaling approach on the C62 is to create a function that returns the necessary amount of scaling on the input. For any multiply/accumulate type of operations, such as filtering or transform, the worst case is the multiplication and addition of all 1's. Then the required amount of scaling would be dependent on the number of additions in the summation. To examine the worst case, it is required to obtain the number of scalings needed for all overflows to disappear. This can be achieved by writing a function to compute the required number of scalings or shiftings of input samples. For the example in this lab, such a function is as follows and is named *getNumberOfScaling()*:

```
#incluce <stdio.h>
#incluce <regs.h.
#define SIZE 16
float Coeff[SIZE] = { /* Test array */
       0,
   0.8311,
  -0.2977,
   0.4961,
   0.6488,
  -0.3401,
  -0.0341,
  -0.2336,
  -0.38C1,
  -0.3984,
  -0.2568,
   0.4884,
   0.1113,
   0.2495,
   0.9999,
  -0.4088} ; /* coefficient */
short safe_add(short A, short B, int *status);
void rescale(short g[]);
void main()
{
   int n;
   n = GetNumberOfScaling(Coeff);
   printf("Numer of times to scale: %d\n", n);
}

int GetNumberOfScaling(float *Coeff)
{
        short sum, g[SIZE];
        int i,bOverFlow, numberOfScaling;

        /* Convert to Q-15, good approximate */
        for(i=0;i<SIZE;i++)
        {
```

```
                g[i]=0x7fff*Coeff[i];
        }
        numberOfScaling = 0;
start:
        sum = 0;
        //Add all values to see if OVERFLOW occurs
        for(i=0;i<SIZE;i++)
        {
                sum = safe_add(sum,g[i],&bOverFlow);
                if(bOverFlow == 1)
                { /* Overflow occurred. */
                        rescale(g);
                        numberOfScaling++;
                        printf("Overflow occurred at summation %d\n", i+1);
                        goto start;
                }
        }
        return numberOfScaling;
}
void rescale(short g[])
{
        int k, temp;
        //Rescale Input since it Overflows
        for(k=0;k<SIZE;k++)
        {
                temp=(0x4000 * g[k]) << 1; // Half it
                g[k] = temp >> 16; // Half it
        }
}
short safe_add(short A, short B, int *status)
{
        int AA,BB,result,SAT_BIT;
        AA=A<<16;
        BB=B<<16;
        result = _sadd(AA,BB);
        SAT_BIT=GET_REG_BIT(CSR,9);
        if(SAT_BIT==1)
        {
                //Overflow Occured
                RESET_REG_BIT(CSR,9); //Reset Sat Bit
                *status = 1;
        }
        else
        {
                *status = 0;
        }
        return (result >> 16);
}
```

The function *GetNumerOfScaling()* first produces a good approximation to Q-15 format by multiplying the input float values by 0x7FFF (effectively scaling by 2^{15}). The summation is then obtained by using the function *safe_add()*, which sets the saturation bit if an overflow occurs. The overflow status is checked after every call to *safe_add()*. If it is 1, indicating an overflow, the function *rescale()* is called to scale down the input. The number of scalings is also counted. After scaling the input, the summation is repeated. If another overflow occurs, the input sample is scaled down further. The process is continued until no overflow occurs. The final number of scalings is then returned. Care must be taken not to scale the input too many times; otherwise, the input signal gets buried in quantization noise.

It should be noted that, in addition to scaling the input, it is also possible to scale the coefficients or constants in a summation (such as filter coefficients or FFT twiddle factors) to force the outcome to stay within the dynamic range. Depending on the values of constants or coefficients, it may not be necessary to do the maximum shift for each value. As far as the preceding program is concerned, it can be seen that an overflow occurs at the fourth summation, and one scaling is required to avoid it. The execution result is displayed in Figure L3-4, and Table L3-1 shows the sum of the coefficients C_k. Notice that in the worst case, the inputs are all 1's, so the sum of the C_k's overflows at the fourth summation, which is highlighted in the table. If the coefficients are scaled down by one-half, this is equivalent to scaling down the input samples by one-half, and the overflows disappear.

TABLE L3-1 Scaling example.

C_k	$\sum C_k$	$\dfrac{C_k}{2}$	$\sum \dfrac{C_k}{2}$
0	0	0	0
0.8311	0.8311	0.41555	0.41555
−0.2977	0.5334	−0.14885	0.2667
0.4961	**1.0295**	0.24805	0.51475
0.6488	**1.6783**	0.3244	0.83915
−0.3401	**1.3382**	−0.17005	0.6691
−0.0341	**1.3041**	−0.01705	0.65205
−0.2336	**1.0705**	−0.1168	0.53525
−0.3801	0.6904	−0.19005	0.3452
−0.3984	0.292	−0.1992	0.146
−0.2568	0.0352	−0.1284	0.0176
0.4884	0.5236	0.2442	0.2618
−0.1113	0.4123	−0.05565	0.20615
0.2495	0.6618	0.12475	0.3309
0.9999	**1.6617**	0.49995	0.83085
−0.4088	**1.2529**	−0.2044	0.62645

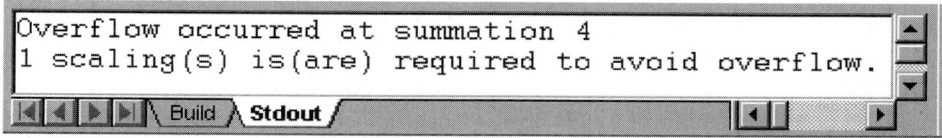

FIGURE L3-4 Overflow program execution result.

L3.3   Simulator   **157**

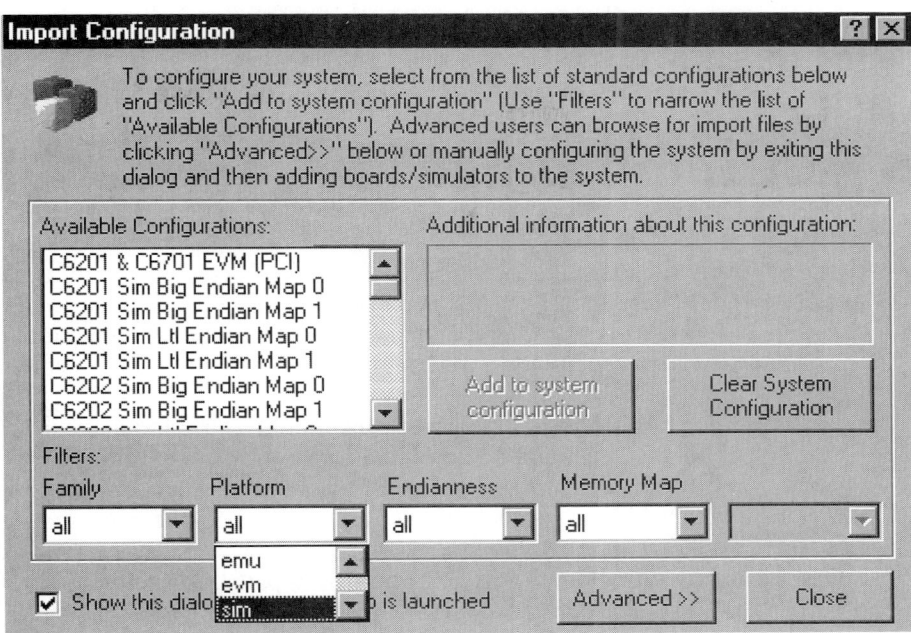

FIGURE L3-5   Simulator installation.

L3.3   SIMULATOR

When no DSP board is available, the CCS simulator can be used to run the lab programs. To install the simulator, simply select sim in the field platform during the installation process of CCS, as shown in Figure L4-9. In the Available configurations window, select the simulator option for one of the specified DSP boards. By clicking the button Add to system configuration, the simulator gets installed and becomes ready to use. Note that although the simulator supports DMA and EMIF operations, operations related to McBSP, HPI, and Timer are not supported. The files for running Lab3 via the simulator are provided on the attached CD under the subdirectory SIM.

CHAPTER 9

Code Optimization

Four relatively simple modifications of assembly code can be done to generate a more efficient code. These modifications make use of the available C6x resources, such as multiple buses, functional units, pipelined CPU, and memory organization. They include (a) using parallel instructions, (b) eliminating delays or NOPs, (c) unrolling loops, and (d) using word-wide data.

Wherever possible, parallel instructions should be used to make maximum use of idle functional units. It should be noted that, whenever the order in which instructions appear is important, care must be taken not to have any dependency in the operands of the instructions within a parallel instruction. It may become necessary to have cross paths when making instructions parallel. There are two types of cross paths: data and address. As illustrated in Figure 9-1(a), in data cross paths, one source part of an instruction on the A or B side comes from the other side. A cross path is indicated by x as part of functional unit assignment. The destination is determined by the unit index 1 or 2. As an example, we might have

```
MPY .M1x A2,B3,A4
MPY .M2x A2,B3,B4
```

In address cross paths, a .D unit gets its data address from the address bus on the other side. There are two address buses: DA1 and DA2, also known as T1 and T2, respectively. Figure 9-1(b) presents an example where a load and a store are done in parallel via the address cross paths.

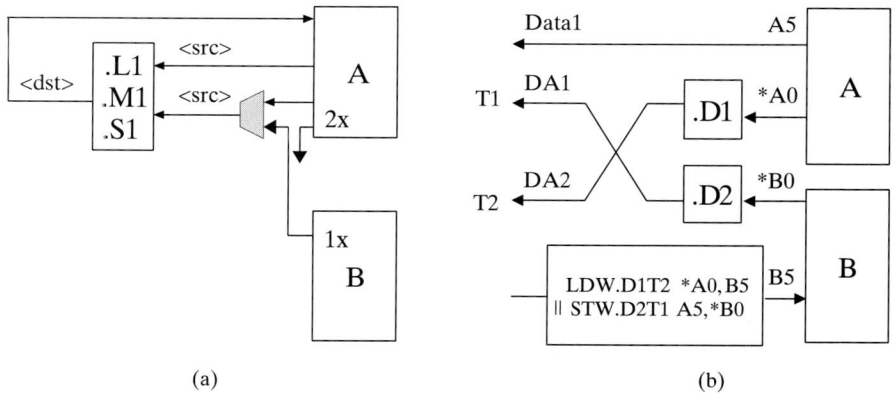

FIGURE 9-1 (a) Data cross-path, and (b) address cross-path.[†]

```
Loop:
         LDH.D1     *A8++,A2        ;load input 1 into A2
         LDH.D2     *B9++,B3        ;load input 2 into B3
   [B0]  SUB.L2     B0,1,B0         ;decrement counter
   [B0]  B.S1       Loop            ;branch to Loop
         NOP        2               ;5 latency slots required
         MPY.M1X    A2,B3,A4        ;A4=A2*B3, crosspath
         NOP
         ADD.L1     A4,A6,A6        ;A6 += A4
```

FIGURE 9-2 Optimized dot product example.

Wherever possible, branches should be placed five places ahead of where they are intended to appear. This would create a delayed branch, minimizing the number of NOPs. The approach should also be applied to load and multiply instructions that involve four delays and one delay, respectively. If the code size is of no concern, loops should be repeated or copied. By copying or unrolling a loop, fewer clock cycles would be needed, primarily due to deleting branches. Figure 9-2 shows the optimized version of the dot-product loop incorporating the preceding steps.

Considering that there exists a delay associated with getting information from off-chip memory, program codes should be run from the on-chip program RAM whenever possible. In situations where program codes would not fit into the on-chip RAM, faster execution can be achieved by placing the most time-consuming routine or function in the on-chip memory. The C6x has a cache feature that can be enabled to turn the program RAM into cache memory. This is done by setting the Program Cache Control (PCC) bits of the CSR to 010. For repetitive operations or loops, it is recommended that this feature be enabled, since there is then a good chance that the cache will contain the needed fetch packet and the EMIF will be left unused, speeding up code execution. Figure 9-3 shows the code for enabling the cache feature. The instructions CLR and SET are used to clear and set bits, respectively, from the second argument position to the third argument position. For more detailed operation of cache and its options, refer to the *CPU Reference Guide* [14].

```
              .def    _enable_cache

_enable_cache:
         b      .s2     B3
         mvc    .s2     CSR, B0
         clr    .s2     B0, 5, 7, B0
         set    .s2     B0, 6, 6, B0
         mcv    .s2     B0, CSR
         nop
```

FIGURE 9-3 Enabling cache feature.

9.1 WORD-WIDE OPTIMIZATION

If data are in halfwords (16 bits), it is possible to perform two loads in one instruction, since the CPU registers are 32 bits wide. In other words, as shown in Figure 9-4, one datum can be loaded into the lower part of a register and another one into the upper part.

However, this would require that the programmer be aware of the location of data during their manipulation. For example, to do a multiplication, two multiplication instructions, MPY and MPYH, should be used, one taking care of the lower part and the other of the upper part, as shown in Figure 9-5. Note that A5 and B5 appear as arguments in both MPY and MPYH instructions. This does not pose any conflict, since, on the C62x, up to four reads of a register in one cycle are allowed.

Figure 9-6 provides the word-wide optimized version of the dot-product function, *DotP()*. When the looping is finished, register A2 would contain the sum of even terms and register B2 the sum of odd terms. To obtain the total sum, these registers are added outside the loop.

Out of the preceding modifications, it is possible to do the last one, word-wide optimization, in C. This demands using an appropriate datatype in C. Figure 9-7 shows the word-wide C code optimized by using the _mpy() and _mpyh() intrinsics.

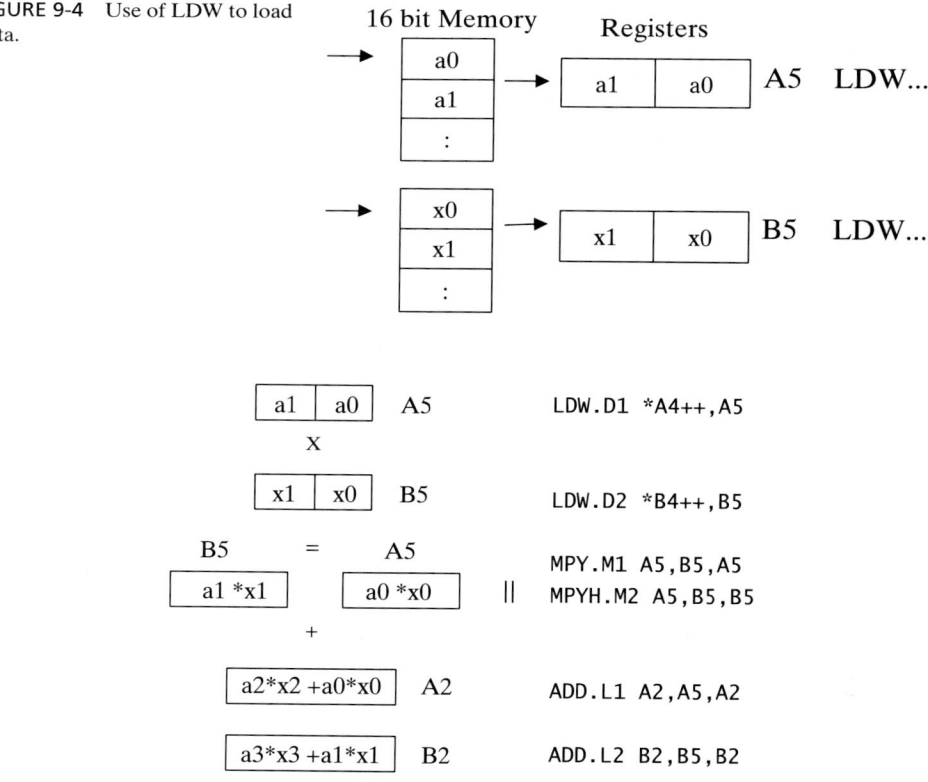

FIGURE 9-4 Use of LDW to load data.

FIGURE 9-5 Word-wide optimization.†

```
            .def   DotP
;A4 = &a, B4 = &x, A6 = 20, B3 = return address
     DotP:   zero   A2                 ;A2=0
                    B2                 ;B2=0
             mv     A6,B0              ;set B0 to argument passed in A6
loop:
             ldw    .d1    *A4++,A5    ;input word
             ldw || .d2    *B4++,B5    ;input word
       [B0]  sub    .l2    B0,1,B0     ;decrement loop counter
       [B0]  b      .s1    loop        ;branch to loop (5 delay slots filled below)
             nop           2
             mpy    .m1    A5,B5,A5    ;A5=A5(low)*B5(low)
             mpyh|| .m2    A5,B5,B5    ;B5=A5(high)*B5(high)
             nop
             add    .l1    A2,A5,A2    ;A2 += A5
             add || .l2    B2,B5,B2    ;B2 += B5
     rtn:    b      .s2    B3          ;branch back to calling address
             add    .l1    A2,B2,A4    ;A4 = A2 + B2 return value
             nop           4
```

FIGURE 9-6 Word-wide optimized version of dot product code.

9.2 MIXING C AND ASSEMBLY

To mix C and assembly, it is necessary to know the register convention used by the compiler to pass arguments. This convention is illustrated in Figure 9-8. DP, the base pointer, points to the beginning of the .bss section, containing all global and static variables. SP, the stack pointer, points to local variables. The stack grows from higher memory to lower memory, as indicated in Figure 9-8. The space between even registers (odd registers) is used when passing 40-bit or 64-bit values.

9.3 SOFTWARE PIPELINING

Software pipelining is a technique for writing highly efficient assembly loop codes on the C6x processor. Using this technique, all functional units on the processor are fully utilized within one cycle. However, to write hand-coded software pipelined assembly code, a fair amount of coding effort is required, due to the complexity and number of steps involved in writing such code. In particular, for complex algorithms encountered in many communications, and signal-image-processing applications, hand-coded software pipelining considerably increases coding time. The C compiler at the optimization levels 2 and 3 (-o2 and -o3) performs software pipelining to some degree. (See Figure 6-1.) Compared with linear assembly, the increase in code efficiency when writing hand-coded software pipelining is relatively slight.

```
//Prototype
short DotP(int *m, int *n, short count);

//Declarations
short a[40] = {40,39,...1};
short x[40] = {1,2,...40};
short y = 0;
main()
{
        y = DotP((int *)a, (int *)x, 20);
}

short DotP(int *m, int *n, short count)
{
        short i;
        short productl;
        short producth;
        short suml = 0;
        short sumh = 0;

        for(i=0, i<count; i++)
        {
                productl = _mpy(m[i],n[i]);
                producth = _mpyh(m[i],n[i]);
                suml += productl;
                sumh += producth;
        }
        suml += sumh;
        return(suml);
}
```

FIGURE 9-7 Word-wide optimized code in C.

9.3.1 Linear Assembly

Linear assembly is a coding scheme that allows one to write efficient codes (compared with C) with less coding effort (compared with hand-coded software pipelined assembly). The assembler optimizer is the software tool that makes linear assembly code parallel across the eight functional units. It attempts to achieve a compromise between code efficiency and coding effort.

In a linear assembly code, it is not required to specify any functional units, registers, or NOP's. Figure 9-9 shows the linear assembly code version of the dot-product function. The directives .proc and .endproc define the beginning and end, respectively, of the linear assembly procedure. The symbolic names p_m, p_n, m, n, count, prod, and sum are defined by the .reg directive. The names p_m, p_n, and count are respectively associated with the registers A4, B4, and A6 by using the assignment MV instruction.

9.3 Software Pipelining

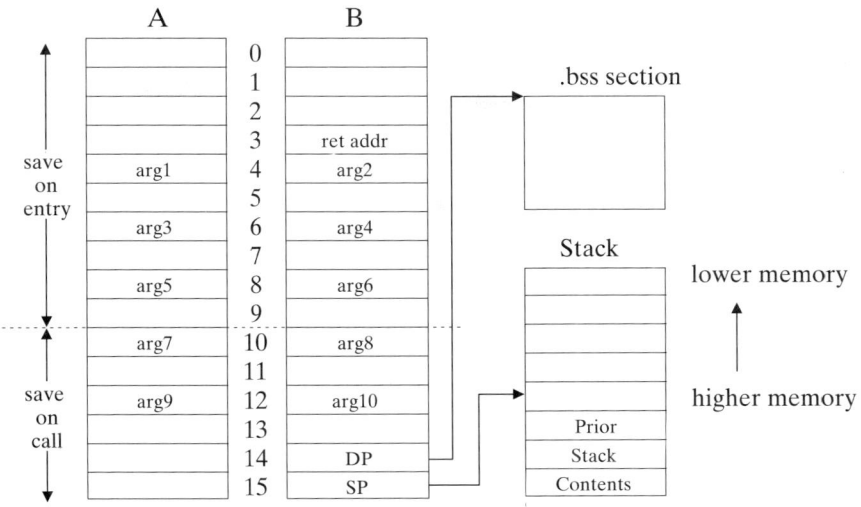

FIGURE 9-8   Passing arguments convention.†

```
        .title "dotp.sa"
        .def   dotp
        .sect  "code"
dotp:   .proc  A4, B4, A6, B3
        .reg   p_m, m, p_n, n, prod, sum, count
        mv     A4, p_m                  ;p_m now has the address of m
        mv     B4, p_n                  ;p_n now has the address of n
        mv     A6, count                ;count = the number of iterations
        mvk    0, sum                   ;sum=0

loop:   .trip  40                       ;minimum 40 iterations through loop
        ldh    *p_m++, m                ;load element of m, postincrement pointer
        ldh    *p_n++, n                ;load element of n, postincrement pointer
        mpy    m, n, prod               ;prod=m*n
        add    prod, sum, sum           ;sum += prod
[count] sub    count, 1, count          ;decrement counter
[count] b      loop                     ;branch back to loop

        mv     sum, A4                  ;store result in return register A4
        .endproc A4, B3
```

FIGURE 9-9   Linear assembly code for dot product example.

As per the register convention, the arguments are passed into and out of the procedure via registers A4, B4, A6, and B3. A4 is used to pass the address of m (arg1), B4 the address of n (arg2), and A6 the address of sum (arg3). Register B3, referred to as a preserved register, is passed in and out with no modification. This is done to prevent it from being used by the procedure. Here, this register is used to contain the return address reached by the branch instruction outside of the procedure. Preserved registers must be specified in both input and output arguments even when they are not used within the procedure. Table 9-1 provides a list of linear assembly directives.

TABLE 9-1 Linear assembly directives.†

Directive	Description	Restrictions
.call	Calls a function	Valid only within procedures
.cproc	Start a C/C++ callable procedure	Must use with .endproc
.endproc	End a C/C++ callable procedure	Must use with .cproc
.endproc	End a procedure	Must use with .proc; cannot use variables in the register parameter
.mdep	Indicates a memory dependence	Valid only within procedures
.mptr	Avoid memory bank conflicts	Valid only within procedures; can use variables in the register parameter
.no_mdep	No memory aliases in the function	Valid only within procedures
.proc	Start a procedure	Must use with .endproc; cannot use variables in the register parameter
.reg	Declare variables	Valid only within procedures
.reserve	Reserve register use	
.return	Return value to procedure	Valid only within .cproc procedures
.trip	Specify trip count value	Valid only within procedures

If the number of iterations is known, a .trip directive should be used for the assembly optimizer to generate the pipelined code. For n iterations of a loop, in a pipelined code, the loop is repeated n' times, where $n' = n - prolog\ length$. (Prolog will be explained later in the chapter.) The number of iterations, n', is known as the *minimum trip count*. If .trip is greater than or equal to n', only the pipelined code is created. Otherwise, both the pipelined and the non-pipelined code are created. If .trip is not specified, only the non-pipelined code is created. In C, the function *_n_assert()* can be used to provide the same information as .trip.

To further optimize a linear assembly code, partitioning information can be added. Such information consists of the assignment of data paths to instructions.

9.3.2 Hand-coded Software Pipelining

First let us review the pipeline concept. Figures 9-10(b) and 9-10(c) show a nonpipelined and a pipelined version of the loop code shown in Figure 9-10(a). As can be seen from this figure, the functional units in the pipelined version are not fully utilized, leading to more cycles compared with the pipelined version. There are three stages to a pipelined code, named prolog, loop kernel, and epilog. Prolog corresponds to instructions that are needed to build up a loop kernel or loop cycle, and epilogue to instructions that are needed to complete all loop iterations. When a loop kernel is established, the entire loop is executed in one cycle via one parallel instruction using the maximum number of functional units. This parallelism is what causes a reduction in the number of cycles.

9.3 Software Pipelining

```
loop:        ldh
      ||     ldh
             mpy
             add
```

(a)

cycle\unit	.D1	.D2	.M1	.M2	.L1	.L2	.S1	.S2
1	ldh	ldh						
2			mpy					
3					add			
4	ldh	ldh						
5			mpy					
6					add			
7	ldh	ldh						
8			mpy					
9					add			

(b)

	cycle\unit	.D1	.D2	.M1	.L1
Prolog	1	ldh	ldh		
loop buildup	2	ldh	ldh	mpy	
Loop Kernel	3	ldh	ldh	mpy	add
	4	ldh	ldh	mpy	add
	5	ldh	ldh	mpy	add
Epilog	6			mpy	add
Completing final operations	7				add

(c)

FIGURE 9-10 (a) A loop example, (b) non-pipelined code, and (c) pipelined code.[†]

Three steps are needed to produce a hand-coded software pipelined code from a linear assembly loop code: (a) drawing a dependency graph, (b) setting up a scheduling table, and (c) deriving the pipelined code from the scheduling table.

In a dependency graph (see Figure 9-11 for the terminology), the nodes denote instructions and symbolic variable names. The paths show the flow of data and are annotated with the latencies of their parent nodes. To draw a dependency graph for the loop part of the dot-product code, we start by drawing nodes for the instructions and symbolic variable names.

FIGURE 9-11 Dependency graph terminology.

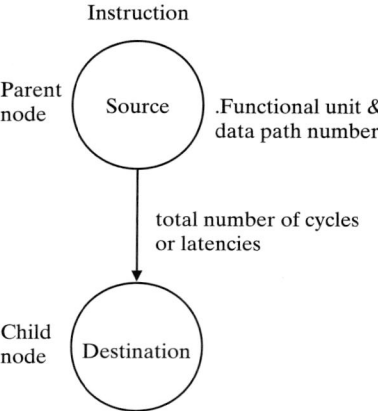

After the basic dependency graph is drawn, a functional unit is assigned to each node or instruction. Then, a line is drawn to split the workload between the A- and B-side data paths as equally as possible. It is apparent that one load should be done on each side, so this provides a good starting point. From there, the rest of the instructions need to be assigned in such a way that the workload is equally divided between the A- and B-side functional units. The dependency graph for the dot-product example is shown in Figure 9-12.

The next step for handwriting a pipelined code is to set up a scheduling table. To do so, the longest path must be identified in order to determine how long the table should be. Counting the latencies of each side, we see that the longest path is 8. This means that 7 prolog columns are required before entering the loop kernel. Thus, as shown in Table 9-2, the scheduling table consists of 15 columns (7 for prolog, 1 for loop kernel, 7 for epilog) and eight rows (one row for each functional unit). Epilog and prolog are of the same length.

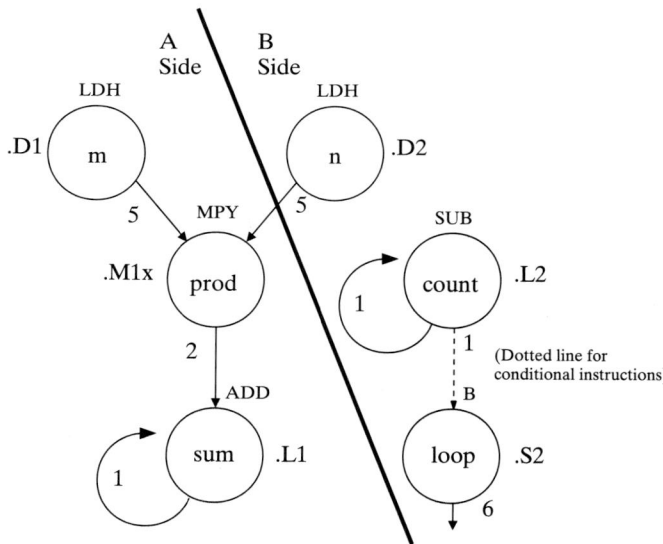

FIGURE 9-12 Dot product dependency graph.[†]

9.3 Software Pipelining

TABLE 9-2 Dot product scheduling table.†

cycle unit	1	2	3	4	5	6	7	8	9	10	11	12	13	14	15
	\multicolumn{7}{c}{PROLOG}		LOOP	\multicolumn{7}{c}{EPILOG}						
.D1	①LDH	LDH	LDH	LDH	LDH	LDH	LDH	LDH							
.D2	④LDH	LDH	LDH	LDH	LDH	LDH	LDH	LDH							
.L1								③ADD	ADD	ADD	ADD	ADD	ADD	ADD	ADD
.L2		⑥SUB	SUB	SUB	SUB	SUB	SUB	SUB							
.S1															
.S2			⑤B	B	B	B	B	B							
.M1						②MPY	MPY	MPY	MPY	MPY	MPY	MPY	MPY		
.M2															

The scheduling is started by placing the load instructions in parallel in cycle 1. These instructions are repeated every cycle thereafter. The multiply instruction must appear five cycles after the loads (1 cycle for loads + 4 load delays), so it is scheduled into slot or cycle 6. The add must appear two cycles after the multiply (1 cycle for multiply + 1 multiply delay), requiring it to be placed in slot or cycle 8, which is the loop kernel part of the code. The branch instruction is scheduled in slot or cycle 3 by reverse counting 5 cycles back from the loop kernel. The subtraction must occur before the branch, so it is scheduled in slot or cycle 2. The completed scheduling table appears in Table 9-2.

Next, the code is handwritten directly from the scheduling table as 7 prolog parallel instructions, 40 − 7 = 33 loop kernel parallel instructions, and 7 epilog parallel instructions. This hand-coded software pipelined code is shown in Figure 9-13. It can be seen that it requires only 47 cycles to perform the dot product 40 times. Note that, as

```
cycle 1:
         ldh  .D1   *A1++,A2
      || ldh  .D2   *B1++,B2
cycle 2:
         ldh  .D1   *A1++,A2
      || ldh  .D2   *B1++,B2
      || [B0] sub .L2  B0,1,B0
cycles 3, 4 and 5:
         ldh  .D1   *A1++,A2
      || ldh  .D2   *B1++,B2
      || [B0] sub .L2  B0,1,B0
      || [B0] B   .S2  loop
cycles 6 and 7:
         ldh  .D1   *A1++,A2
      || ldh  .D2   *B1++,B2
      || [B0] sub .L2  B0,1,B0
      || [B0] B   .S2  loop
      || mpy  .M1x  A2,B2,A3

cycles 8 to n: Single-cycle loop
loop:    ldh  .D1   *A1++,A2
      || ldh  .D2   *B1++,B2
      || [B0] sub .L2  B0,1,B0
      || [B0] B   .S2  loop
      || mpy  .M1x  A2,B2,A3
      || add  .L1   A4,A3,A4

cycles n+1 to n+5:
         mpy  .M1x  A2,B2,A3
      || add  .L1   A4,A3,A4

cycles n+6 and n+7:
         add  .L1   A4,A3,A4
```

FIGURE 9-13 Hand-coded software pipelined dot product code.

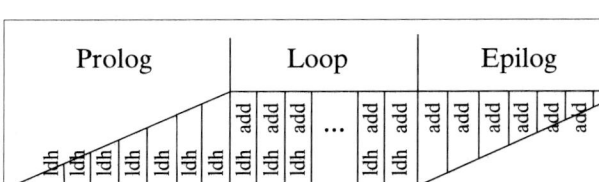

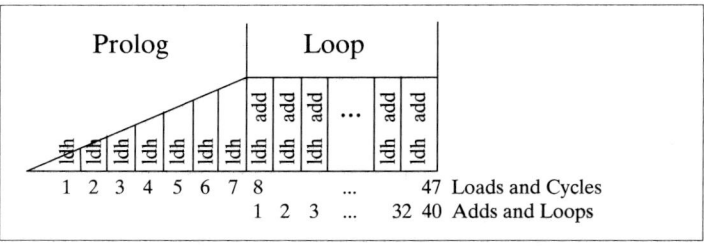

FIGURE 9-14 Elimination of epilogue instructions.†

shown in Figure 9-14, it is possible to eliminate the epilog instructions by performing the loop kernel instruction 40 times instead of 33 times, leading to a lower code size and a higher number of loads.

Figures 9-15(b) and 9-15(c) show the dependency graph and the scheduling table, respectively, of the word-wide optimized dot-product code appearing in Figure 9-15(a). The corresponding hand-coded pipelined code is shown in Figure 9-16. This time, 28 cycles are required: 7 prolog instructions, 40/(2 word datatype) = 20 loop kernel instructions, and one extra add to sum the even and odd parts. Table 9-3 provides the number of cycles for different optimizations of the dot-product example discussed throughout the book. The interested reader is referred to the *TI TMS320C6x Programmer's Guide* [19] for more details on how to handwrite software pipelined assembly code.

Multicycle loops. Let us now examine an example of a weighted vector sum $c = a + r * b$, where a and b denote two arrays or vectors of size 40 and r designates a constant or scalar. Figure 9-17 shows the linear assembly code and the corresponding dependency graph used to compute c. A problem in this dependency graph is that there are more than two loads/stores operations (i.e., the .D1 unit is assigned to two nodes). This, of course, is not possible in a single-cycle loop. Consequently, we must have two cycle loops instead of one. In other words, two parallel instructions per iteration are needed to compute the vector sum c.

9.3 Software Pipelining

```
; for (i=0;i < count;i++)
; prod = m[i] * n[i];
;sum += prod;     count becomes 20

loop:    ldw      *p_m++, m
         ldw      *p_n++, n
         mpy      m, n, prod
         mpyh     m, n, prodh
         add      prod, sum, sum
         add      prodh, sumh, sumh
[count]  sub      count, 1, count
[count]  b        loop

; Outside of Loop
         add      sum, sumh, sum
```

(a)

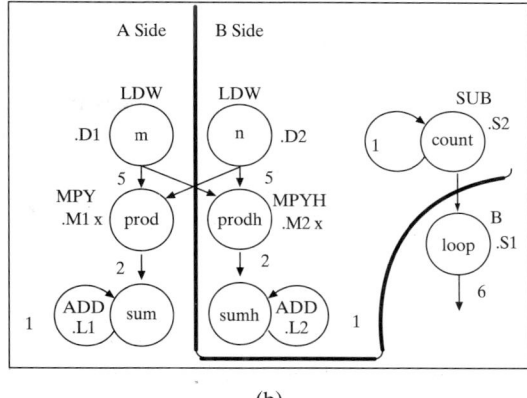

(b)

unit \ cycle	1	2	3	4	5	6	7	8
			PROLOG					LOOP
.D1	① ldw m	ldw	ldw	ldw	ldw	ldw	ldw	ldw
.D2	④ ldw n	ldw	ldw	ldw	ldw	ldw	ldw	ldw
.L1								③ add
.L2								⑥ add
.S1			⑧ B	B	B	B	B	B
.S2		⑦ sub	sub	sub	sub	sub	sub	sub
.M1						② mpy	mpy	mpy
.M2						⑤ mpyh	mpyh	mpyh

(c)

FIGURE 9-15 (a) Linear assembly dot product code, (b) corresponding dependency graph, and (c) scheduling table.†

This time, the scheduling table consists of two sets of functional units arranged as shown in Figure 9-17. In this example, the length of the longest path is 10, which corresponds to the load–multiply–shift–add–store path. This means that there should be 10 cycle columns. The cycle number is set up by alternating between the two sets of functional units. The scheduling is started by entering the instructions for the longest path. The load LDH bi is placed in slot 1. MPY is placed in slot 6, 5 slots after slot 1, to accommodate for the load latencies. SHR is placed in slot 8, two slots after slot 6, to accommodate for the multiply latency. ADD is placed in slot 9, one slot after slot 8, and STH ci in the last slot 10. The other path is then scheduled. The loading LDH ai for this path must be done 5 slots or cycles before the ADD instruction. This would place LDH ai in slot 4. However, notice that the .D1 unit as part of the second loop cycle has already been used for STH ci and cannot be

```
cycle 1:
            ldw   .D1   *A4++,A5
     ||     ldw   .D2   *B4++,B5
cycle 2:
            ldw   .D1   *A4++,A5
     ||     ldw   .D2   *B4++,B5
     || [B0] sub  .S2   B0,1,B0
cycles 2,3 and 4:
            ldw   .D1   *A4++,A5
     ||     ldw   .D2   *B4++,B5
     || [B0] sub  .S2   B0,1,B0
     || [B0] B    .S1   loop
cycles 5 and 6:
            ldw   .D1   *A4++,A5
     ||     ldw   .D2   *B4++,B5
     || [B0] sub  .S2   B0,1,B0
     || [B0] B    .S1   loop
     ||     mpy   .M1x  A5,B5,A6
     ||     mpyh  .M2x  A5,B5,B6
```

```
cycles 8 to n+7: Single-cycle loop
loop:       ldw   .D1   *A4++,A5
     ||     ldw   .D2   *B4++,B5
     || [B0] sub  .S2   B0,1,B0
     || [B0] B    .S1   loop
     ||     mpy   .M1x  A5,B5,A6
     ||     mpyh  .M2x  A5,B5,B6
     ||     add   .L1   A7,A6,A7
     ||     add   .L2   B7,B6,B7
```

FIGURE 9-16 Hand-coded pipelined code for word wide dot product loop.

TABLE 9-3 Optimization methods cycles.

No optimization	16 cycles * 40 iterations =	640
Parallel optimization	15 cycles * 40 iterations =	600
Filling delay slots	8 cycles * 40 iterations =	320
Word wide optimizations	8 cycles * 20 iterations =	160
Software pipelined - LDH	1 cycle * 40 loops + 7 prolog =	47
Software pipelined - LDW	1 cycle * 20 loops + 7 prolog + 1 epilog =	28

used at the same time. This creates a conflict that must be resolved. As indicated by the shaded boxes in Figure 9-17, the conflict is resolved either by using the .D2 unit in the second cycle loop or by using the .D1 unit in the first cycle loop.

9.4 C64X IMPROVEMENTS

This section illustrates how the additional features of the C64x DSP can be used to further optimize the dot-product example. Figure 9-18(b) shows the C64x version of the dot-product loop kernel for multiplying two 16-bit values. The equivalent C code appears in Figure 9-18(a).

Now, by using the DOTP2 instruction of the C64x, we can perform two 16*16 multiplications on a single functional unit, reducing the number of cycles by one-half. This requires accessing two 32-bit values every cycle. As shown in Figure 9-19(a), in C, these

9.4 C64x Improvements

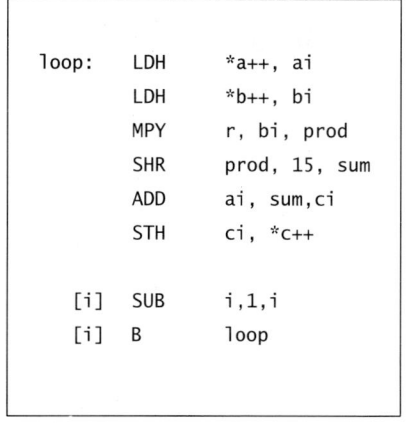

```
loop:   LDH   *a++, ai
        LDH   *b++, bi
        MPY   r, bi, prod
        SHR   prod, 15, sum
        ADD   ai, sum, ci
        STH   ci, *c++

 [i]    SUB   i,1,i
 [i]    B     loop
```

(a)

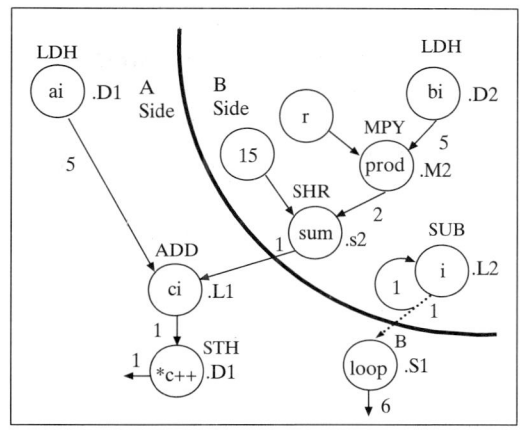

(b)

Unit\cycle	1	3	5	7	9
.L1					ADDci
.L2					
.S1			B		
.S2					
.M1					
.M2					
.D1		LDHai			
.D2	LDHbi				

Unit\cycle	2	4	6	8	10
.L1					
.L2		SUB i			
.S1					
.S2				SHRsum	
.M1					
.M2			MPYbi		
.D1		~~LDHai~~			STHci
.D2		LDHai			

conflict

(c)

FIGURE 9-17 Multi-cycle loop: (a) loop code, (b) dependency graph, and (c) scheduling table.[†]

```
main()
{
   y = DotP(a,x,40);
}
int DotP(short *m, short *n, int count)
{
   int i;
   int product;
   int sum = 0;
   for(i=0;i<count;i++)
   {
      product = m[i] * n[i];
      sum += product;
   }
   return(sum);
}
```

(a)

```
;PIPED LOOP KERNEL
LOOP:
         [A0]   SUB    .L1    A0,1,A0
   ||    [!A0]  ADD    .S1    A6,A5,A5    ;keep running sum
   ||           MPY    .M1X   B4,A4,A6    ;multiply two 16-bit values
   ||    [B0]   BDEC   .S2    LOOP, B0    ;decrement loop counter and branch if > 0
   ||           LDH    .D1T1  *A3++,A4    ;load 16-bit value
   ||           LDH    .D2T2  *B5++,B4    ;load 16-bit value
```

(b)

FIGURE 9-18 C64x pipelined code: (a) C, and (b) assembly.[†]

can be achieved by using the intrinsic _dotp2() and by casting shorts as integers. The equivalent loop kernel code generated by the compiler is shown in Figure 9-19(b). This is a double-cycle loop containing four 16*16 multiplications. The instruction LDW is used to bring in the required 32-bit values.

Considering that the C64x can bring in 64-bit data values by using the double-word loading instruction LDDW, the foregoing code can be further improved by performing four 16*16 multiplications via two DOTP2 instructions within a single-cycle loop, as shown in Figure 9-20(b). This way the number of operations is reduced fourfold, since four 16*16 multiplications are done per cycle. To do this in C, we need to cast short datatypes as doubles and to specify which 32 bits of a 64-bit data a DOTP2 is supposed to operate on. This is done by using the _lo() and _hi() intrinsics to specify the lower and the upper 32 bits, respectively, of 64-bit data. Figure 9-20(a) shows the equivalent C code.

9.4 C64x Improvements

```
main()
{
   y=DotP((int )a, (int *)x,20);
}
int DotP(int *m, int *n, int count)
{
   int i;
   int product;
   int sum=0;
   for(i=0;i<count;i++)
   {
      product=_dotp2(m[i], n[i]);
      sum=product + sum;
   }
   return(sum);
}
```

(a)

```
;PIPED LOOP KERNEL
LOOP:
      [!A1]  ADD     .L2    B8,B4,B4       ;running sum 0
||           DOTP2   .M2X   B7,A6,B8       ;2 16x16 multiplies+add; prod 0
||    [A0]   BD EC   .S1    LOOP, A0       ;decrement loop counter and branch if > 0
||           LDW     .D1T1  *+A4(4),A3     ;load a 32-bit value
||           LDW     .D2T2  *+B5(4),B6     ;load a 32-bit value
      [A1]   SUB     .L1    A1,1,A1
||    [!A1]  ADD     .S1    A7,A5,A5       ;running sum1
||           DOTP2   .M1X   B6,A3,A7       ;2 16x16 multiplies+add; prod 1
||           LDW     .D1T1  *++A4(8), A6   ;load a 32-bit value
||           LDW     .D2T2  *++B5(8), B7   ;load a 32-bit value
```

(b)

FIGURE 9-19 C64x packed datatype code: (a) C, and (b) assembly.[†]

```
int DotP(const short * restrict m, const short * restrict n, int count)
{
   int i;
   int sum=0;
   const double * restrict m_dbl=(const double *) m;
   const double * restrict n_dbl=(const double *) n;

   count>>2;/*count is divided by two if using same main function ↵
                                        to call this subroutine */
   for(i=0;i<count;i++)
   {
      sum +=_dotp2(_lo(m_dbl[i]), _lo(n_dbl[i]))+
            _dotp2(_hi(m_dbl[i]), _hi(n_dbl[i]));

  }
    return  sum ;
}
```

(a)

```
;PIPED LOOP KERNEL
LOOP:
        [ B0]   SUB     .L2     B0,1,B0         ;decrement running sum counter
||      [!B0]   ADD     .S2     B8,B6,B6        ; running sum 0
||      [!B0]   ADD     .L1     A7,A6,A6        ; running sum 1
||              DOTP2   .M2X    B4,A4,B8        ; 2 16x16 multiplies+add; prod 0
||              DOTP2   .M1X    B5,A5,A7        ; 2 16x16 multiplies+add; prod 1
||      [A0]    BDEC    .S1     LOOP,A0         ;branch to loop & decrement loop count
||              LDDW    .D1T1   *A3++,A5:A4     ;load a 64-bit value
||              LDDW    .D2T2   *B7++,B5:B4     ;load a 64-bit value
```

(b)

FIGURE 9-20 C64x double-word packed datatype code: (a) C, and (b) assembly.[†]

Lab 4: Real-time Filtering

The purpose of this lab is to design and implement a finite impulse response filter on the C62 and C67. The design of the filter is done by using Matlab. Once the design is completed, the filtering code is inserted into the sampling EVM shell program as an ISR to process live signals in real-time.

L4.1 DESIGN OF FIR FILTER

Matlab or filter design packages can be used to obtain the coefficients for a desired FIR filter. To have a more realistic simulation, a composite signal may be created and filtered in Matlab. A composite signal consisting of three sinusoids, as shown in Figure L4-1, can be created by the following Matlab code:

```
Fs=8e3;
Ts=1/Fs;
Ns=512;

t=[0:Ts:Ts*(Ns-1)];

f1=750;
f2=2500;
f3=3000;
x1=sin(2*pi*f1*t);
x2=sin(2*pi*f2*t);
x3=sin(2*pi*f3*t);

x=x1+x2+x3;
```

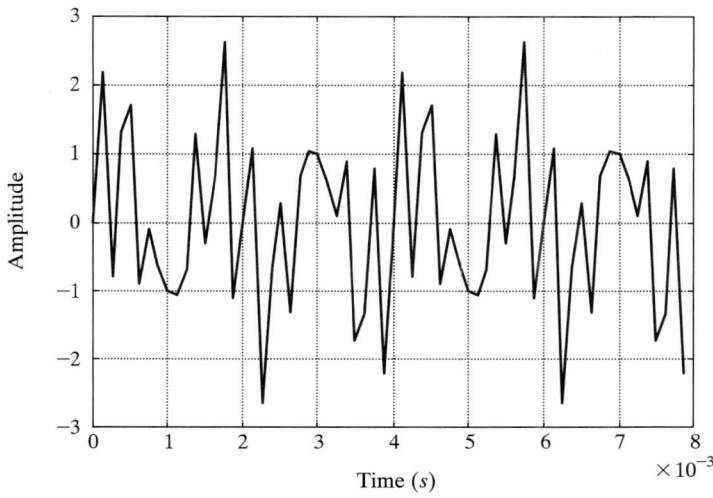

FIGURE L4-1 Two cycles of composite signal.

The signal frequency content can be plotted by using the Matlab 'fft' function. Three spikes should be observed, at 750 Hz, 2500 Hz, and 3000 Hz. The frequency leakage observed on the plot is due to windowing caused by the finite observation period. A lowpass filter is designed here to filter out frequencies greater than 750 Hz and retain the lower components. The sampling frequency is chosen to be 8 kHz, which is common in voice processing. The following code is used to get the frequency plot shown in Figure L4-2:

```
X=(abs(fft(x,Ns)));
y=X(1:length(X)/2);
f=[1:1:length(y)];
plot(f*Fs/Ns,y);
grid on;
```

To design an FIR filter with passband frequency = 1600 Hz, stopband frequency = 2400 Hz, passband gain = 0 dB, stopband attenuation = 20 dB, and sampling rate = 8000 Hz, the Parks–McClellan method is used via the 'remez' function of Matlab [2]. The magnitude and phase response are shown in Figure L4-3, and the coefficients are given in Table L4-1. The Matlab code is as follows:

```
N=10;
F=[0 0.4 0.6 1];
M=[1 1 0 0];
B=remez(N,F,M);
A=1;
freqz(B,A);
```

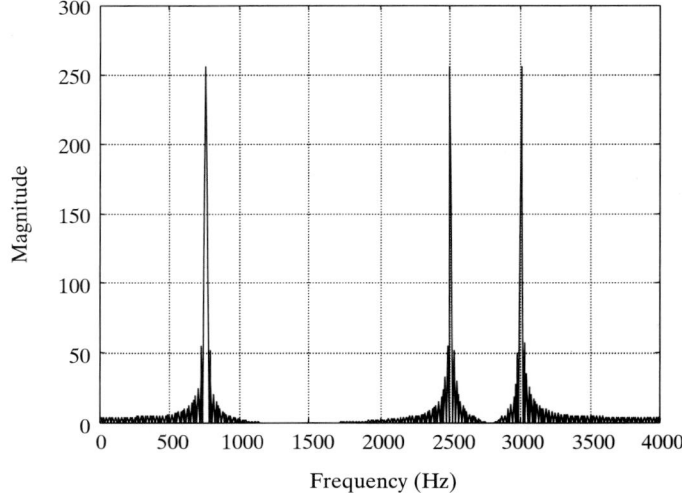

FIGURE L4-2 Frequency components of composite signal.

L4.1 Design of FIR Filter

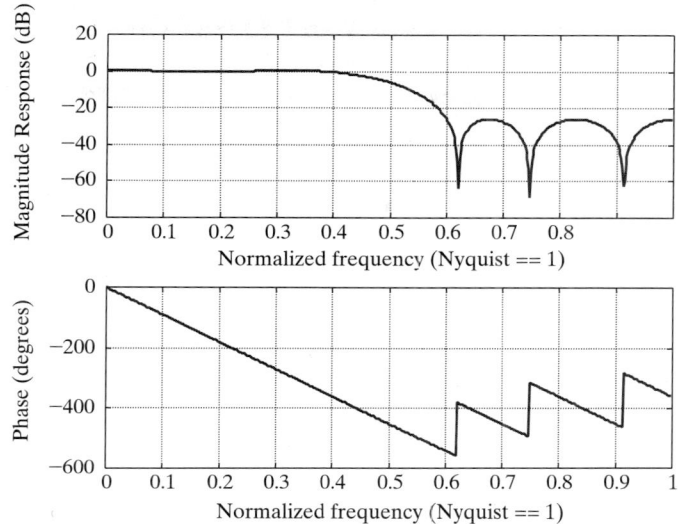

FIGURE L4-3 Filter magnitude and phase response.

TABLE L4-1 FIR filter coefficients.

Coefficient	Values	Q-15 Representation
B0	0.0537	0x06DF
B1	0.0000	0x0000
B2	−0.0916	0xF447
B3	−0.0001	0xFFFD
B4	0.3131	0x2813
B5	0.4999	0x3FFC
B6	0.3131	0x2813
B7	−0.0001	0xFFFD
B8	−0.0916	0xF447
B9	0.0000	0x0000
B10	0.0537	0x06DF

(Note: Do not confuse B coefficients with B registers!)

With these coefficients, the 'filter' function of Matlab is used to verify that the FIR filter is actually able to filter out the 2.5 kHz and 3 kHz signals. The following Matlab code allows one to visually inspect the filtering operation:

```
subplot(3,1,1);
va_fft(x,1024,8000);
subplot(3,1,2);
[h,w]=freqz(B,A,512);
plot(w/(2*pi),10*log(abs(h)));
```

```
grid on;
subplot(3,1,3);
y = filter(B,A,x);
va_fft(y,1024,8000);

function va_fft(x,N,Fs)
X=fft(x,N);
XX=(abs(X));
XXX=XX(1:length(XX)/2);
y=XXX;
f=[1:1:length(y)];
plot(f*Fs/N,y);
grid on;

n=128
subplot(2,1,1);
plot(t(1:n),x(1:n));
grid on;
xlabel('time(s)');
ylabel('Amplitude');
title('Original and Filtered Signals');
subplot(2,1,2);
plot(t(1:n),y(1:n));
grid on;
xlabel('time(s)');
ylabel('Amplitude');
```

Looking at the plots appearing in Figures L4-4 and L4-5, we see that the filter is able to remove the desired frequency components of the composite signal. Observe that the time response has an initial setup time causing the first few data samples to be inaccurate. Now that the filter design is complete, let us consider the implementation of the filter.

L4.2 FIR FILTER IMPLEMENTATION

An FIR filter can be implemented on the C62 in C or assembly. The goal of the implementation is to have a minimum cycle time algorithm [i.e., to do the filtering as fast as possible, in order to achieve the highest sampling frequency (the smallest sampling time interval)]. Initially, the filter is implemented in C, since this demands the least coding effort. Once a working algorithm in C is obtained, the compiler optimization levels (i.e., -o2, -o3) are activated to reduce the number of cycles. An implementation of the filter is then done in hand-coded assembly, which can be software pipelined for optimum performance. A final implementation of the filter is performed in linear assembly, and the timing results are compared.

L4.2 FIR Filter Implementation

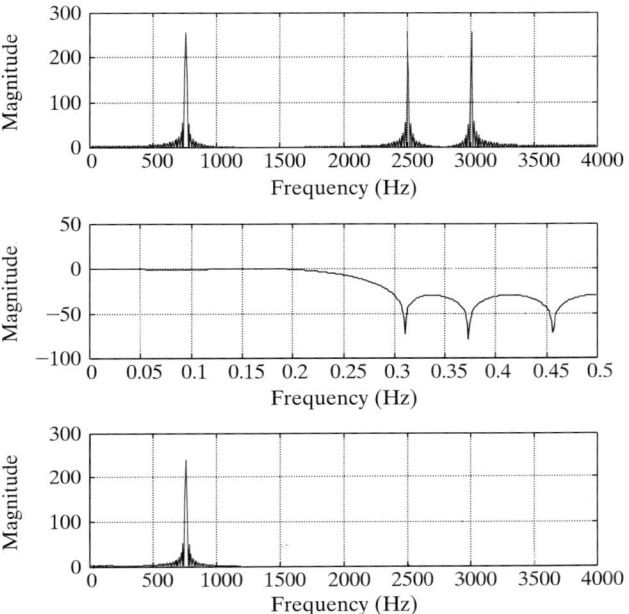

FIGURE L4-4 Frequency representation of filtering operation.

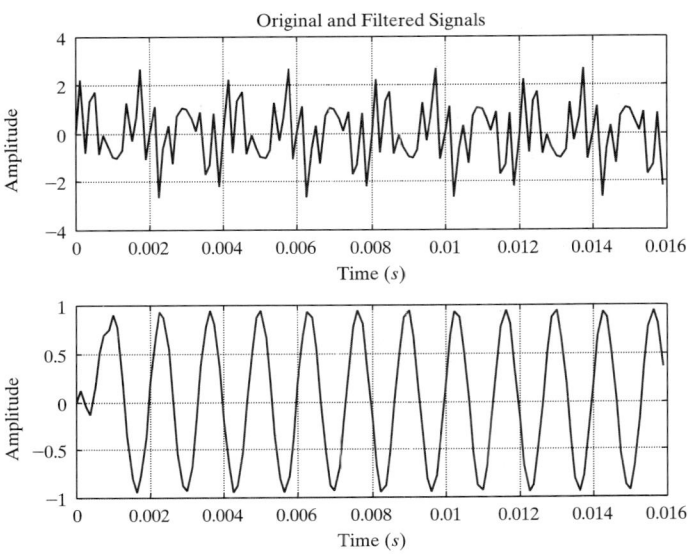

FIGURE L4-5 Time domain representation of filtering operation.

The difference equation $y[n] = \sum_{k=0}^{N-1} B_k * x[n-k]$ is implemented to realize the filter. Since the filter is implemented on the EVM, the coding is done by modifying the sampling program in Lab 2, which uses an ISR that is able to receive a sample from the serial port and send it back out without any modification.

Since Q-15 representation is used here, the MPY instruction cannot be utilized to multiply a 32-bit sample by a 16-bit coefficient, as this will not produce the correct product. Instead, the upper part of the samples must be multiplied by the coefficients, which are only 16 bits wide. After the multiplication, in order to store the product in 32 bits, it has to be left shifted by 1 to get rid of the extended sign bit. The product may be stored in 16 bits if a right shift of 15 is used.

To implement the algorithm in C, the _mpyhl() intrinsic and the binary shift operator << should be used as follows:

```
result = ( _mpyhl(sample,coefficient) ) << 1;
```

Here, result and sample are 32 bits wide, while coefficient is 16 bits wide. The intrinsic _mpyhl() multiplies the upper 16 bits of the first argument by the lower 16 bits of the second argument. Therefore, the upper 16 bits of sample is used in the multiplication. Alternatively, the sample in the DRR can be right shifted by 16 and stored as a short, and then the MPY instruction can be used to multiply the 16-bit sample with the 16-bit coefficient.

For the proper operation of the FIR filter, it is required that the current sample and $N - 1$ previous samples be processed at the same time, where N is the number of coefficients. Hence, the N most current samples have to be stored and updated with each incoming sample. This can be done easily via the following code:

```
interrupt void serialPortRcvISR (void)
{
        int i,temp;
        temp = MCBSP_READ(0);

        //update array samples
        for(i=N-2;i>=0;i--)
                samples[i+1]=samples[i];
        samples[0] = temp;
        MCBSP_WRITE(0, temp);
}
```

In this program, as a new sample comes in, each of the previous samples is moved into the next cell in the array. As a result, the oldest sample, sample[N], is discarded, and the newest sample, temp, is put into sample[0].

This approach adds some overhead to the ISR, but for now it is acceptable, since at a sampling frequency of 8 kHz, there is a total of 16,625 available cycles [1/(8 kHz/133 MHz) = 16,625] between consecutive samples, considering that the EVM here is configured to run at 133 MHz. The total overhead for this manipulation is 659 cycles without any optimization. It should be noted that the proper way of doing this type of filtering is by using circular buffering. The circular buffering approach will be discussed in Lab 5.

Now that the *N* most current samples are in the array, the filtering operation may get started. All that needs to be done, according to the difference equation, is to multiply each sample by the corresponding coefficient and sum the products. This is achieved by the following code:

```
interrupt void serialPortRcvISR (void)
{
        int i,temp,result = 0;
        temp = MCBSP_READ(0);

        //update array samples
        for(i=N-2;i>=0;i--)
                samples[i+1]=samples[i];
        samples[0] = temp;

        //Filtering
        for(i=0;i<N;i++)
                result += ( _mpyhl(samples[i],coefficients[i]) ) << 1;

        MCBSP_WRITE(0, result);
}
```

To complete the FIR filter implementation, we need to incorporate the filter coefficients previously designed into the C program. This is done by modifying the sampling program in Lab2 as follows:

```
.
.
.
#define N 11

// FIR filter coefficients
short coefficients[] =
{0x6DF,0x0,0xF447,0xFFFD,0x2813,0x3FFC,0x2813,0xFFFD,0xF447,0x0,0x6DF};
int samples[N];
        .
        .
int main()
{
```

```
    for(i=0;i<N;i++)
        samples[i]=0;
    .
    .
    .
}
```

The filtering program can now be built and run. Using a function generator and an oscilloscope, it is possible to verify that the filter is working as expected. The output of the function generator should be connected to the line-in jack of the EVM board, and the line-out jack of the board to the input of the oscilloscope. As the input frequency is increased, it is seen that the signal attenuation starts at 1.6 kHz and dies out at 2.4 kHz.

Given that a working design is at hand, it is time to start the optimization of the filtering algorithm. The first step in optimization is to use the compiler optimizer. The optimizer can be invoked by choosing Project->Options from the CCS menu bar. This option will invoke a dialog box, as shown in Figure L4-6. In this dialog box, select Basic in the Category field, and then choose the desired optimization level from the pull-down

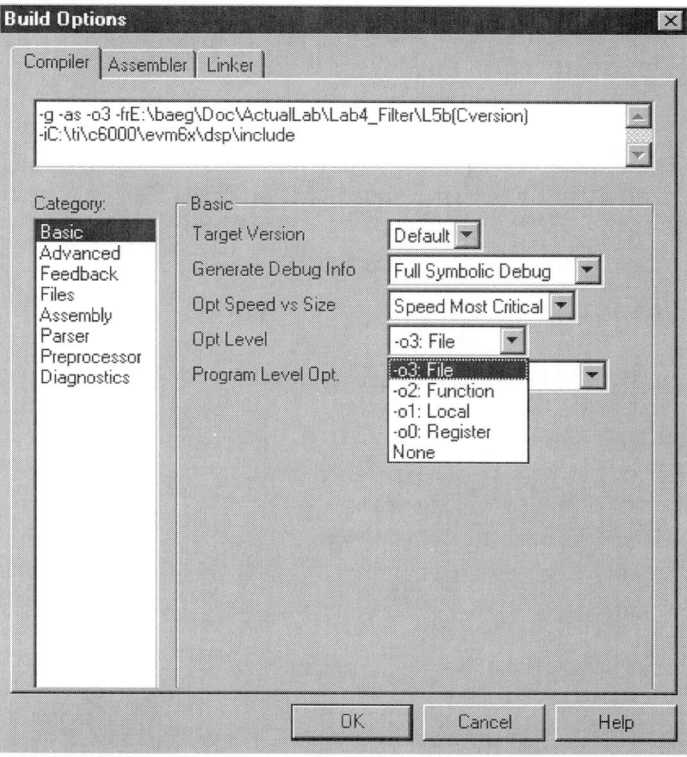

FIGURE L4-6 Selection of different optimization levels.

list in the Opt Level field. Table L4-2 summarizes the timing results for different optimization levels.

TABLE L4-2 Timing cycles for different builds.

Build Type	Number of Cycles
Compile without optimization	857
Compile with -o0	708
Compile with -o1	690
Compile with -o2/-o3	315

As can be seen from Table L4-2, the number of cycles diminishes as the optimization level is increased. It is important to remember that because the compiler optimizer changes the flow of a program, the debugger may not work in some cases. Therefore, it is advised that one make sure that a program works correctly before performing any compiler optimization.

Before doing the linear assembly implementation, the code is written in assembly to see how basic optimization methods such as placing instructions in parallel, filling delay slots, loop unrolling, and word-wide optimization affect the timing cycle of the code.

To perform the operation of multiplying and adding N coefficients, a loop has to be set up. This can be done by using a branch instruction. A counter is required to exit the loop once N iterations have been performed. For this purpose, one of the conditional registers (A1, A2, B0, B1 or B2) is used. No other register allows for conditional testing. Adding [A2] in front of an instruction permits the processor to execute the instruction if the value in A2 does not equal zero. If A2 contains zero, the instruction is skipped, noting that an instruction cycle is still consumed. The S1 unit may be used to perform the move constant and branch operations. The value in the conditional register A2 decreases by using a subtract instruction. Since the subtract operation should stop if the value drops below zero, this conditional register is mentioned in the SUB instruction to execute it only if the value is not equal to zero. The programmer should remember to add five delay slots for the branch instruction. The code for this loop is as follows:

```
        MVK    .S1    11, A2          ;move 11 into A2 count register
Loop1:
            .
            .
            .
   [A2] SUB   .L1    A2,1,A2          ;decrement counter
   [A2] B     .S1    Loop1            ;branch back to Loop1
        NOP   5
```

We can now start adding instructions to perform the multiplication and accumulation of the values. First, those values that are to be multiplied need to be loaded from their memory locations into the CPU registers. This is done by using load word (LDW) and load half-word (LDH) instructions. Upon executing the load instructions, the pointer is postincremented so that it is pointed to the next memory location. Once the values have appeared in the registers (four cycles after the load instruction), the MPYHL instruction is used to multiply them and store the product in another register. Then, the summation is performed by using the ADD instruction. The completed assembly program is as follows:

```
        .global _fir_simple
_fir_simple:
        MV      .S1     A6,A2           ;Count register
        ZERO    .S1     A9              ;Sum register

loop:   LDW     .D1     *A4++,A7        ;Load data from samples
        LDH     .D2     *B4++,B7        ;Load data from coefficients
        NOP     4
        MPYH_   .M1x    A7,B7,A8        ;A7 is 32 bit sample, B7 is Q-15
                                        ;  representation coefficient
        NOP
        SHL     A8,1,A8                 ;Eliminate sign extension bit
        ADD     .S1     A8,A9,A9        ;Accumulate result
[A2]    SUB     .S1     A2,1,A2         ;Decrement counter
[A2]    B       .S1     loop
        NOP     5

        MV      .S1     A9,A4           ;Move result to return register
        B       .S2     B3              ;Branch back to calling address
        NOP     5
```

The preceding code is a C callable assembly function. To call this function from C, a function declaration must be added as external (extern) without any arguments. The arguments to the function are passed via registers A4, B4, and A6. The return value is stored in A4. Here, the pointers to the arrays are passed in A4 and B4 as the first two arguments and the number of iterations in A6 as the third argument. The return address from the function is stored in B3. Therefore, a final branch to B3 is required to return from the function. For a complete explanation of calling assembly functions from C, see the *TI TMS320C6x Optimizing C Compiler User's Guide* [17].

Running this code from the external SBSRAM memory takes a total of 590 cycles for the assembly function to complete. To move the code so that it runs in the internal program memory, an appropriate section in the command file should be defined. The assembly file must also contain the directive .sect with the same name so that the linker would know which part of the code to place in the internal memory. By adding the directive .sect "fir" at the beginning of the assembly file, it is possible to move the code into the internal memory space. Running the code from there results in 230 cycles. Notice that only the assembly function is running in the internal program memory; the rest of the ISR is still slow taking a total of 982 cycles. It is possible to move the complete ISR

L4.2 FIR Filter Implementation

into the internal memory space to allow a faster execution of the interrupt. This can be performed by going through the steps just mentioned.

To optimize the foregoing function, basic optimization methods, such as placing instructions in parallel, filling delay slots, and loop unrolling, are applied. Examining the code, one sees that some of the instructions can be placed in parallel. Because of operand dependencies, care must be taken not to schedule parallel instructions that use previous operands as their operands. The two initial load instructions are independent and can be made to run in parallel. Looking at the rest of the program, we can see that the operands are dependent on the previous operands; hence, no other instructions can be placed in parallel.

To reduce the number of cycles taken by the NOP instructions, we can use the delay slot filling technique. For example, as the load instructions are executed in parallel, it is possible to schedule the subtraction of the loop counter in place of their NOPs. The branch instruction takes five cycles to execute. It is therefore possible to slide the branch instruction four slots up to get rid of its NOPs. Incorporating these optimizations, we can rewrite the function as follows:

```
            .global _fir_filled
            .sect    ".fir_filled"           ;used to load into internal
                                             ;   program memory
_fir_filled:
            MV       .S1    A6,A2           ;Count register
            ZERO     .S1    A9              ;Sum register

loop:       LDW      .D1    *A4++,A7        ;Load data from samples
     ||     LDH      .D2    *B4++,B7        ;Load data from coefficients
            NOP
     [A2]   SUB      .S1    A2,1,A2         ;Decrement counter
     [A2]   B        .S1    loop            ;branch back to loop
            NOP
            MPYHL    .M1x   A7,B7,A8        ;A7 is 32 bit sample, B7 is Q-15
                                            ;   representation coefficient
            NOP
            SHL             A8,1,A8         ;Eliminate sign extension bit
            ADD      .S1    A8,A9,A9        ;Accumulate result

            MV       .S1    A9,A4           ;Move result to return register
            B        .S2    B3              ;Branch back to calling address
            NOP      5
```

By filling delay slots, the number of cycles is reduced. In repetitive loops such as this one, it is seen that the branch instruction takes up extra cycles that can be eliminated. As just mentioned, one method to perform this elimination is to fill the delay slots by sliding the branch instruction higher in the execution phase, thus filling the latencies associated with branching. Another method for reducing the effect of the latencies is to unroll the loop. However, notice that loop unrolling eliminates only the last

186   Chapter 9   Lab 4:   Real-time Filtering

latency of the branch. Since, in the preceding delay filled version, the branch latency has no effect on the number of cycles, loop unrolling does not achieve any further improvement in timing.

To perform word optimization, the ISR has to be modified to store a sample into 16 bits rather than 32. This is achieved by simply right shifting the input by 16 bits during the read process. The following code stores a sample into a short variable, assuming that the result is 32 bits.

```
interrupt void serialPortRcvISR (void)
{
        int i,result = 0;
        short temp;
        temp = MCBSP_READ(0) >> 16;
        //Filtering
        MCBSP_WRITE(0, result);
}
```

Using word optimization, it is possible to load two consecutive 16-bit values in memory with a single load-word instruction. This way, the input register contains two values, one in the lower and the other in the upper part. The instructions MPY and MPYH can then be used to multiply the upper and lower parts, respectively. The following assembly code shows how this is done for the FIR filtering program:

```
        .global _fir_wordoptimized
_fir_wordoptimized:
        MV      .S1     A6,A2                   ;Count register
        ZERO    .S1     A9                      ;Sum register
||      ZERO    .S2     B9

loop:   LDW     .D1     *A4++,A7                ;Load data from samples (here the
                                                ; input data is in 16 bit format)
||      LDW     .D2     *B4++,B7                ;Load data from coefficients
[A2]    SUB     .S1     A2,1,A2                 ;Decrement counter
[A2]    B       .S1     loop
        NOP     2
        MPY     .M2     A7,B7,B8                ;B8 is the lower part product
||      MPYH    .M1     A7,B7,A8                ;A8 is the higher part product
        NOP
        ADD     .S1     A8,A9,A9                ;Accumulate result
||      ADD     .S2     B8,B9,B9                ;Accumulate result

        LDH     .D1     *A4++,A7                ;Load the final elements
||      LDH     .D2     *B4++,B7                ;Load the final elements
        NOP     4
        MPY     .M1     A7,B7,A8                ;Final multiply
        NOP
        ADD     .L1     A8,A9,A9                ;Final add
```

```
            ADD     .S1    A9,B9,A4      ;Move result to return register
            SHL            A4,1,A4       ;Eliminate sign extension bit

            B       .S2    B3            ;Branch back to calling address
            NOP            5
```

Notice here that since two loads are done consecutively, it takes half the amount of time to loop through the program. When calling this function, the value passed in A6 must be the truncated $N/2$, where N is the number of coefficients of the FIR filter. In our case, we have 11 coefficients requiring five iterations plus an additional multiply and accumulate. With this code, it is possible to bring down the number of cycles to 92. The timing cycles for the aforementioned optimizations are listed in Table L4-3.

TABLE L4-3 Timing cycles for different optimizations.

Optimization	Number of Cycles
Un-optimized assembly	230
Delay slot filled assembly	149
Word optimized assembly	92

L4.2.1 Handwritten Software-pipelined Assembly

To produce a software-pipelined version of the code, it is required to first write it in symbolic form without any latency or register assignment. The following code shows how to write the FIR program in a symbolic form:

```
            LDW     *p_sample++,sample    ;load sample word
            LDH     *p_coef++,coef        ;load coef half-word
            MPYHL   sample,coef,temp      ;temp = sample(high)*coef
            SHL     temp,1,temp           ;shift left to remove sign extended bit
            ADD     sum,temp,sum          ;sum += temp
    [count] SUB     count,1               ;decrement counter
    [count] B       loop                  ; branch back to loop
```

To handwrite software-pipelined code, a dependency graph of the loop must be drawn and a scheduling table be created from it. The software-pipelined code is then derived from the scheduling table. To draw a dependency graph, we start by drawing nodes for the instructions and symbolic variable names. Then we draw lines or paths that show the flow of data between nodes. The paths are marked by the latencies of the instructions of their parent nodes.

After the basic dependency graph is drawn, functional units have to be assigned. Then, a line is drawn between the two sides of the CPU so that the workload is split as equally as possible. In the preceding FIR program, the loads should be done one on

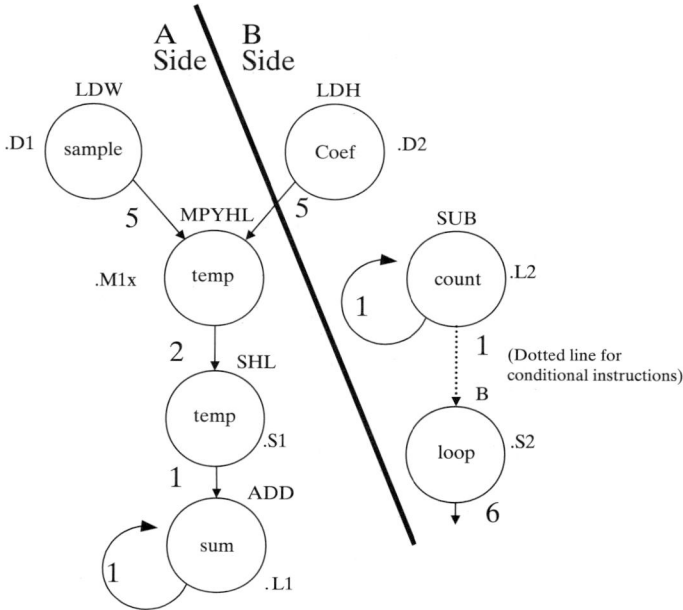

FIGURE L4-7 FIR dependency graph.†

each side, so that they run in parallel. It is up to the programmer on which side to place the rest of the instructions to divide the workload equally between the A-side and B-side functional units. The completed dependency graph for the FIR program is shown in Figure L4-7.

The next step for handwriting a pipelined code is to set up a scheduling table. To do so, the longest path must be identified to determine how long the table should be. Counting the latencies of each side, one sees that the longest path located on the left side is 9. Thus, eight prolog columns are required in the table before entering the main loop. There need to be eight rows (one for each functional unit) and nine columns in the table. The scheduling is started by placing the parallel load instructions at slot 1. The instructions are repeated at every loop thereafter. The multiply instruction must appear five slots after the loads, so it is scheduled into slot 6. The shift must appear one slot after the multiply, and the add must appear after the shift instruction, placing it in slot 9, which is the loop kernel of the code. The branch instruction is scheduled in slot 4 by reverse counting five cycles back from the loop kernel. The subtraction must occur before the branch, so it is scheduled in slot 3. The completed scheduling table appears in Table L4-4.

The software-pipelined code is handwritten directly from the scheduling table as eight parallel instructions before entering a loop that completes all the adds. The resulting code, with which the number of cycles is reduced to 74, is as follows:

L4.2 FIR Filter Implementation

```
        .global _fir_pipelined

_fir_pipelined:
        ZERO    A8
        ZERO    A9
        MV      A6,B2

        LDW     .D1     *A4++,A7
||      LDH     .D2     *B4++,B7

        LDW     .D1     *A4++,A7
||      LDH     .D2     *B4++,B7

        LDW     .D1     *A4++,A7
||      LDH     .D2     *B4++,B7
|| [B2] SUB     .L2     B2,1,B2

        LDW     .D1     *A4++,A7
||      LDH     .D2     *B4++,B7
|| [B2] SUB     .L2     B2,1,B2
|| [B2] B       .S2     loop10

        LDW     .D1     *A4++,A7
||      LDH     .D2     *B4++,B7
|| [B2] SUB     .L2     B2,1,B2
|| [B2] B       .S2     loop10

        LDW     .D1     *A4++,A7
||      LDH     .D2     *B4++,B7
|| [B2] SUB     .L2     B2,1,B2
|| [B2] B       .S2     loop10
||      MPYHL   .M1     A7,B7,A8
```

TABLE L4-4 FIR scheduling table.[†]

			PROLOG						LOOP
	1	2	3	4	5	6	7	8	9
.L1									ADD
.L2			SUB	SUB	SUB	SUB	SUB	SUB	SUB
.S1								SHL	SHL
.S2				B	B	B	B	B	B
.M1						MPYHL	MPYHL	MPYHL	MPYHL
.M2									
.D1	LDW	LDW	LDW	LDW	LDW	LDW	LDW	LDW	LDW
.D2	LDH	LDH	LDH	LDH	LDH	LDH	LDH	LDH	LDH

```
            LDW     .D1     *A4++,A7
||          LDH     .D2     *B4++,B7
|| [B2]     SUB     .L2     B2,1,B2
|| [B2]     B       .S2     loop10
||          MPYHL   .M1     A7,B7,A8

            LDW     .D1     *A4++,A7
||          LDH     .D2     *B4++,B7
|| [B2]     SUB     .L2     B2,1,B2
|| [B2]     B       .S2     loop10
||          MPYHL   .M1     A7,B7,A8
||          SHL     .S1     A8,1,A8

loop10:     LDW     .D1     *A4++,A7
||          LDH     .D2     *B4++,B7
|| [B2]     SUB     .L2     B2,1,B2
|| [B2]     B       .S2     loop10
||          MPYHL   .M1     A7,B7,A8
||          SHL     .S1     A8,1,A8
||          ADD     .L1     A8,A9,A9

            MV      .L1     A9,A4
            B       .S2     B3
            NOP     5
```

L4.2.2 Assembler Optimizer Software-Pipelined Assembly

Since handwritten pipelined codes are time consuming to write, linear assembly is usually used to generate pipelined codes. In linear assembly, latencies, functional units and register allocations do not need to be specified. Instead, symbolic variable names are used to write a sequential code with no delay slots (NOPs). The file extension for a linear assembly file is .sa. The assembly optimizer is automatically invoked if a file in a CCS Project has a .sa extension. The assembly optimizer turns a linear assembly code into a pipelined code. Notice that the optimization level option in C also affects the optimization of linear assembly.

A code line in linear assembly consists of five fields: label, mnemonic, unit specifier, operand list, and comment. The general syntax of a linear assembly code line is

[label[:]] [[register]] mnemonic [unit specifier] [operand list] [;comment]

Fields in square brackets are optional. A label must begin in column 1 and can include a colon. A mnemonic is an instruction such as MPY or an assembly optimizer directive such as .proc. Notice that a mnemonic will be interpreted as a label if it begins in column 1. At least one blank space should be placed in front of a mnemonic when there is no label. A mnemonic becomes a conditional instruction when it is preceded by a register or a symbolic variable within two square brackets. A unit specifier specifies the functional unit performing the mnemonic. Operands can be symbols, constants, or expressions and should be separated by commas. Comments begin with a semicolon. Comments beginning in column 1 can begin with either an asterisk or a semicolon.

L4.2 FIR Filter Implementation

In writing code in linear assembly, the assembler optimizer must be supplied with the right kind of information for performing optimizations. The first such piece of information is where it should optimize. The optimizer considers only code between the directives .proc and .endproc. Symbolic variable names are used to allow the optimizer to select which registers to use. This is done by using the .reg directive together with the names of the variables. Also, registers that contain input arguments, such as variables passed to a function, must be specified. The registers declared to contain input arguments cannot be modified and have to be declared as operands of the .proc statement. To connect the input register arguments to the symbolic variable names, the move instruction, mv, is used. Registers that contain output values upon exiting the procedure must be declared as arguments to the .endproc directive.

To write the FIR code in linear assembly, we start by creating the main loop and then add the load, multiply, and add instructions. Since two pointers to two arrays and an integer are passed to the function, we must declare registers A4, B4, and A6 as part of the .proc directive. Also, register A4 is used for returning values, so it needs to appear as part of the .endproc directive. The preserved register B3 is indicated as an argument in both of these directives. To connect the symbolic variable names to the input registers, the mv instruction is used. And finally, the optimizer is told that the loop is to be performed a minimum of 11 times by inserting the .trip directive. The final code is as follows:

```
          .global _fir_la
          .sect ".fir_la"

_fir_la:   .proc  A4,B4,A6,B3
           .reg   p_m,m,p_n,n,prod,sum,cnt

           mv     A4, p_m           ;move argument in A4 to p_m
           mv     B4, p_n           ;move argument in B4 to p_n
           mv     A6, cnt           ;set up counter (third argument)
           zero   sum               ;sum=0

loop8:     .trip  11                ;minimum 11 times through loop
           ldw    *p_m++, m         ;load m
           ldh    *p_n++, n         ;load n
           mpy    m,n,prod          ;prod = m * n
           shl    prod,1,prod       ;prod << 1
           add    prod,sum,sum      ;sum += prod

[cnt]      sub    cnt,1,cnt         ;decrement counter
[cnt]      b      loop8             ;branch back to loop8

           mv     sum,A4            ;move result into return register A4
           .endproc A4,B3

           B      B3                ;branch back to address stored in B3
           NOP    5
```

Using this code, we obtain a timing outcome of 74 cycles, which is the same as the timing obtained by the handwritten software-pipelined assembly.

To summarize the programming approach, start writing your code in C, and then use the optimizer to achieve a faster code. If the code is not as fast as expected, you may write it in assembly and incorporate the aforementioned simple optimization techniques. However, it is usually easier and more efficient to rewrite your code in linear assembly, since the assembly optimizer attempts to create pipelined code for you. Figure L4-8

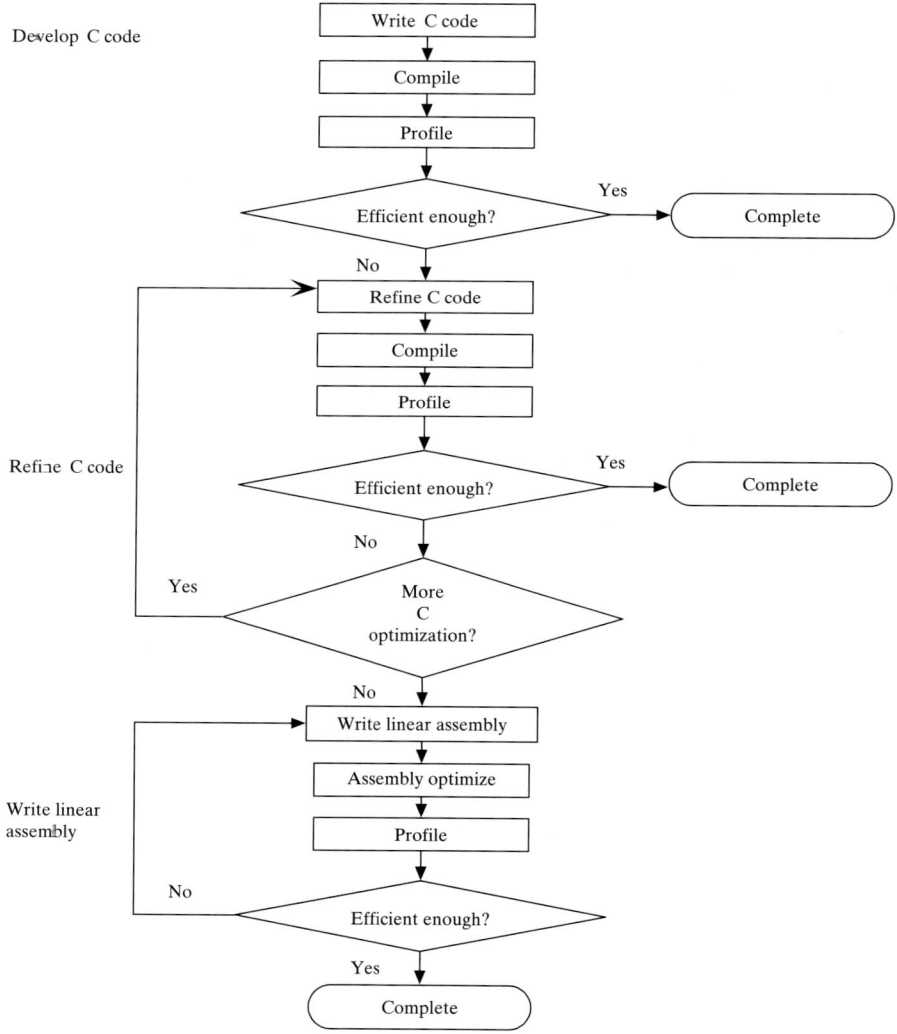

FIGURE L4-8 Code development flow.[†]

illustrates the code development flow to get an optimum performance on the C6x. If, at the end, none of these approaches provide a satisfactory timing cycle, you are left no choice but to rewrite your code in hand-coded pipelined assembly. Appendix A includes an optimization checklist for writing DSP application programs.

L4.3 FLOATING-POINT IMPLEMENTATION

Implementing the FIR filter on the C67 takes relatively less effort. Since the hardware is capable of multiplying and adding floating-point numbers, Q-format number manipulation is not required. However, in general, the floating-point code is slower, because floating-point operations have more latencies than their fixed-point counterparts. As is shown shortly, the FIR filter interrupt and function are modified to run on a C67 EVM board. The code written in C is fairly simple. The coefficients are entered directly as float. The data buffer is declared as float, and a sample is initially read as an integer and then typecast to a float.

You can view your C source code interleaved with disassembled code in Code Composer Studio. To view this mixed mode, after loading the program into the DSP, open the source file by double clicking on it from the Project View panel and then right click in the source window. Select Mixed Mode from the pop-up menu. Using this mixed mode, it can be verified that the compiler is actually using the MPYSP and ADDSP instructions to perform the floating-point multiply and add, rather than calling a separate function to do them in software. The code is as follows:

```
float dotp1(const float a[], const float b[])
{
        int i;
        float sum = 0.0;
        for(i=0; i<11;i++)
                sum += a[i] * b[i];
        return sum;
}
interrupt void serialPortRcvISR (void)
{
int i,sample;
        float temp, sum;
        sample = MCBSP_READ(0);
        temp = (float)sample;
        for(i=10;i>=0;i--)
                x[i]=x[i-1];
        x[0]=temp;
        sum = dotp1(coef, x);
        MCBSP_WRITE(0, sum);
}
```

All the programs associated with this lab can be loaded from the attached CD.

C H A P T E R   1 0

Circular Buffering

In many DSP algorithms, such as filtering, adaptive filtering, or spectral analysis, we need to shift data or update samples (i.e., we need to deal with a moving window). The direct method of shifting data is inefficient and uses many cycles. Circular buffering is an addressing mode by which a moving-window effect can be created without the overhead associated with data shifting. In a circular buffer, if a pointer pointing to the last element of the buffer is incremented, it is automatically wrapped around and pointed back to the first element of the buffer. This provides an easy mechanism for excluding the oldest sample while including the newest sample, creating a moving-window effect as illustrated in Figure 10-1.

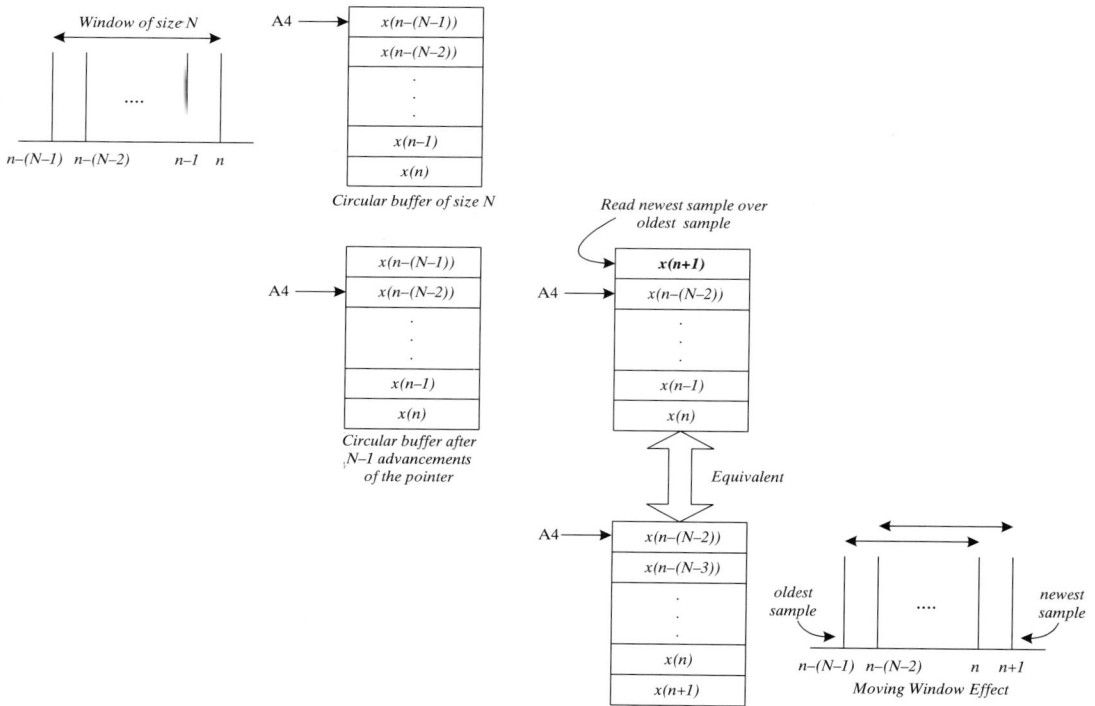

FIGURE 10-1 Moving-window effect.

Some DSPs have dedicated hardware for doing this type of addressing. On the C6x processor, the arithmetic logic unit has the circular buffer addressing mode capability built into it. To use circular buffering, first the circular buffer sizes need to be writ-

Introduction

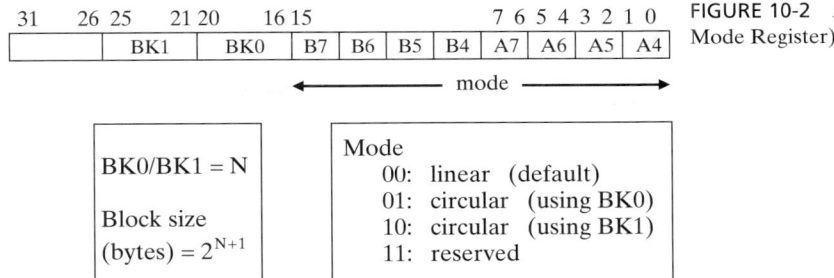

FIGURE 10-2 AMR (Address Mode Register).†

ten into the BK0 and BK1 block size fields of the Address Mode Register (AMR), as shown in Figure 10-2. The C6x allows two independent circular buffers of powers of 2 in size. Buffer size is specified as $2^{(N+1)}$ bytes, where N indicates the value written to the BK0 and BK1 block size fields.

Then, the register to be used as the circular buffer pointer needs to be specified by setting appropriate bits of AMR to 1. For example, as shown in Figure 10-2, using A4 as a circular buffer pointer, bits 0 and 1 are set to 1. Of the 32 registers on the C6x, 8 can be used as circular buffer pointers: A4 through A7 and B4 through B7. Note that linear addressing is the default mode of addressing for these registers.

Figure 10-3 shows the code to set up the AMR register for a circular buffer of size 8, together with a load example. To set up such a circular buffer in C, one must use so called in-line assembly as follows:

```
asm ("MVK.S2     0001h,B2");
asm ("MVKLH.S2   0002h,B2");
asm ("MVC.S2     B2,AMR");
```

When using circular buffers, care must be taken to align data on the buffer size boundary. In C, this can be achieved by using so-called pragma directives, which indicate what kinds of preprocessing are done by the compiler. The DATA_ALIGN pragma can be used to align symbol with a power of 2 alignment boundary constant (in bytes) as follows:

```
#pragma DATA_ALIGN (symbol,constant)
```

```
; Blk size = 8, use A4/BK0
        MVK.S2      0001H , B2
        MVKLH.S2    0002H , B2
        MVC.S2      B2, AMR

        LDH.D1      * A4++[2], A1 ; A1= 0, A4=&s[2]
        LDH.D1      * A4++[3], A1 ; A1= 2, A4=&s[1]
```

FIGURE 10-3 AMR setup example.†

Lab 5: Adaptive Filtering

Adaptive filtering is used in many applications ranging from noise cancellation to system identification. In most cases, the coefficients of an FIR filter are modified according to an error signal in order to adapt to a desired signal. In this lab, a system identification example is implemented wherein an adaptive FIR filter is used to adapt to the output of a seventh-order IIR bandpass filter. The IIR filter is designed in Matlab and implemented in C. The adaptive FIR is first implemented in C and later in assembly using circular buffering.

In system identification, the behavior of an unknown system is modeled by accessing its input and output. An adaptive FIR filter can be used to adapt to the output of the system based on the same input. The difference in the output of the system, $d[n]$, and the output of the adaptive filter, $y[n]$, constitutes the error term $e[n]$, which is used to update the coefficients of the FIR filter. Figure L5-1 illustrates this process.

The error term calculated from the difference of the outputs of the two systems is used to update each coefficient of the FIR filter according to the formula (least mean square (LMS) algorithm [1]):

$$h_n[k] = h_{n-1}[k] + \delta * e[n] * x[n - k], \qquad \textbf{(L5.1)}$$

where the h's denote the unit sample response or FIR filter coefficients. The output $y[n]$ is required to approach $d[n]$. The term δ indicates the step size. A small step size will ensure convergence, but results in a slow adaptation rate. A large step size, though faster, may lead to skipping over the solution.

L5.1 DESIGN OF IIR FILTER

A seventh-order bandpass IIR filter is used as the unknown system. The adaptive FIR is designed to adapt to the response of the IIR system. Using a sampling frequency of 8 kHz, let the IIR filter has a passband from $\pi/3$ to $2\pi/3$ (radians), with a stopband attenuation

FIGURE L5-1 Adaptive filtering.

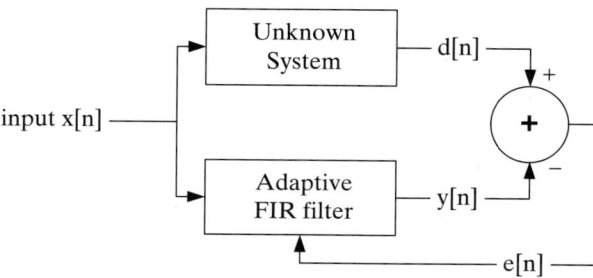

of 20 dB. The design of the filter can be easily achieved with the Matlab function 'yulewalk' [2]. The following Matlab code may be used to obtain the coefficients of the filter:

```
Nc=7;
f=[0 0.32 0.33 0.66 0.67 1];
m=[0 0 1 1 0 0];
[B,A]=yulewalk(Nc,f,m);
freqz(B,A);

%Create A sample signal
Fs=8000;
Ts=1/Fs;
Ns=128;
t=[0:Ts:Ts*(Ns-1)];
f1=750;
f2=2000;%The one to keep
f3=3000;

x1=sin(2*pi*f1*t);
x2=sin(2*pi*f2*t);
x3=sin(2*pi*f3*t);

x=x1+x2+x3;
%Filter it
y=filter(B,A,x);
```

It can be verified that the filter is working by deploying a simple composite signal. Using the Matlab function 'filter', verify the design by observing that the frequency components of the composite signal falling in the stopband are removed. (See Figure L5-2 and Table L5-1.)

TABLE L5-1 IIR filter coefficients

A's	B's
1.0000	0.1191
0.0179	0.0123
0.9409	−0.1813
0.0104	−0.0251
0.6601	−0.1815
0.0342	0.0307
0.1129	−0.1194
0.0058	−0.0178

Note: Do not confuse A&B coefficients with the CPU A&B registers!

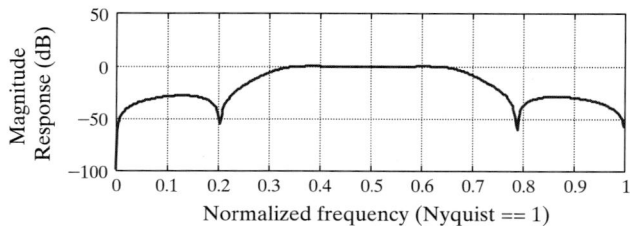
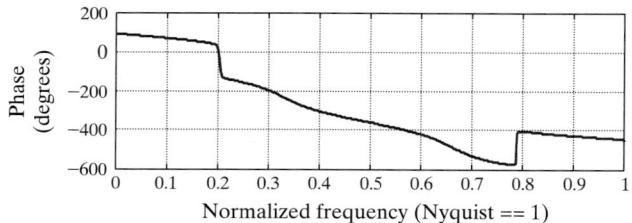
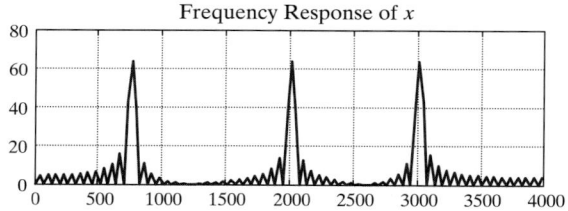
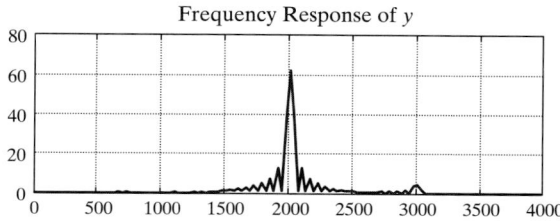

FIGURE L5-2 IIR Filter response.

L5.2   IIR FILTER IMPLEMENTATION

The implementation of the IIR filter is first done in C, using the difference equation

$$y[n] = -\sum_{k=1}^{N} a_k * y[n-k] + \sum_{k=0}^{N} b_k * x[n-k], \quad \textbf{(L5.2)}$$

where the a_k's and b_k's denote the IIR filter coefficients. Two arrays are required, one for the input samples $x[n]$ and the other for the output samples $y[n]$. Given that the filter is of order 7, an input array of size 8 and an output array of size 7 are considered. The arrays are used to simulate a circular buffer, since in C this property of the CPU cannot be accessed. As a new sample comes in, all elements in the input array are shifted down by one. In this manner, the last element is lost and the last eight samples are always kept. The input array is used to calculate the resulting output, and then the output is used to modify the output array. A simple implementation of this scheme is shown in the following code, which is a modification of the sampling program in Lab 2:

```
interrupt void serialPortRcvISR (void)
{
  int temp,n,ASUM,BSUM;
  short input,IIR_OUT;

    temp = MCBSP_READ(0);
    input = temp >> S;                          //Scaling factor

    for(n=7;n>0;n--)                            //Input buffer
       IIRwindow[n] = IIRwindow[n-1];

    IIRwindow[0] = input;

    BSUM = 0;


    for(n=0;n<=7;n++)
    {                                           //Multiplication of Q-15 with Q-15
       BSUM += ((BS[n] * IIRwindow[n]) << 1);   //Results in Q-30. Shift by one to
    }                                           //Eliminate Sign Extension bit

    ASUM = 0;
    for(n=0;n<=6;n++)
    {
       ASUM += ((AS[n] * y_prev[n]) << 1);
    }

    IIR_OUT = (BSUM - ASUM) >> 16;
    for(n=6;n>0;n--)                            //Output buffer
       y_prev[n] = y_prev[n-1];

    y_prev[0] = IIR_OUT;

    MCBSP_WRITE(0, IIR_OUT << S);               // Scaling factor S
}
```

By running this program while connecting a function generator and an oscilloscope to the line-in and line-out of the EVM board, the functionality of the IIR filter can be easily verified. Whenever the DRR receives a new incoming sample from the function generator, the ISR is invoked. Then, the new sample is right shifted by the scaling factor S. This factor is included for the correction of any possible overflow, but here using the original shift of 16 is adequate. Once the new sample is scaled, the last eight samples are kept by discarding the oldest sample and adding the new sample to the input buffer IIRwindow. This operation is done by shifting the data in the input array. Note that this array is global and is initialized to zero in the main function.

Now that the last eight samples are ready to be used, it is time to compute BSUM (the b coefficient terms) and ASUM (the a coefficient terms). Attention needs to be paid to the datatype of BSUM, ASUM, the coefficient arrays, and the input array. The datatype of the coefficient arrays is short, so the coefficients are converted to Q-15 format by multiplying them by 0x7FFF in the main function. The datatype of the input array IIRwindow is also short. However, the datatype of ASUM or BSUM is int (32 bits). Therefore, ASUM and BSUM need to be left shifted by 1 to eliminate the extended sign bit, since the multiplication of Q-15 by Q-15 results in Q-30 representation. In order to obtain the IIR output IIR_OUT, the partial output ASUM is subtracted from BSUM. Note that the difference (BSUM - ASUM) is right shifted by 16 to convert it to a short datatype. The IIR output is then used to compute ASUM in the next iteration. Finally, the IIR output is scaled back and sent to the data transmit register (DXR).

L5.3 ADAPTIVE FIR FILTER

By replacing the following piece of code with MCBSP_WRITE(0,IIR_OUT<<S) in the previous IIR function, we can make an FIR filter to adapt to the output of the IIR filter:

```
//Simulate Circular buffer for FIR
for(n=31;n>0;n--)
        FIRwindow[n] = FIRwindow[n-1];

FIRwindow[0] = input;

//Perform Filtering with current coefficients
temp = 0;
for(n=0;n<32;n++)
{
        temp += ((h[n]*FIRwindow[n]) << 1);
}

y = temp >> 16;

//Calculate Error Term

e = IIR_OUT - y;

//Update Coefficients

stemp = (DELTA*e)>>15;

for(n=0;n<32;n++)
{
        stemp2 = (stemp*FIRwindow[n])>>15;
        h[n] = h[n] + stemp2;
}
MCBSP_WRITE(0,y<<S);
```

In this program, a 32-coefficient FIR filter is used to adapt to the output of the IIR filter. To do this in C, an additional buffer of length 32 is needed: one for the input buffer FIRwindow and the other for the coefficients h of the FIR filter. Initially, all the data in both arrays are zero. The order of processing is as follows: First, the last 32 samples are shifted. The shift discards the oldest sample and adds the newly read sample into the input buffer. Next, the FIR filtering is done by performing a dot product between the coefficients h and the input buffer. The dot product is converted to Q-15 format by left shifting it by 1. Then, the error term between the IIR and the FIR filter output is computed. The coefficients of the FIR filter are then updated to match the IIR and FIR filter outputs. Finally, the FIR filter output is sent to the DXR. By using a function generator and an oscilloscope, the adaptation process can be observed by scanning through different frequencies.

It is worth mentioning a point about the step size δ. In a floating-point processor, δ is usually chosen to be in the range of e^{-7}. However, the precision on the C62 does not allow for such a small number. We can at most use 0x0001, which is $1/(2^{15}) \approx 0.0000305$. When a multiplication is done with this number, any positive number will be defaulted to 0 and any negative number to -1. This is due to the nature of multiplication of Q-15 format numbers, where the product is right shifted by 15. However, the contribution of negative numbers to the coefficients is sufficient for the LMS algorithm to converge. Using a larger δ for this adaptive filtering example results in faster adaptation, but convergence is not guaranteed. Satisfactory results can be observed with δ in the range of 0x0100 to 0x0001.

The reason for implementing the LMS algorithm in assembly is to make use of the circular buffering capability of the C6x. Of the 32 registers on the C6x, 8 can perform circular addressing. These registers are A4 through A7 and B4 through B7. Since linear addressing is the default mode of addressing, each of these registers must be specified as circular using the AMR register. The lower 16 bits of the AMR register are used to select the mode for each of the 8 registers. The upper 10 bits (6 are reserved) are used to set the length of the circular buffer. The buffer size is determined by $2^{(N+1)}$ bytes, where N is the value appearing in the block size fields of the AMR register.

Since we are using both C and assembly, we have to initialize the circular buffer when we enter the assembly part of the program. When the assembly code is executed, the register that is used in the circular mode allows a certain location in memory to always contain the newest sample. As the assembly code is completed and returns to the calling C program, the location of the pointer to the buffer must be saved, and the AMR register must be returned to the linear mode, since leaving it in the circular mode disrupts the flow of the program.

To do such a task, a section of memory not used by the compiler must be set aside for the buffer and the coefficients. A simple way to do this is to reserve 64 bytes for the coefficients, 64 bytes for the buffer, and 1 byte for the pointer. Since the data and coefficients are short formatted here, 64 bytes are used to provide 32 locations. The following memory representation is employed for this purpose:

8000_0000	64 Bytes	Circular Buffer
8000_0040	64 Bytes	Coefficients
8000_0080	1 Byte, Pointer	

The command file must also be modified. A simple assembly file is needed to initialize the memory locations with zeros. The following command file defines a new memory section called MMEM in the internal data memory and uses it for the code section .mydata:

```
MEMORY
{
  INT_PROG_MEM   (RX)   : origin = 0x00000000 length = 0x00010000
  SBSRAM_PROG_MEM (RX)  : origin = 0x00400000 length = 0x00014000
  SBSRAM_DATA_MEM (RW)  : origin = 0x00414000 length = 0x0002C000
  SDRAM0_DATA_MEM (RW)  : origin = 0x02000000 length = 0x00400000
  SDRAM1_DATA_MEM (RW)  : origin = 0x03000000 length = 0x00400000
  INT_DATA_MEM   (RW)   : origin = 0x80000100 length = 0x0000FF00
  MMEM                  : origin = 0x80000000 length = 0x00000100
}

SECTIONS
{
  .vec:      load = 0x00000000
  .text:     load = INT_PROG_MEM
  .const:    load = INT_DATA_MEM
  .bss:      load = INT_DATA_MEM
  .data:     load = INT_DATA_MEM
  .cinit     load = INT_DATA_MEM
  .pinit     load = INT_DATA_MEM
  .stack     load = INT_DATA_MEM
  .far       load = INT_DATA_MEM
  .sysmem    load = SDRAM0_DATA_MEM
  .cio       load = INT_DATA_MEM
  .int       load = INT_PROG_MEM
  .mydata    load = MMEM
  sbsbuf     load = SBSRAM_DATA_MEM
         {_SbsramDataAddr = .; _SbsramDataSize = 0x0002C000; }
}
```

The file *initmem.asm* appearing next is used to initialize the memory locations with zeros and set the pointer to the first free location, which is 0x80000000:

initmem.asm

```
.sect ".mydata"
      .short
0,0,0,0,0,0,0,0,0,0,0,0,0,0,0,0,0,0,0,0,0,0,0,0,0,0,0,0,0,0,0,0
      .short
0,0,0,0,0 0,0,0,0,0,0,0,0,0,0,0,0,0,0,0,0,0,0,0,0,0,0,0,0,0,0,0
      .field    0x80000000,32
```

L5.3 Adaptive FIR Filter

With the command file and the brief assembly code just shown, it is ensured that 129 bytes of space, starting at 0x80000000, will not be used for anything other than the adaptive filter. Now, as mentioned before, the circular buffer must be initialized upon entering the assembly part. To do this, it is necessary to modify the AMR register. Since a buffer of length 32 is needed, one must have 5 in the block fields (block size = $2^{(5+1)}$ = 64 bytes). With register A5 as the circular buffer pointer, the value to set the AMR register becomes 0x00050004. Entering the assembly function, the last pointer location is read from 0x80000000. The last free location of the buffer is saved to the same location upon exit. The following code shows how this is achieved:

```
;Initialize the Circular buffer for the FIR filter
MVK        .S2      0x0004,B10           ;A5 is selected as circular
MVKLH      .S2      0x0005,B10           ;2^(5+1)=64
MVC        .S2      B10,AMR

;Load the pointer to A0
;Assume that the current location of the circular buffer
;is pointed to

MVK        .S1      0x0080,A0
MVKLH      .S1      0x8000,A0            ;A0=0x80000080 (has last pointer)

LDW        .D1      *A0,A5
NOP        4                             ;A5 now points to the first free
                                         ;Location of the circular buffer
;Load the current sample to the Circular buffer
STH        .D1      A4,*A5               ;A4 has sample passed from calling

//FIR FILTERING HERE

;Now save the Last location of A0 to memory

MVK        .S1      0x0080,A0
MVKLH      .S1      0x8000,A0            ;This address has the pointer to x

LDH        .D1      *A5++,A13            ;Dummy Load
STW        .D1      A5,*A0               ;Saved last
;Restore Linear Addressing
MVK        .S2      0x0000,B10
MVKLH      .S2      0x0004,B10
MVC        .S2      B10,AMR

;return the result y

MV         .S1      A9,A4
B          .S2      B3
NOP        5
```

Chapter 10 Lab 5: Adaptive Filtering

Upon entering the assembly function, the AMR register is loaded with 0x00050004 for the desired circular buffer operation. Then, the memory location (or address) to which register A5 was last pointing is loaded into A5. Hence, A5 points to the first free location of the circular buffer, and the content of register A4 (the newest sample passed from the C program) is stored in this location. After adaptive filtering, the address pointed to by A5 is stored at the location 0x80000000. Note that here a dummy load is performed to increment the pointer so that it points to the last element (the next free location). This is needed because only a load or store operation increments the pointer in a circular fashion.

The following adaptive FIR filter assembly code resides in the section of the foregoing code labeled 'FIR FILTERING HERE':

```
            ;Do the filtering
            MVK    .S2         0x0040,B1
            MVKLH  .S2         0x8000,B1      ;This is the address of h[n]

            MVK    .S2         32,B2          ;Set up a counter
            ZERO   .S1         A9             ;Accumulator
loop:
            LDH    .D2         *B1++,B7       ;load h_k
            LDH    .D1         *A5-,A7
            NOP    4
            MPY    .M1x        A7,B7,A7       ;A7 is Q-30
            NOP
            SHL    .S1         A7,1,A7
            ADD    .S1         A7,A9,A9
    [B2]    SUB    .S2         B2,1,B2        ;Decrement Counter
    [B2]    B      .S2         loop
            NOP    5

            SHR    .S1         A9,16,A9       ;Make Short, Eliminate Sign extension bit
                                              ;A9 is now short Y

            ;Calculate Error Term
            MV     .S1         B4,A13
            SUB    .S1         A13,A9,A8      ;A13=d(IIR_OUT),A9=y,A8=e

            ;Update Coefficients
            MVK    .S1         0x0100,A10     ;A10=DELTA
            MPY    .M1         A8,A10,A10     ;A10=DELTA*e this is actually ineffective
            NOP
            SHR    .S1         A10,15,A10     ;A10=DELTA*e is now short Q-15
```

```
            MVK    .S2        32,B2           ;Loop Counter
            MVK    .S2        0x0040,B1
            MVKLH  .S2        0x8000,B1       ;This is the address of h[n]

loop2:
            LDH    .D1        *A5-,A8         ;Load x[n-k]
            LDH    .D2        *B1,A12         ;Load h[n]
            NOP    4
            MPY    .M1        A10,A8,A8       ;A10 = DELTA*e*x in Q-31
            NOP
            SHR    .S1        A8,15,A8        ;A10 is now Q-15
            ADD    .S1        A8,A12,A8       ;Updated h
            STH    .D2        A8,*B1++        ;Update the coefficient
   [B2]     SUB    .S2        B2,1,B2         ;Decrement Counter
   [B2]     B      .S2        loop2
            NOP    5
```

In this code, the adaptive filtering process is done via two separate loops. The first loop calculates a dot product between the coefficients and the samples. The error term is then calculated and used in the second loop for updating the coefficients, based on the input samples. Notice that a circular buffer is not used for the coefficients, since they do not change in a time-windowed manner, as do the input samples.

We now have two versions of the adaptive filter. One is written entirely in C, and the other is a mix of C and assembly. When the entire C program runs in the external memory, the output does not adapt to the IIR filter output. Only when the entire C program runs in the internal memory does the output adapt to the IIR filter output. Also, when the assembly part of the mixed C/assembly program runs in the external memory, the output adapts to the IIR filter output. Of course, the assembly part of the program can be configured to run in the internal memory space, where the number of cycles is noticeably reduced. The main reason for running the C code from the internal memory space is that running it from the external memory is too slow and samples get missed. Table L5-2 gives a summary of the timing cycles for different memory options of the adaptive filtering program. All the programs for this lab are included on the attached CD.

TABLE L5-2 Timing cycles for different memory options.

Type of build	Number of cycles
C program in external memory	7880
C program in internal memory	2882
Non-optimized assembly in external memory	3577
Non-optimized assembly in internal memory	1166

CHAPTER 11

Frame Processing

When it comes to processing frames of data (for example, in doing FFT and block convolution), triple buffering is an efficient data frame handling mechanism. While samples of the current frame are being collected by the CPU in an input array via an ISR, samples of the previous frame in an intermediate array can get processed during the time left between samples. At the same time, the DMA can be used to send out samples of a previously processed frame residing in an output array. In this manner, the CPU is used to set up the input array and process the intermediate array while the DMA is used to move processed data from the output array. At the end of each frame or the start of a new frame, the roles of these arrays are interchanged. The input array is reassigned as the intermediate array to be processed, the processed intermediate array is reassigned as the output array to be sent out, and the output array is reassigned as the input array to collect incoming samples for the current frame. This process is illustrated in Figure 11-1.

11.1 DIRECT MEMORY ACCESS

Many DSP chips are equipped with a Direct Memory Access resource acting as a coprocessor to move data from one part of memory into another without interfering with the CPU operation. As a result, the chip throughput is increased since, in this manner, the CPU can process and the DMA can move data without interfering with each other. The C6x

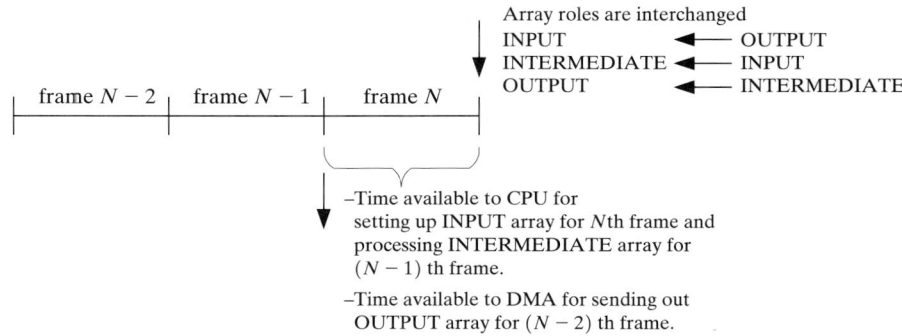

FIGURE 11-1 Triple buffering technique.

11.1 Direct Memory Access

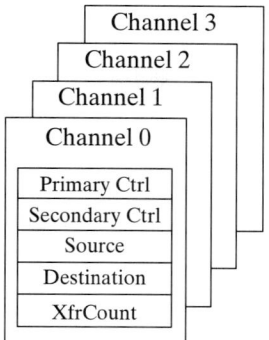 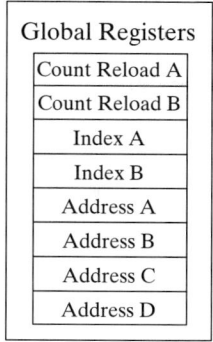

FIGURE 11-2 DMA channels control registers.[†]

DSPs possess a number of DMA channels (e.g., the C6201 has four channels). Each DMA channel has its own memory-mapped control registers which can be set up to move data from one place to another place in memory. Figure 11-2 shows the DMA control registers, consisting of the Primary Control, Secondary Control, Source Address, Destination Address, and Transfer Counter Registers. These registers contain the information regarding the source and destination locations in memory, number of transfers, and format of transfers. As shown in Figure 11-2, in addition to the DMA control registers, there are several global registers shared by all DMA channels.

An example is presented here to show how some of the fields of the DMA registers are set for block or frame processing. More details on all the fields are available in the *TI TMS320C6x Peripherals Reference Guide* [18]. It is possible to transfer a block of data consisting of a number of frames, which in turn contain a number of elements. Elements here mean the smallest piece of data. The example shown in Figure 11-3 illustrates the DMA register setup for transferring a block of data (this can be viewed as image data) consisting of four frames (rows), where each frame contains four elements (16-bit pixels). Figure 11-3 also provides the options for some of the fields of the Primary Control Register, Transfer Counter Register, and Global Index Register. Source/destination address fields can be incremented or decremented by an element size or by an index as specified in the global register. The element size field is used to indicate the datatype. The Transfer Counter Register contains the number of frames and elements. In this example, the element size field of the Primary Control Register is set to 01 (halfwords), the source address field to 11 (index option for accessing next value), and the destination to 01 (increment option for writing consecutively). As specified in the Transfer Counter Register, the data transfer in this example has four frames, each consisting of four elements. The global register A (chosen by the index field) contains the element as well as the frame index. In this example, the element index is set to 2 to increase addresses by 2 bytes or element size, and the frame index to 6 bytes to get to the next frame, since at the end of a frame the pointer points to the beginning of the last element of that frame.

208   Chapter 11   Frame Processing

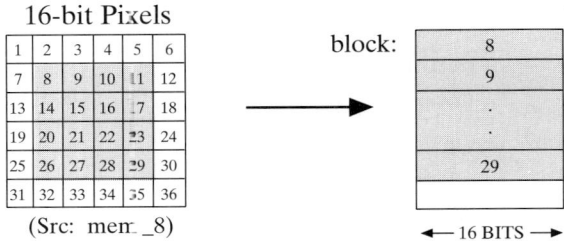

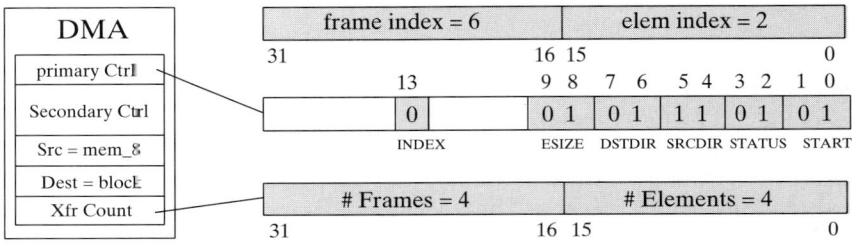

ESIZE		SRC/DSTDIR		STATUS		START	
00	32-bits	00	no modifications	00	Stopped	00	Stop
01	16-bits	01	inc by element size	01	Running (without auto-init)	01	Start without auto-init
10	8-bits	10	dec by element size	10	Paused	10	Pause
11	reserved	11	index	11	Running (with auto-init)	11	Start with auto-init

FIGURE 11-3   DMA transfer example.[†]

11.2 DSP–HOST COMMUNICATION

There are two 32-bit FIFO (first in first out) registers, called FIFO-read and FIFO-write, in the PCI controller of the EVM board through which the host PC and the C6x may communicate. The FIFO-read is used for data transfer from the host to the DSP and the FIFO-write for data transfer from the DSP to the host. The initialization of the FIFO is done through the memory-mapped FIFO status register. The communication can be simply achieved by using the supplied *WriteFIFO_DMA()* function as part of Lab 6.

Lab 6: Fast Fourier Transform

Operations such as DFT or FFT require that a frame of data be present at the time of processing. Unlike filtering, where operations are done on every incoming sample, in frame processing N samples are captured first and then operations are performed on all N samples.

To perform frame processing, a proper method of gathering data and of processing and sending data out is required. The processing of a frame of data is not usually completed within the sampling time interval, rather it is spread over the duration of a frame before the next frame of data is gathered. Hence, incoming samples must be stored to a separate buffer other than the one being processed. Also, another buffer is needed to send out a previously processed frame of data. As explained earlier, this can be achieved by triple buffering involving three buffers: input, intermediate, and output.

To do triple buffering on the C6x, the sampling shell program in Lab 2 is modified to incorporate an endless loop revolving around the rotation of three buffers. The buffers rotate every time the input buffer is full so that a new frame of N sampled data is passed to the intermediate buffer for processing and a previously processed frame is passed to the output buffer for transmission. The following modifications of the shell program achieve this:

```
short *output;          /* POINTER TO DATA ARRAY FOR OUTPUT   */
short *input;           /* POINTER TO DATA ARRAY FOR INPUT    */
short *intermediate;    /* POINTER TO DATA ARRAY FOR DMA ACCESS */
int index=0;

main()
{
        evm_init();
        mcbsp_drv_init();
        dev = mcbsp_open(0);
        init_serial();
        init_arrays();

        /* Main Loop, wait for Interrupt */
        for(;;)
        {
                wait_buffer();     /* WAIT FOR A NEW BUFFER OF DATA*/
        }
}
```

The preceding code shows how an endless loop is added to the shell program. Here, most of the initializations for the codec and McBSP have not been shown to make the code easier to follow. Once the serial port is initialized, the three arrays are allocated in memory and initialized to zero. The program then goes into an endless loop where the function *wait_buffer()*, shown next, is executed endlessly:

```
void wait_buffer(void)
{
        short *p;
        /* WAIT FOR ARRAY INDEX TO BE RESET TO ZERO BY ISR */
        while(index);

        /* ROTATE DATA ARRAYS */
        p = input;
        input = output;
        output = intermediate;

   //Function call here...

        intermediate = p;
        WriteFIFO_DMA();
        while(!index);
}
```

This function checks on the global variable `index` to do the rotation of the arrays and to start processing. When the `input` array becomes full (indicated by `index`), the arrays are rotated and the `intermediate` array gets set for processing. The comment //Function call here... indicates where the processing function such as FFT should be placed.

The ISR is also modified as shown in the next code block. Note that index is incremented within the ISR.

```
interrupt void serialPortRcvISR (void)
{
        int temp;
        temp = MCBSP_READ(0);

        input[index] = temp >> 16;

        MCBSP_WRITE(0, output[index] << 16);

        if (++index == BUFFLENGTH)
                index = 0;
}
```

This ISR reads an input sample from the DRR and shifts it by 16 bits since the representation of numbers are considered to be in Q-15 format. After shifting, the input samples are placed into the `input` array to build a frame of length BUFFLENGTH. The variable `index` is incremented every time a new input sample is read. This variable is reset to zero when the `input` buffer becomes full, that is, `index` reaches BUFFLENGTH. This reset causes the program to go out of the while(index) loop in the function *wait_buffer()*. Then, the rest of the program in *wait_buffer()* gets executed.

Now let us go back to the function *wait_buffer()*. After `index` is reset to zero, the `input` buffer is reassigned to a pointer `p`. This reassignment is necessary for the `//Function call here...` part to avoid any conflict with the ISR. Note that the ISR uses the `input` buffer to receive new samples. If say the FFT function processes the data in the `input` buffer, wrong results may be produced, because the ISR may change the content of the `input` buffer anytime. This malfunction may occur because the ISR runs on a higher priority basis, while the FFT function runs on a lower priority one. Following the `p=input` statement, the `output` buffer is reassigned to the `input` buffer. This reassignment allows the ISR to use the `output` buffer to receive and store new incoming samples. Notice that the data in the `output` buffer has been sent out by the DMA. The next reassignment `output=intermediate` is necessary in order to send the processed data in the `intermediate` buffer to the DXR.

After the data is processed in `//Function call here...`, the pointer `p` is reassigned to the `intermediate` buffer for it to point to the processed samples. The data in the `intermediate` buffer is sent out by the function *WriteFIFO_DMA()* as part of the *Wait_buffer()* function. The loop `while(!index)` at the end of *wait_buffer()* ensures that a frame is processed only once. In the absence of a new sample in the DRR, `index` remains zero and the program waits at `while(!index)` because `!index` is TRUE. When a new sample arrives in the DRR, `index` is incremented and the program gets out of *wait_buffer()* and falls into the loop in *main()*. There it waits for a new frame.

The function *WriteFIFO_DMA()* uses the C6x's DMA capability to send the intermediate array to the host through the FIFO on the EVM. The following is the code for doing so:

```
void WriteFIFO_DMA(void)
{
    dma_reset();
    dma_init(2,  //Channel
            0x0A000110u,                          //Primary Control Register
                                                  //(Peripherals pp4-9)
            0x0000000Au,                          //Secondary Control Register
            (unsigned int) intermediate,          //Source Address
            0x01710000u,                          //Destination Address
            0x00010080u);                         //Transfer Counter Register
    DMA_START(DMA_CH2);
}
```

The two DMA API functions *dma_reset()* and *dma_init()* are used to initialize the DMA for data transfer between the `intermediate` array and the FIFO. A single frame of samples of length 128 is transferred as indicated by the content of the Transfer Counter Register (TCR). The *dma_reset()* API resets all the DMA registers to their power-on reset states. The *dma_init()* API initializes a selected DMA channel by assigning appropriate

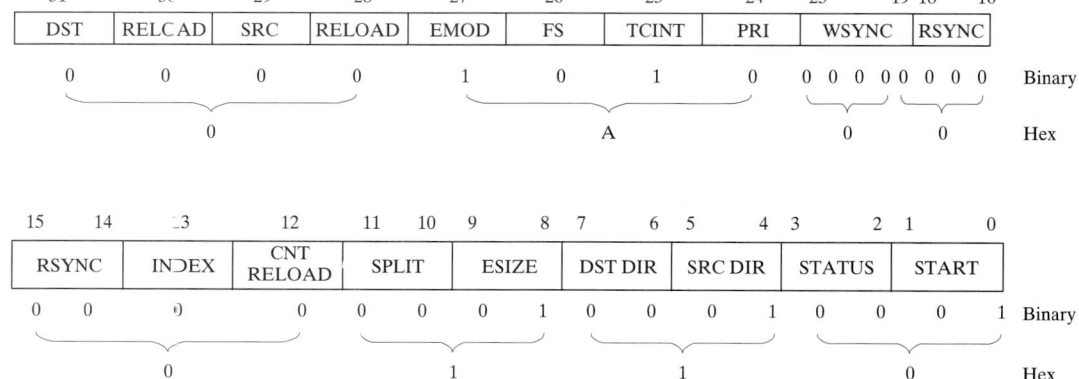

FIGURE L6-1 Primary Control Register.

values to its control registers. Here, the first argument of the *dma_init()* API is used to select the DMA channel 2. The second argument sets the Primary Control Register to 0x0A000110u, as shown in Figure L6-1. This value is specified based upon the following: EMOD bit is set to 1 to pause the DMA channel during an emulation halt. TCINT bit is set to 1 to enable the transfer controller interrupt. The element size is 16 bits, so ESIZE bits are set to 01. SRC DIR bits are set to 01 to increment the source address by element size in bytes. The third argument of the *dma_init()* API sets the Secondary Control Register. SX IE and FRAME IE bits of this register are set to 1 to enable the DMA channel interrupt. The fourth argument of the *dma_init()* API assigns the intermediate array as the source address. The fifth argument is set to 0x01710000u, which is the address of the EVM PCI interface, through which the host program reads the output. The last argument sets the Transfer Counter Register to 0x00010080u. The upper 16 bits specify the number of frames. Here, these bits are set to 0x0001u in order to send one frame out at a time. The lower 16 bits are set to 0x0080u for having 128 elements in a frame. Finally, the function *WriteFIFO_DMA()* calls the macro DMA_START() to activate the DMA channel 2. This macro sets the START field of the Primary Control Register to 01.

By using a function generator and an oscilloscope, the operation of the modified sampling program can be verified. A simple program called *gui.exe* is written utilizing the *evm6x_read()* API in the EVM host support library to get data from the FIFO on the host side and display it on the screen. This program, placed on the attached CD, is written in Microsoft Visual C++ using a Dialog wizard. It basically starts a thread that continuously reads data in the FIFO on the host side and plots it on the screen.

L6.1 DFT IMPLEMENTATION

DFT can be simply calculated from the equation

$$X[k] = \sum_{n=0}^{N-1} x[n] * W_N^{nk}, \quad k = 0, 1, \ldots, N-1, \tag{L6.1}$$

where $W_N = e^{-j2\pi/N}$. This equation requires N complex multiplications and $N - 1$ complex additions for each term. For all N terms, N^2 complex multiplications and $N^2 - N$ complex additions are needed. As is well known, this method is not efficient, since the symmetry properties of the transform are not utilized. However, it is useful to implement this equation on the C6x as a comparison to the FFT implementation. The graphing capability of CCS is used here for this purpose. This is carried out in an offline manner because the amount of time required to do the DFT exceeds the duration of a frame capture.

First, a simple composite signal is generated in Matlab with the frequency components located at 750 Hz, 2500 Hz, and 3000 Hz. Saving two periods of this signal sampled at 8000 Hz results in a 64-point signal. Figure L6-2 shows the signal read into CCS and plotted using its graphing capability. The frequency content of the signal is also plotted based on a built-in FFT option.

The DFT code used is the one appearing in the *TI Application Report SPRA291* [8]. Here is the code:

```
#include <math.h>
#include "params.h"
void dft(int N, COMPLEX *X){
        int n, k;
        double arg;
        int Xr[1024];
        int Xi[1024];
        short Wr, Wi;
        for(k=0; k<N; k++){
                Xr[k] = 0;
```

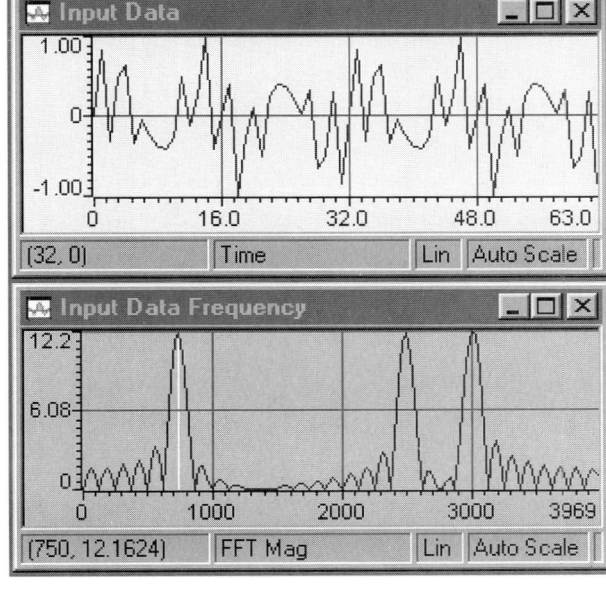

FIGURE L6-2 Input signal in time and frequency domains.

```
                    Xi[k] = 0;
                    for(n=0; n<N; n++){
                            arg =(2*PI*k*n)/N;
                            Wr = (short)((double)32767.0 * cos(arg));
                            Wi = (short)((double)32767.0 * sin(arg));
                            Xr[k] = Xr[k] + X[n].real * Wr + X[n].imag * Wi;
                            Xi[k] = Xi[k] + X[n].imag * Wr - X[n].real * Wi;
                    }
            }

            for (k=0;k<N;k++){
                    X[k].real = (short)(Xr[k]>>15);
                    X[k].imag = (short)(Xi[k]>>15);
            }
}
```

In order to use this code, the input has to be represented as complex numbers. This is done using a structure definition to create a complex variable with components `real` and `imag`. The main program used to perform DFT is as follows:

```
main()
{
        int i,j;
        COMPLEX x[128];
        int mag[128];

        /*Change input to Q-15*/
        for(i=0;i<128;i++)
        {
                x[i].real=0x7FFF * input_data[i];
                x[i].imag=0;
        }
        dft(128, x);

        for(i=0;i<128;i++)
                mag[i]=(x[i].real*x[i].real + x[i].imag*x[i].imag) << 1;
        return(0);
}
```

In this program, the input is converted to Q-15 format and is stored in the complex structure, which is then used to call the DFT function. The magnitude of the DFT outcome is shown in Figure L6-3. As expected, there are three spikes, at 750Hz, 2500Hz, and 3000Hz. Notice that this code is quite inefficient, as it calculates each twiddle factor using the math library at every iteration. Running this code from the external SBSRAM results in an execution time of about 16×10^9 cycles for a 128-point frame.

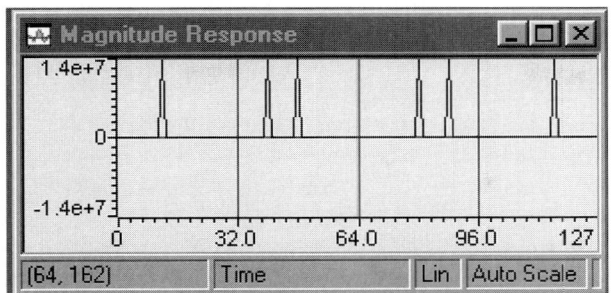

FIGURE L6-3 Magnitude response of DFT.

L6.2 FFT IMPLEMENTATION

The above DFT code will not run in real-time on the EVM, since there are only $16{,}625 \times N$ cycles to perform an N-point transform. For a 128-point signal, the limit is $\sim 2 \times 10^6$ cycles, and the preceding timing exceeds this limit.

To make use of the symmetry properties of the transform, the approach of computing a $2N$-point FFT as mentioned in the *TI Application Report SPRA291* [8] is adopted here. This approach involves forming two new N-point signals $x_1[n]$ and $x_2[n]$ from the original $2N$-point signal $g[n]$ by splitting it into even and odd parts as follows:

$$x_1[n] = g[2n], \quad 0 \leq n \leq N - 1;$$
$$x_2[n] = g[2n + 1]. \qquad (\text{L6.2})$$

From the two sequences $x_1[n]$ and $x_2[n]$, a new complex sequence is defined as

$$x[n] = x_1[n] + jx_2[n], \quad 0 \leq n \leq N - 1. \qquad (\text{L6.3})$$

To get $G[k]$, the DFT of $g[n]$, the equation

$$G[k] = X[k]A[k] + X[N - k]B[k],$$
$$k = 0, 1, \ldots, N - 1, \quad \text{with } X[N] = X[0], \qquad (\text{L6.4})$$

is used, where

$$A[k] = \frac{1}{2}(1 - jW_{2N}^k)$$

and

$$B[k] = \frac{1}{2}(1 + jW_{2N}^k). \qquad (\text{L6.5})$$

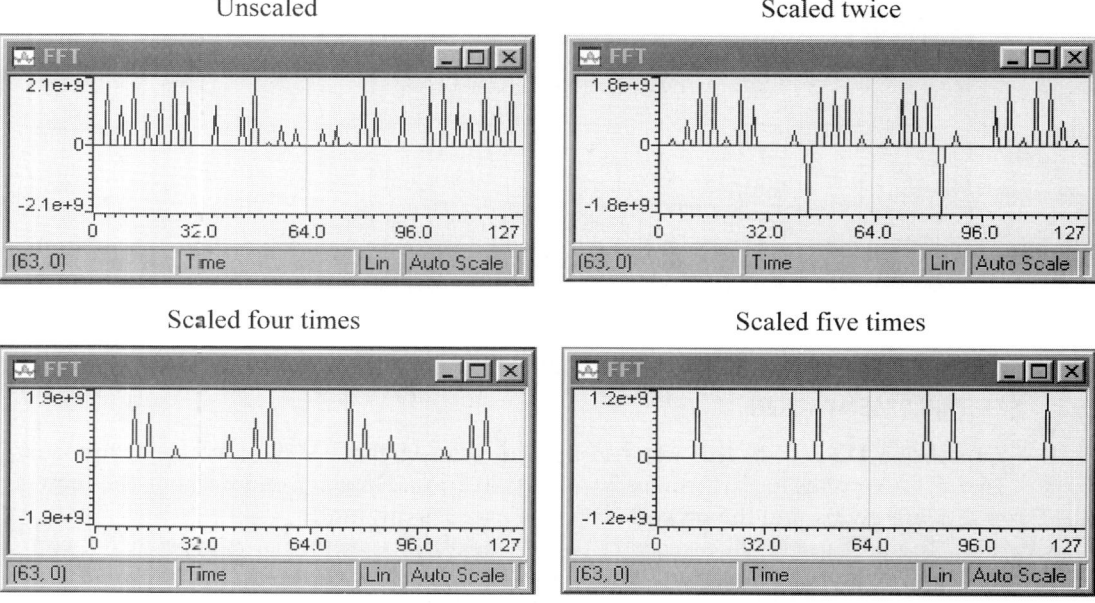

FIGURE L6-4 Scaling to get correct FFT magnitude response.

Only N points of $G[k]$ are computed from Eq. (L6.4). The remaining points are found by using the complex conjugate property of $G[k]$, $G[2N - k] = G^*[k]$. As a result, a $2N$-point transform is calculated based on an N-point transform, leading to a reduction in the number of cycles. The codes for the functions (*split1*, *R4DigitRevIndexTableGen*, *digit_reverse*, and *radix4*) implementing this approach are provided in the *TI Application Report* [8] and appear on the attached CD.

Figure L6-4 shows the FFT outcome where the signal has been scaled down 0, 2, 4, and 5 times, respectively. The scaling is done to get rid of overflows, which are present for the scale factors 0, 2, and 4. As revealed by these figures, the input signal has to be scaled down five times to eliminate overflows. When the signal is scaled down five times, the expected peaks appear. The total number of cycles for this FFT is 56,383. Since this is less than the capture time for a 128-point data frame at a sampling frequency of 8 kHz, it is expected that this algorithm would run in real time on the EVM.

L6.3 REAL-TIME FFT

To perform FFT in real-time, the triple buffering program is used. A frame length of 128 is considered here. The output is observed by halting the processor through CCS. The animate feature of CCS cannot be used here since it slows down the processing and causes frames to overlap.

The following modifications are made to the triple buffering program to run the aforementioned FFT algorithm in real-time:

```
void wait_buffer(void)
{
        int n,k;
        short *p;

        while(index);

        p = input;
        input = output;
        output = intermediate;

        for (n=0; n<NUMPOINTS; n++)
        {
                x[n].imag = p[2*n + 1]; // x2(n) = g(2n + 1)
                x[n].real = p[2*n]; // x1(n) = g(2n)
        }

        radix4(NUMPOINTS, (short *)x, (short *)W4);
        digit_reverse((int *)x, IIndex, JIndex, count);
        x[NUMPOINTS].real = x[0].real;
        x[NUMPOINTS].imag = x[0].imag;

        split1(NUMPOINTS, x, A, B, G);
        G[NUMPOINTS].real = x[0].real - x[0].imag;
        G[NUMPOINTS].imag = 0;
        for (k=1; k<NUMPOINTS; k++){
                G[2*NUMPOINTS-k].real = G[k].real;
                G[2*NUMPOINTS-k].imag = -G[k].imag;
        }
        for (k=0; k<NUMDATA; k++){
                mag1[k] = (G[k].real*G[k].real) << 1;
                mag2[k] = (G[k].imag*G[k].imag) << 1;
                mag[k] = mag1[k] + mag2[k];
        }
        intermediate = p;
        WriteFIFO_DMA();
        while(!index);
}
```

The *wait_buffer()* function is modified with the appropriate function calls so that when the input buffer is full, the transform is calculated and sent out to the host through the FIFO.

The functionality of the program can be verified by connecting a function generator to the line-in. The graphing capability of CCS can be used to plot the FFT outcome. By changing the frequency of the input, the spikes in the frequency response would move to the left or right, accordingly. Figure L6-5 illustrates the output for a 1 kHz and a 2 kHz sinusoidal signal. These snapshots are captured by halting the processor. The input here is scaled by shifting it right 20 bits. All the files associated with this lab can be downloaded from the attached CD.

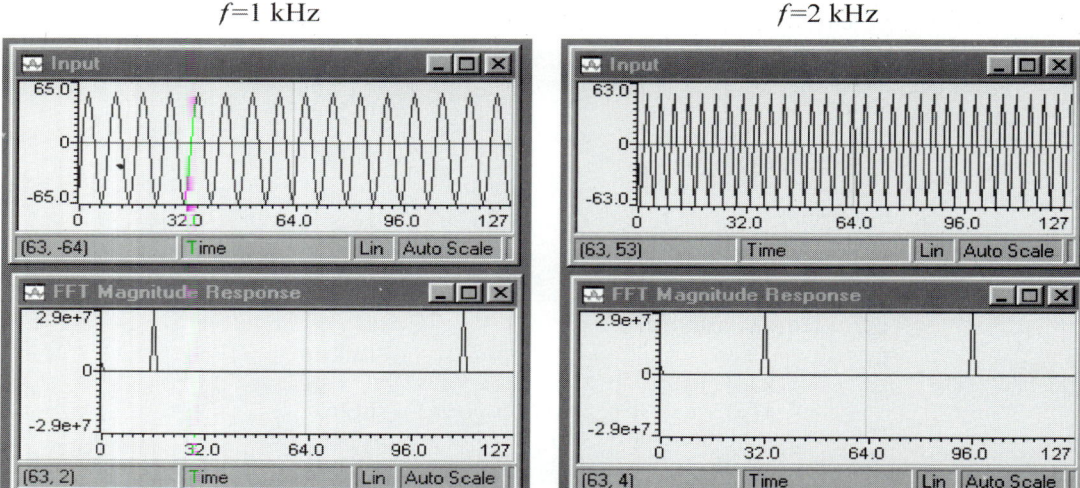

FIGURE L6-5 Real-time FFT magnitude response.

CHAPTER 12

Real-time Analysis and Scheduling

Figure 12-1 provides an overview of the conventional CCS debugging techniques based on breakpoints, probe points, and profile points. Although these debugging tools are very useful to see whether an application program is logically correct or not, when it comes to making sure that real-time deadlines are met, they have limitations. The DSP/BIOS feature of CCS complements the traditional debugging techniques by providing mechanisms to analyze an application program as it runs on the target DSP without stopping the processor. In traditional debugging, the target DSP is normally stopped and a snapshot of the DSP state is examined. This is not an effective way to test for real-time glitches.

DSP/BIOS consists of a number of software modules that get glued to an application program to provide real-time analysis and scheduling capabilities. A listing of all available modules is provided in Figure 12-2. CCS provides an easy-to-use way to glue

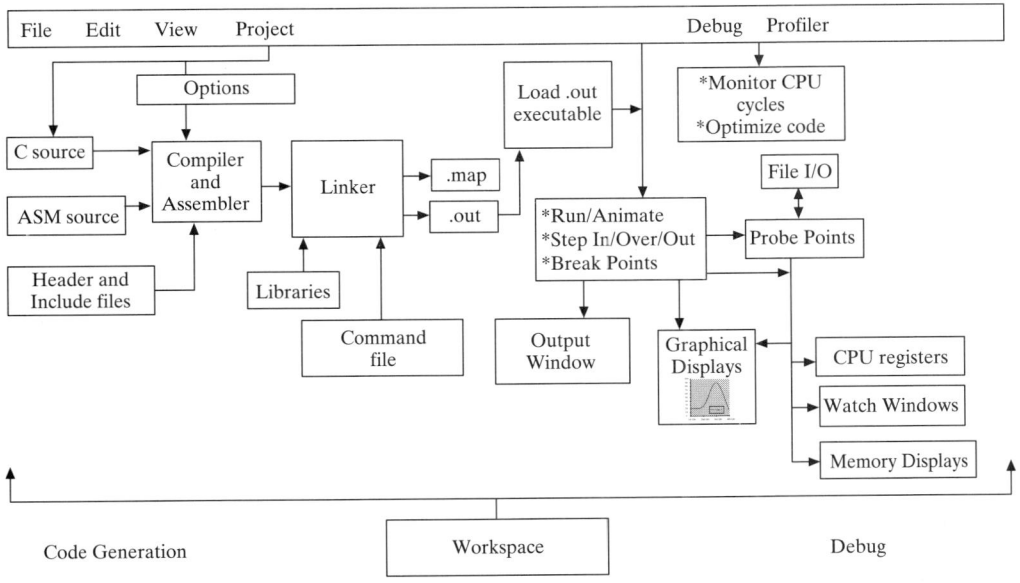

FIGURE 12-1 Code Composer Studio IDE.†

220 Chapter 12 Real-time Analysis and Scheduling

Instrumentation/Real-Time Analysis	
LOG	Message log manager
STS	Statistics accumulator manager
TRC	Trace manager
RTDX	Real-Time Data eXchange Manager
Thread Types	
HWI	Hardware interrupt manager
SWI	Software interrupt manager
TSK	Multitasking manager
IDL	Idle function & process loop manager
Clock and Periodic Functions	
CLK	System clock manager
PRD	Periodic function manager

Comm/Synch Between Threads	
SEM	Semaphores manager
MBX	Mailboxes manager
LCK	Resource lock manager
Input/Output	
PIP	Data pipe manager
HST	Host Input/Output manager
SIO	Stream I/O manager
DEV	Device driver interface
Memory and Low-level Primitives	
MEM	Memory manager
SYS	System service manager
QUE	Queue manager
ATM	Atomic function
GBL	Global setting manager

FIGURE 12-2 DSP/BIOS API modules.†

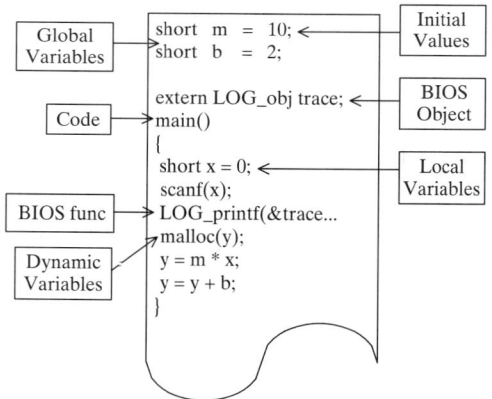
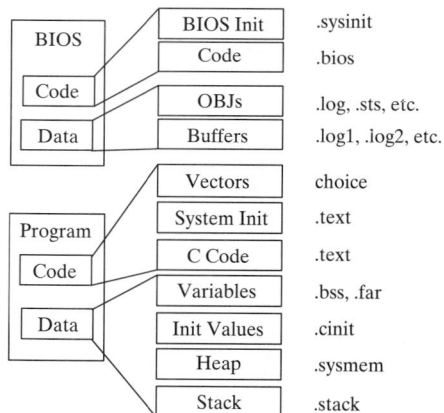

FIGURE 12-3 BIOS sections.†

these modules to the application program. Figure 12-3 shows a sample C file with a BIOS object, a BIOS function, and its corresponding section names. The size of the DSP/BIOS portion of an application program is limited to a maximum of 2K words and is proportional to the number of modules and objects used.

Figure 12-4 shows a listing of the DSP/BIOS modules accessed by the Configuration Tool feature of CCS. This figure also shows the files generated by the Configuration Tool. The Configuration Tool is a visual editor which allows one to create module objects and set their properties. An application program can interact with objects by using

Introduction 221

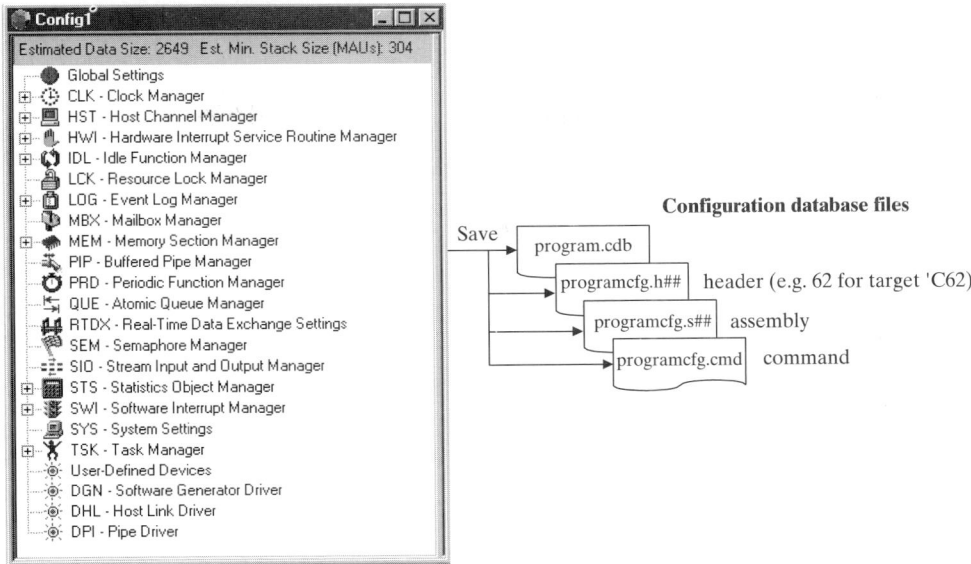

FIGURE 12-4 Files generated by the Configuration Tool.

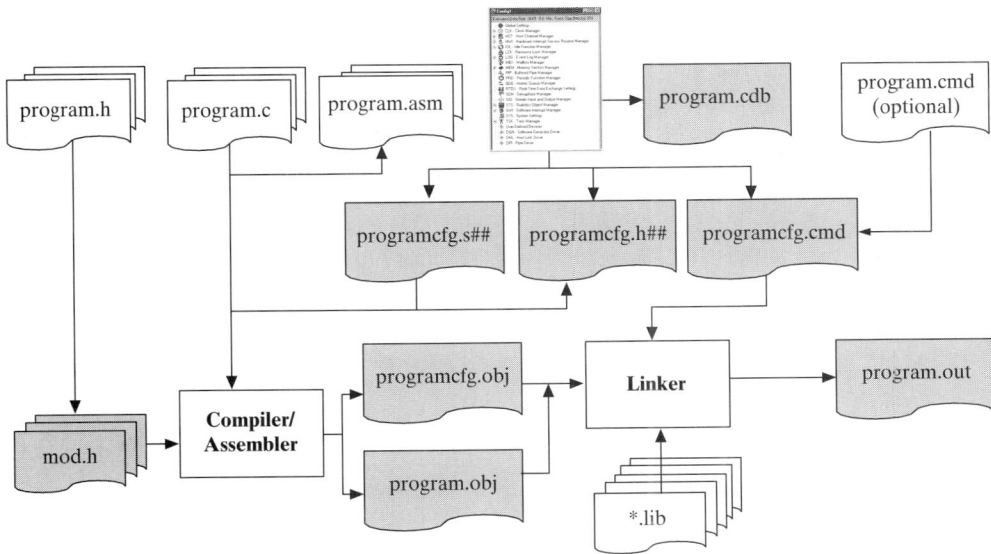

FIGURE 12-5 Files indicated in white created by user and in grey by CCS.[†]

DSP/BIOS API functions. In addition, DSP/BIOS plug-ins can be activated from the CCS environment, providing real-time instrumentation including log and statistics displays.

Figure 12-5 illustrates all the files created within the CCS environment when using DSP/BIOS; the files with white background are created by the user and the ones with grey background by CCS. The naming convention used by the modules is shown in

222 Chapter 12 Real-time Analysis and Scheduling

CATEGORY	CONVENTION	EXAMPLE
Function Calls	MOD_lowercase	LOG_printf
Data Types	MOD_Titlecase	LOG_Obj
Constants	MOD_UPPERCASE	HWI_INT3
Internal Calls	MOD_F_lowercase	FXN_F_nop

FIGURE 12-6 Three-letter prefix module naming conventions; capitalization convention distinguishes functions, types, and constants.[†]

Figure 12-6. The datatypes associated with a module are defined in its header file. Header files of those modules that are used in the real-time analysis of an application program must be included in the program.

In essence, DSP/BIOS provides real-time analysis, real-time scheduling, and real-time data exchange capabilities for debugging application programs. As a result, one can make sure that an application program is meeting its real-time deadlines in addition to being logically correct.

12.1 REAL-TIME ANALYSIS

There are two types of real-time constraints: (a) hard real-time, denoting critical real-time needs (i.e., timing needs that should be met to avoid system failure), and (b) soft real-time, denoting not-so-critical real-time needs that can be performed as time becomes available. For example, as shown in Figure 12-7, the response to incoming samples must be done in a hard real-time manner in order not to lose any information, whereas data transfer from the target DSP to the host PC can be done in soft real-time.

To monitor the status of a program in real-time, it is possible to use the C function *printf()* as defined in the real-time support library. However, this function takes too many cycles to run, an undesirable property as far as real-time performance is concerned. On the other hand, *LOG_printf()* is a LOG module API that creates a buffer. The buffer is

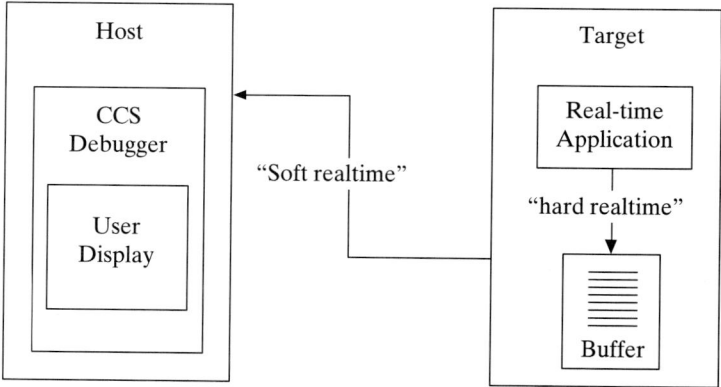

FIGURE 12-7 Hard and soft real time: data buffered in hard real time, and sent to host in soft real time.

then sent to the host in soft real-time. The buffer size *n* is specified as part of a LOG object. A LOG object can be configured in a fixed or circular mode. The fixed mode captures the first *n* occurrences, while the circular mode captures the last *n* occurrences. *LOG_printf()* runs in much fewer clock cycles as compared with *print()*. Figures 12-8(a), 12-8(b), and 12-8(c) illustrate the difference between *printf()* and *LOG_printf()*.

Statistics on a value can be captured by using the *STS_add()* API. Two other statistics API, *STS_set()* and *STS_delta()*, can be used to time a piece of code. The plug-in Statistics View window can be activated on the host as part of CCS to monitor the statistics associated with a variable. These statistics are reconfigured on the host as shown in Figure 12-9.

The trace module through the RTA Control Panel of DSP/BIOS allows various modules to be enabled or turned on so that only a specific or needed portion of the DSP/BIOS kernel is glued to the application program. The RTA Control Panel properties can be set to decide how often the host should poll the target DSP for various logging and statistics data. The CPU Load Graph is another plug-in instrumentation that shows the monitoring of the CPU active time as a program is running on the target DSP.

12.2 REAL-TIME SCHEDULING

Real-time scheduling is done by breaking down the application program into threads each doing a specific function or task. Some of the threads may occur more often than others. Some of them may be subject to hard real-time constraints and some to soft real-time constraints. The real-time need of the entire application program or system is met by appropriately prioritizing threads. This multithreaded real-time scheduling approach is what makes it possible to meet real-time timing deadlines.

Threads can be scheduled using hardware interrupts (ISRs) in a nonpreemptive fashion by disabling hardware interrupts. The scheduling can be performed in a preemptive fashion by prioritizing hardware interrupts. Considering that not all real-time situations can be handled by preemptive hardware interrupts, a more robust scheduling mechanism based on so-called software interrupts is adopted in DSP/BIOS. Software interrupts automatically perform context switching (i.e., storage/retrieval of the DSP status to the time the interrupt occurred).

DSP/BIOS uses a background/foreground scheduling approach where the background consists of noncritical housekeeping threads such as transferring information to the host and instrumentation. These threads or functions are done in a round-robin fashion as part of the idle or background loop IDL. The foreground consists of more critical threads. These threads are implemented via hardware (HWI module) and software (SWI module) interrupts. Hardware interrupts have a higher priority than software interrupts. As indicated in Figure 12-10, normally software interrupts are used for deadlines of 100 microsec or more, and hardware interrupts for more restrictive deadlines of 2 microsec or more. The priority of software interrupts can easily be changed through SWI objects. Software interrupts can be posted unconditionally or conditionally via mailboxes. In essence, the DSP/BIOS scheduling is based on preemptive software interrupts. Figure 12-11 shows an example of how a single-thread ISR is converted to a hardware, software, and idle multi-thread program. In addition to hardware and software

224 Chapter 12 Real-time Analysis and Scheduling

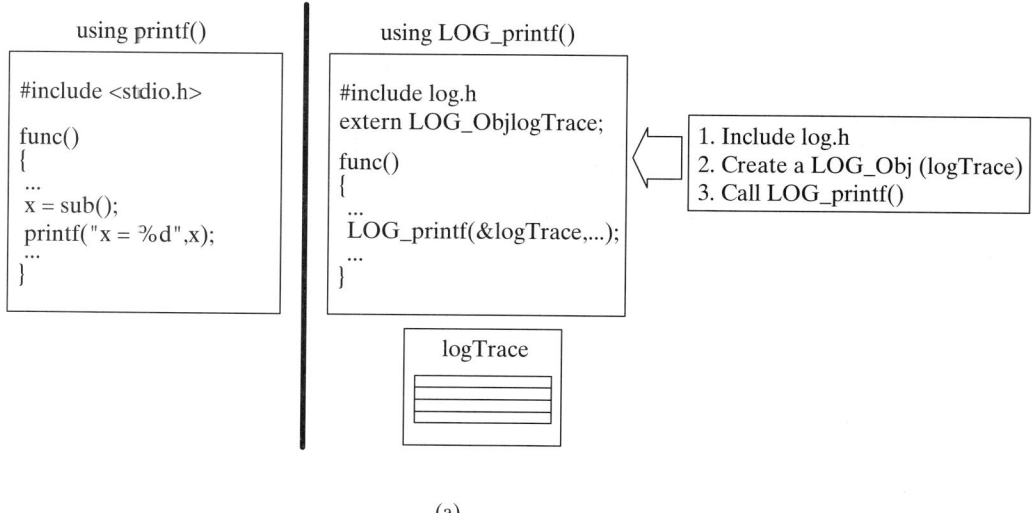

FIGURE 12-8 printf() vs LOG_printf().[†]

Host
Display
Filter=(A*x+B)/C
64
Accumulate
Count
Total
Maximum
Average
Count
(A x total + B)/C
(A x max + B)/C
(A x total + B) / (C x count)
Count
Total
0
Max
Target
32
Previous
Count
Total
Max

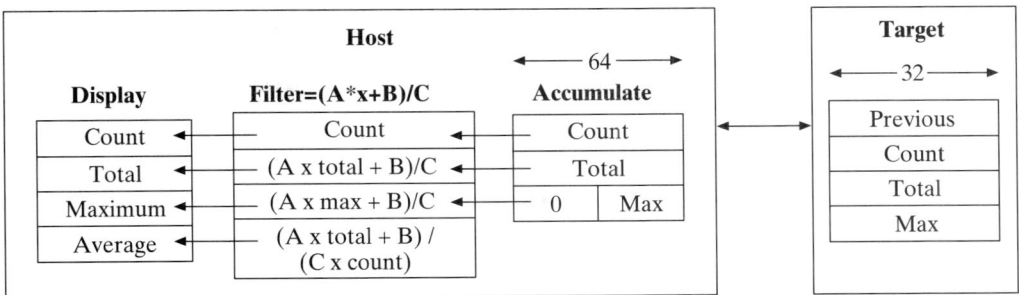

FIGURE 12-9 Reconfiguration of statistics on host.†

real-time deadlines
higher priority
Hardware Interrupts (HWI)
>= 2 microseconds
Software Interrupts (SWI)
>= 100 microseconds
Background threads (IDL)
lower priority

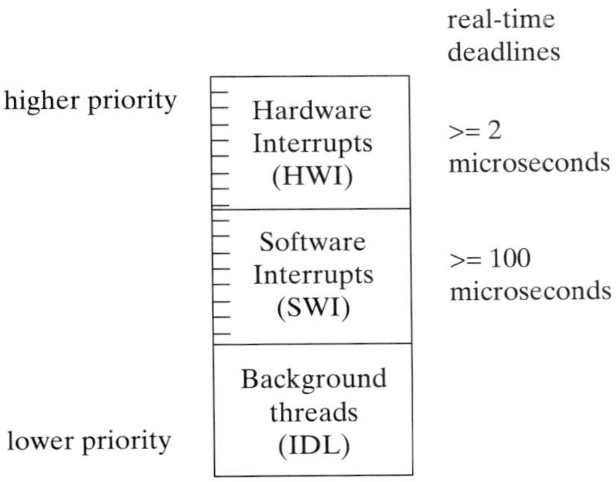

FIGURE 12-10 Thread priorites.†

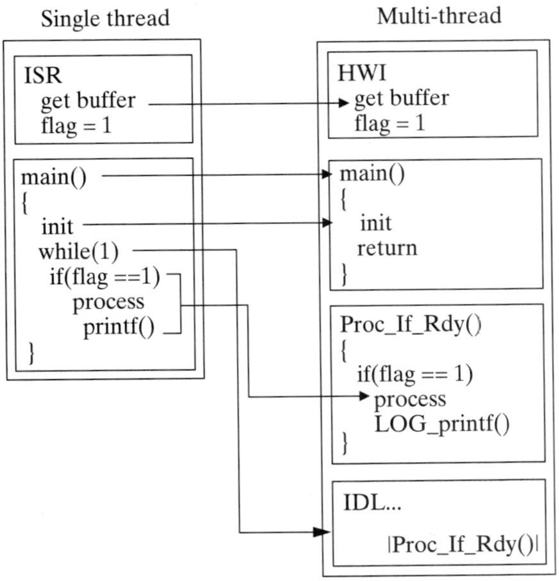

Single thread
Multi-thread
ISR
get buffer
flag = 1
main()
{
init
while(1)
if(flag ==1)
process
printf()
}
HWI
get buffer
flag = 1
main()
{
init
return
}
Proc_If_Rdy()
{
if(flag == 1)
process
LOG_printf()
}
IDL...
|Proc_If_Rdy()|

FIGURE 12-11 Multithread programming.†

threads, the upgraded DSP/BIOS II provides a task TSK module that can be used for posting threads capable of yielding to other threads.

There are basically two thread scheduling rules. The first rule is that if a higher priority thread becomes ready, the running thread is preempted. The second rule is that threads having the same priority are scheduled in a first-in first-out fashion. Three examples are shown in Figure 12-12 to illustrate how various threads run based on the scheduling rules. In this figure, a running thread is shown by shaded blocks and a thread in ready state by white blocks per time tick. The Execution Graph window feature of DSP/BIOS provides a visual display of the execution of threads. It shows which thread is running and which threads are in a ready state. It also provides useful feedback information on errors. Errors are generated when real-time deadlines are missed or when the system log is corrupted. An example of the Execution Graph window is shown in Figure 12-13. It should be noted that in the Execution Graph window, although time intervals between time ticks are the same, they may not be displayed as equal. This provides a more compact way to show all events between two successive time ticks.

There are two timers on the C6x, each is controlled by three memory-mapped registers: the Timer Control Register for setting the operating mode, the Timer Period Register for holding the number of clock cycles to count, and the Timer Counter Register for holding the current clock cycle count. As illustrated in Figure 12-14, the clock module CLK is used to set the on-chip timer registers for low-resolution (determined by the

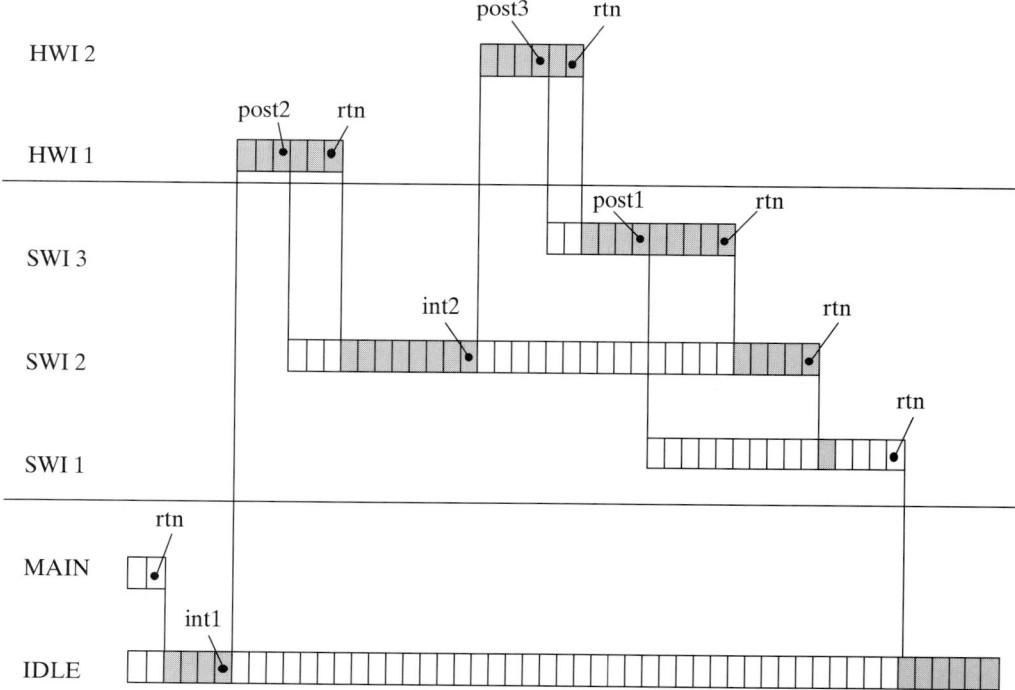

FIGURE 12-12 Examples of threads running based on scheduling rules; gray blocks indicate running status and white blocks ready status.†

12.2 Real-time Scheduling 227

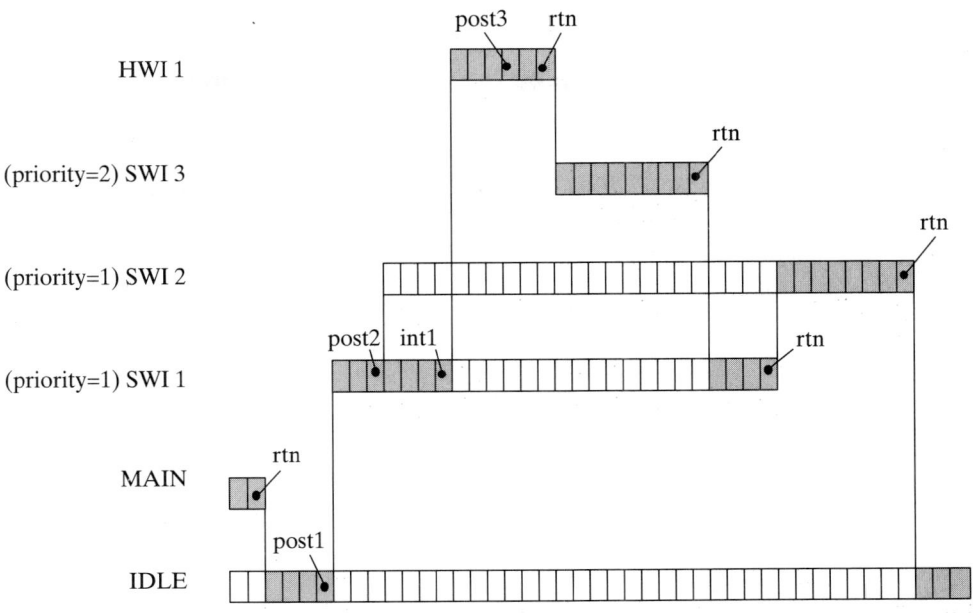

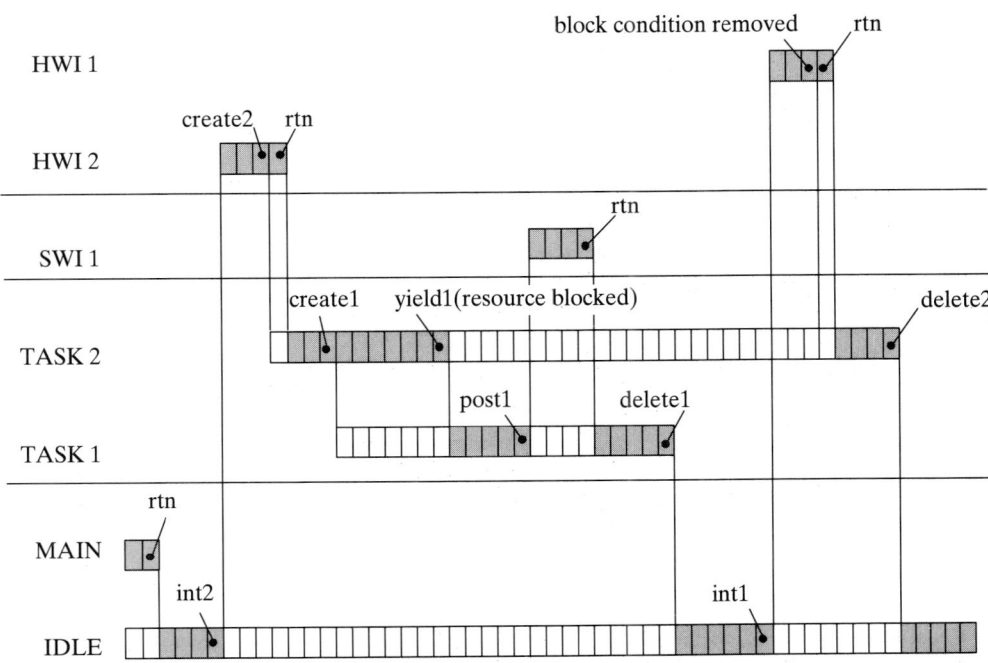

FIGURE 12-12 *(continued)*

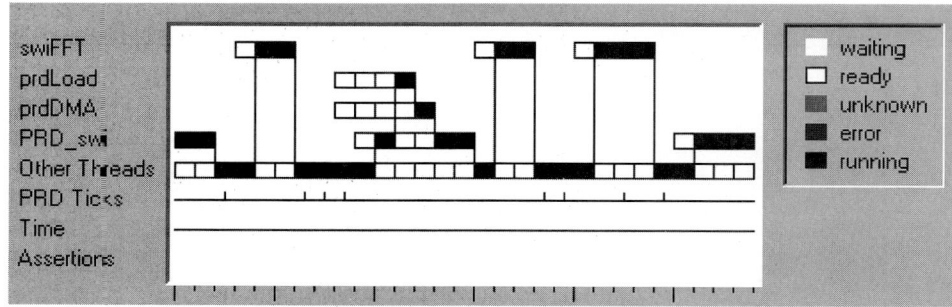

FIGURE 12-13 Execution Graph window.

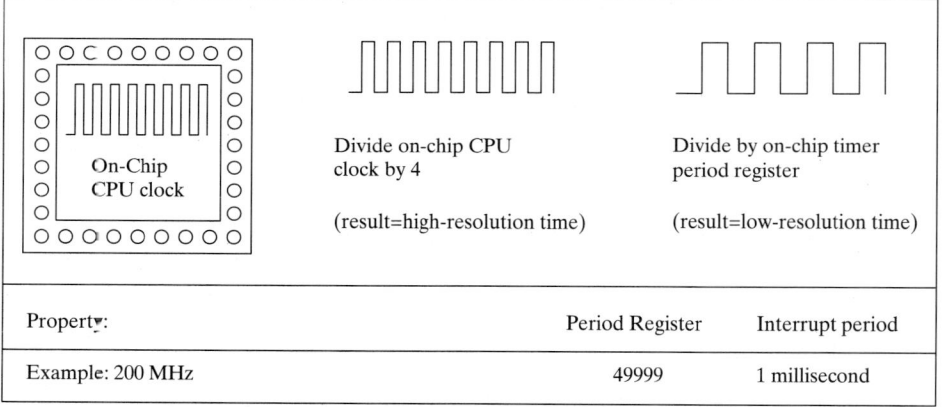

FIGURE 12-14 Low- and high-resolution clock ticks.[†]

Timer Period Register) or high-resolution (the CPU clock divided by 4) ticks. The clock APIs run as hardware functions. The PRD module is used to run threads that are to be executed periodically. The period is specified as part of a PRD object. The period APIs run as software functions. The CLK manager is used to drive the PRD module.

Data frame synchronization and communication can be achieved by using the PIP module. A pipe consists of a specified number of frames having a specified size. It has two ends, a writer end and a reader end. The sequence of operations on the writer side consists of getting a free frame from the pipe via *PIP_alloc*(), writing to it, and putting it back in the pipe via *PIP_put()*, which runs the *notifyReader*() function. The sequence of operations on the reader side consists of getting a full frame from the pipe via *PIP_get()*, reading it, and putting the empty frame in the pipe via *PIP_free()*, which runs the *notifyWriter*() function. Figure 12-15 illustrates this process.

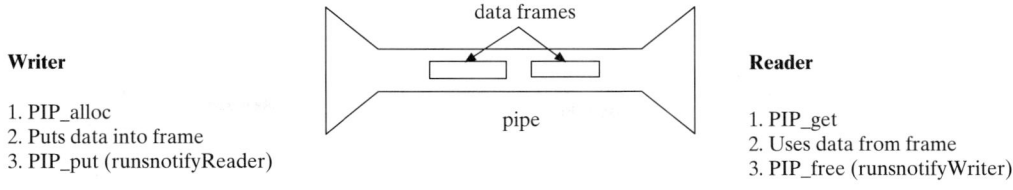

Writer

1. PIP_alloc
2. Puts data into frame
3. PIP_put (runsnotifyReader)

Reader

1. PIP_get
2. Uses data from frame
3. PIP_free (runsnotifyWriter)

FIGURE 12-15 Data pipe.

12.3 REAL-TIME DATA EXCHANGE

The RTDX (Real-time Data Exchange) module can be used to exchange data between the DSP and the host without stopping the DSP. Similar to other modules, this exchange of information between the host and the DSP is done via the JTAG (Joint Test Action Group) connection, an industry-standard connection. The RTDX module provides a useful tool when values need to be modified on the fly as the DSP is running. As shown in Figure 12-16, RTDX consists of both target and host components, each running its own library. On the host side, various displays and analysis OLE (object linking and embedding) automation clients, such as LabVIEW, Visual Basic, and Visual C++, can be used to display and send data to the application program. RTDX can be configured in two modes: noncontinuous and continuous. In the noncontinuous mode, data is written to a log file on the host. This mode is normally used for recording purposes. In the continuous mode, the data is buffered by the RTDX host library. This mode is normally used for continuously displaying data.

Labs 7 and 8 provide a hands-on experience with the DSP/BIOS features of CCS. Lab 7 covers its real-time analysis and scheduling and Lab 8 its data synchronization and communication aspects. More details on the DSP/BIOS modules can be found in the *TMS320C6000 DSP/BIOS User's Guide* manual [15].

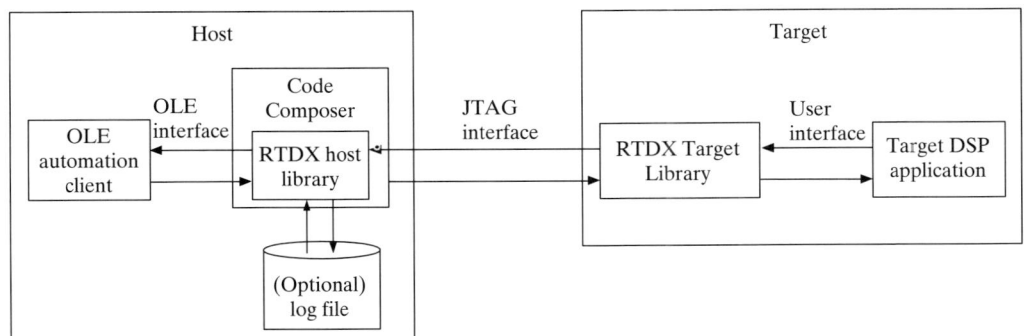

FIGURE 12-16 RTDX target/host dataflow.[†]

Lab 7: DSP/BIOS

The objective of this lab is to become familiar with the DSP/BIOS feature of CCS. DSP/BIOS includes a real-time library which allows one to interact with an application program in real-time as it runs on the target DSP. To build a program based on DSP/BIOS, the Configuration Tool feature of CCS needs to be used to create objects and set their properties. The Configuration Tool is opened by choosing the menu item `File->New->DSP/BIOS Configuration`. The configuration of a program can be saved via `File->Save`, and `Configuration Files(*.cdb)` in the `Save as type` drop-down box. A saved configuration file includes all the necessary files for generating an executable file.

A DSP/BIOS object can be created easily by right clicking on a module displayed in the Configuration Tool window and by selecting `Insert`. For example, as shown in Figure L7-1, to create a PRD object, right click on `PRD-Periodic Function Manager` and select `Insert PRD`. This adds a new object for the PRD module. An object can be renamed by right clicking on its name and choosing `Rename` from the pop-up menu. Properties of an object can be displayed by right clicking on the object icon and selecting `Properties` from the pop-up menu. From the property sheet, property settings can be readily changed.

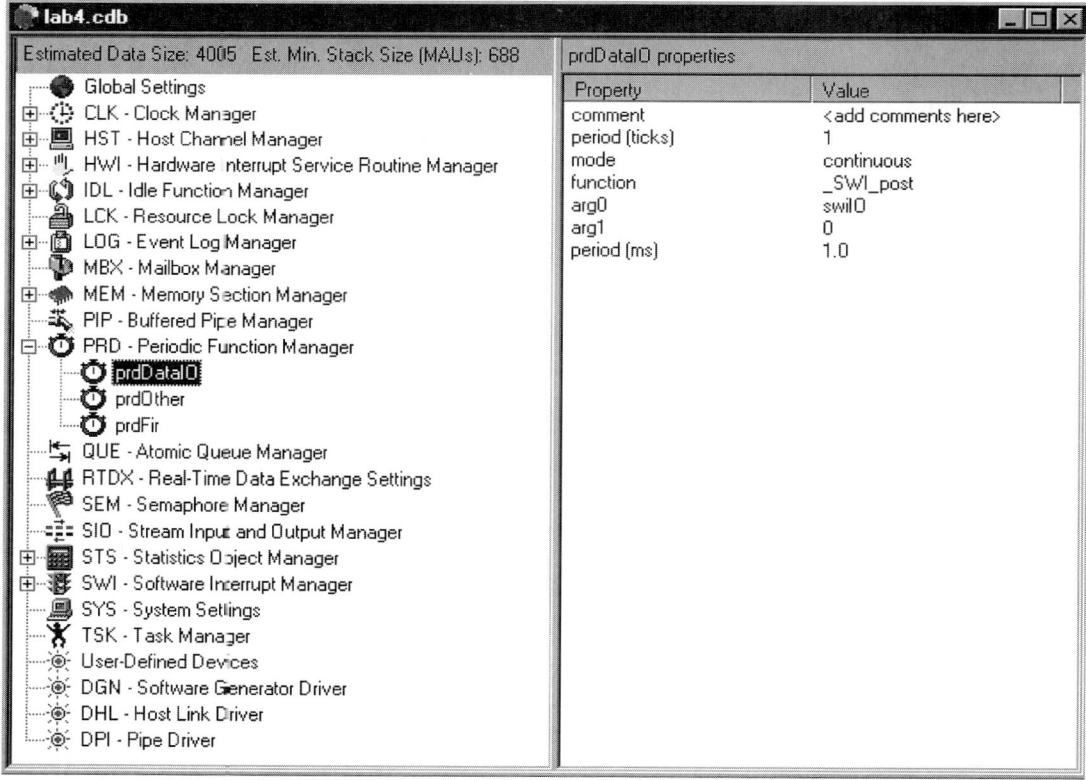

FIGURE L7-1   Configuration Tool.

L7.1 A DSP/BIOS-BASED PROGRAM EXAMPLE

The following C code is an example of a simple DSP/BIOS-based program. This program compares the performance of the DSP/BIOS API *LOG_printf()* to the function *printf()*, which is a part of the run-time support library:

```
#include <stdio.h>         // For printf();
#include <std.h>           // Header files needed for DSP BIOS
#include <log.h>
#include <sts.h>           // Header files added to support statistics
#include <clk.h>
void fun1();               // functions
void fun2();
extern LOG_Obj logTrace1;  // Objects created by the Configuration Tool
extern STS_Obj stsPrintf;
extern STS_Obj stsLogprintf;
void main()                // ======== main ========
{
return;                    // fall into DSP/BIOS idle loop
}

void fun1()
{
  static int i=0;
  i=CLK_gethtime();
  STS_set(&stsPrintf, CLK_gethtime());
  printf("loop: %d\n" , i);           // write a sting to stdio
                                      // using printf
  STS_delta(&stsPrintf, CLK_gethtime());
  return;
}

void fun2()
{
  static int j=0;
  j=CLK_gethtime();
  STS_set(&stsLogprintf, CLK_gethtime());
  LOG_printf(&logTrace1, "loop: %d\n", j);   // write a string using BIOS
                                             // LOG_printf object trace1
  STS_delta(&stsLogprintf, CLK_gethtime());
  return;
}
```

Two functions are declared in this program: *fun1()* and *fun2()*, one using *printf()* and the other *LOG_printf()*. These functions obtain the processing time and print them on the screen. Printing is the most widely used manner in which to view the results of a program. As shown in Figure 12-8, *LOG_printf()* is optimized to take much fewer instruction cycles than *printf()*. It sends buffered data to the host in soft real-time to avoid missing real-time deadlines. On the other hand, *printf()* does not use the background scheduling approach for transferring data to the host and, hence, may cause the program to miss its real-time deadlines. At this point, it is worth mentioning that although it is possible to use Watch Window, this option interrupts the DSP in order to transfer data and does not meet the real-time requirement.

Note that appropriate header files should be included to build a DSP/BIOS-based program. Foremost, the header file *std.h* should be included whenever using any DSP/BIOS API. The header files, *log.h, sts.h,* and *clk.h*, corresponding to the three modules LOG, STS, and CLK, respectively, are included in the aforementioned program. Any created DSP/BIOS objects should also be declared. There are three declared objects here: logTrace1, stsPrintf, and stsLogprintf. The LOG object logTrace1 managed by the LOG module allows real-time logging. The STS objects managed by the STS module store key statistics in real-time. The *STS_set()* and *STS_delta()* APIs use the information stored in the STS objects to compute the required number of instruction cycles to run *printf()* or *LOG_printf()*.

Now let us create the objects declared in the program. First, a configuration file is created by choosing File–>New–>DSP/BIOS Configurations. If a configuration file already exists, it can be activated by double-clicking on it in the Project View panel. To add a LOG object as part of the *LOG_printf()* API, right-click on LOG-Event Log manager in the Configuration Tool and select the option Insert LOG from the pop-up menu. This causes LOG0 to be inserted. Since the name logTrace1 is used here, rename this object by right-clicking on it and then by selecting Rename. Change the name to logTrace1. Right-click on LogTrace1 to change its properties. Select Cancel noting that the default settings are fine for this lab. Next, create two STS objects in a similar manner and rename them as stsPrintf and stsLogprintf. The properties of these objects will be discussed in the next section. Use File–>Save to save the configuration file.

L7.2 DSP/BIOS ANALYSIS AND INSTRUMENTATION

Real-time analysis allows one to determine whether an application program is operating within its real-time deadlines and whether its timing can be improved. DSP/BIOS instrumentation APIs and DSP/BIOS plug-ins enable real-time data gathering and monitoring as an application program is running. For example, when using the instrumentation API *LOG_printf()*, the communication between the DSP and the host is performed during the idle state or in the background. The idle thread has the lowest priority. As a result, the real-time behavior of an application program is not affected.

In the preceding program, the instrumentation APIs *STS_set()* and *STS_delta()* are used to benchmark the functions *printf()* and *LOG_printf()*. *STS_set()* saves the value specified by *CLK_gethtime()* as the previous value in the STS object. *STS_delta()* subtracts this saved value from the value it is passed. Consequently, *STS_delta()* in conjunction with *STS_set()* provide the difference between the start and completion of the function in between. However, to obtain an accurate benchmarking outcome, the overhead associated with the instrumentation APIs should be subtracted. To calculate this overhead, the program should be run again by leaving out *LOG_printf()* and *printf()*.

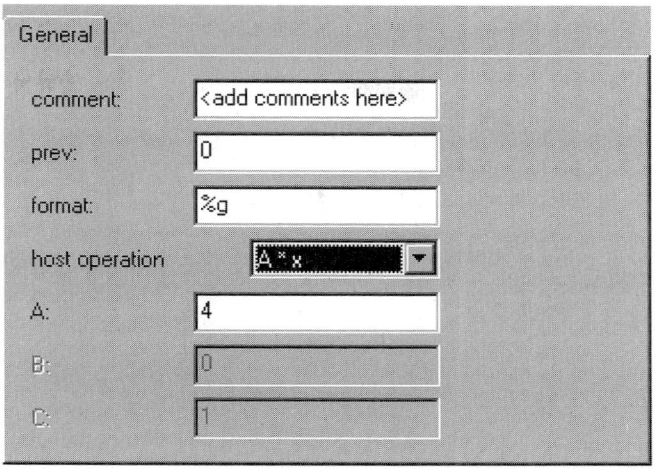

FIGURE L7-2 stsPrintf object properties.

Before calculating the overhead, let us examine how the STS objects should be used during benchmarking. Since the STS objects count system ticks, they do not provide the actual CPU instruction cycles. A filtering operation on the host is normally performed to show the actual CPU instruction cycles. This is done by changing the properties of the STS objects via right-clicking and selecting `Properties` from the pop-up menu. In the properties box, go to the `host operation` field and choose `A * x` from the drop-down menu. Then, enter 4 in the text box next to `A`, as shown in Figure L7-2.

One mechanism to come out of the idle state in *main()* is to use PRD objects to call *fun1()* and *fun2()*. In this lab, such an approach is adopted by activating PRD objects every 50msec. These objects are created by right-clicking on `PRD-Periodic Function Manager` and selecting `Insert PRD`. The objects need to be renamed as `prdPrintf` and `prdLogprintf`. For the object `prdPrintf`, change the properties as illustrated in Figure L7-3. Since the property, `period(ticks)`, is set to 50, this object calls the function mentioned in the property field `function` every 50 msec. This is because 1 tick (or timer interrupt) is set to 1000 microsec (or 1 msec) in the `CLK-Clock Manager` module,

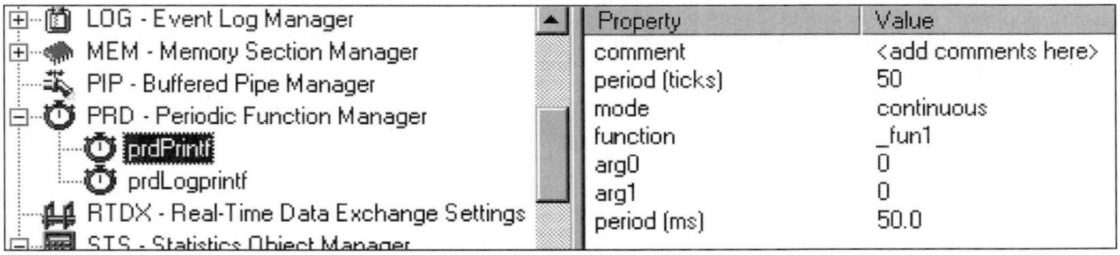

FIGURE L7-3 Property of prdPrintf object.

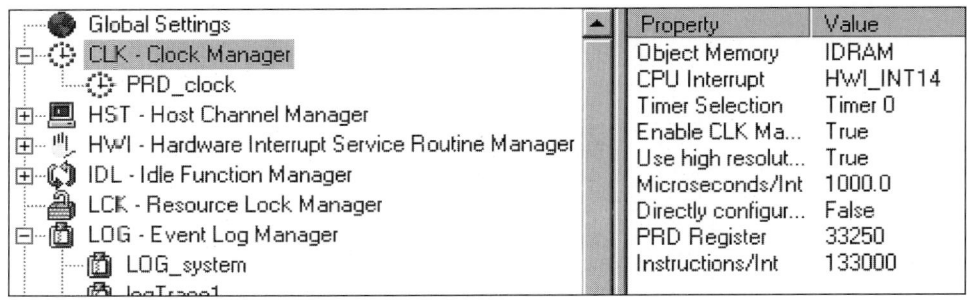

FIGURE L7-4 Property of CLK module.

as shown in Figure L7-4. Notice that when specifying the function *fun1()* in the property field function, an underscore should be added before it. This rule holds for a C function to be run by DSP/BIOS objects. The underscore prefix is necessary because the Configuration Tool creates an assembly source, and the C calling convention requires the underscore when calling C from assembly. For the prdLogprintf object, similarly enter _fun2 in the property field function.

After building the program, in order to view the statistics information captured by the STS objects, choose the menu item Tools->DSP/BIOS->Statistics View. Then, right-click in this window and select Property Page. In the Statistics View Dialog Box, hold down the control key on the keyboard and click on the objects stsPrintf and stsLogprintf. Select all four of the statistics for these objects. Then click OK. You may wish to resize the window so that all of the statistics can be viewed. Run the program. Without *printf()* or *LOG_printf()*, the average number of instruction cycles captured by the STS objects is 68, which is the overhead to call *STS_set()* and *STS_delta()*. Next, rebuild with *printf()* and *LOG_printf()*. To eliminate the overhead, change the properties of the STS objects as follows: host operation = (A * x + B)/C, A = 4, B = −68, and C = 1. Run the program again. The Statistics View window displays 4776 instruction cycles for *printf()* and 36 for *LOG_printf()*, as shown in Figure L7-5.

The output of *LOG_printf()* can be seen via a message log window. Select the menu item Tools->DSP/BIOS->Message Log. A new window will appear. This window should then be linked to the LOG object. In the Message Log window, right-click and select property page. In the drop down box Name, select logTrace1.

	Count	Total	Max	Average
stsPrintf	76	363040	5452	4776.84
stsLogprintf	76	2736	36	36.00

FIGURE L7-5 Statistics View.

L7.3 MULTITHREAD SCHEDULING

Real-time scheduling involves breaking a program into multiple threads in order to meet a specified real-time throughput. The Lab 4 program is used here to study the real-time scheduling issues. First, the ISR in Lab 4 is modified as follows:

```
int cnt=0;
interrupt void serialPortRcvISR(void)
{
        ... ...
        if(cnt++ == 3)
        {
                otherProcessing(400);
                cnt = 0;
        }
        ... ...
}
```

The function *otherProcessing()* does no specific processing and merely consumes CPU time. This function is shown next:

```
        .def _otherProcessing
        .sect ".otherProcessing"
N       .set 1000

;       void otherProcessing(int loopCount)
_otherProcessing:
        mv a4, b0       ; use b0 as loop counter
        mvk N,b1
        mpy b1,b0,b0
        nop
        shru b0,3,b0    ; (loop counter)= (# loops)/8
loop:
        sub b0,1,b0
        nop
  [b0]  b loop
        nop 5
        b b3
        nop 5           ; return
        .end
```

This function sets up a counter using the value passed to it. Let this value be 400. The counter is decreased one at a time in a loop, thus consuming CPU time. This function, of course, can be replaced with an actual processing code.

236 Chapter 12   Lab 7: DSP/BIOS

After building the project, connect the function generator and oscilloscope to the EVM board. Then, run the program. It is observed that the ISR does not meet real-time deadlines due to the extra processing required by the function *otherProcessing()*. In other words, the ISR misses input samples and fails to produce the desired output signal.

Now, let us perform real-time scheduling by breaking up the ISR into three functions or threads, *dataIO()*, *fir()*, and *otherProcessing()*, as follows:

```
/* Include header files here */
... ...
#define N 10
/* Declare global variables here */
... ...
Void main()
{
        index = 0;
        return;       /* fall into DSP/BIOS idle loop */
}

Void dataIO()
{
            temp = inputBuffer[index++];  /* read a sample */
            index = (index == N) ? 0: index;     /* Restart */
}

void fir()
{
  int i,result;
  result = 0;
  for(i=N-1;i>=0;i--)          /* Update array samples */
        samples[i+1] = samples[i];
  samples[0] = temp;

  for(i=0;i<=N;i++)             /* Filtering */
        result += ( _mpyhl(samples[i],b[i]) ) << 1;

}
```

The thread *dataIO()* reads one sample from `inputBuffer` whenever it is called. This thread simulates the operation of the macro `MCBSP_READ()`. The thread *fir()* performs FIR filtering. Let us now use three PRD objects to run these threads. Since *otherProcessing()* is called after every three input samples, it is not necessary to run all three threads or functions at the same period. The `prdDataIO` object runs the function *dataIO()* and `prdFir` object runs the function *fir()* every 1 msec. The `prdOther` object runs the function *otherProcessing()* every 4 msec. The property settings of these PRD objects are shown in Figure L7-6.

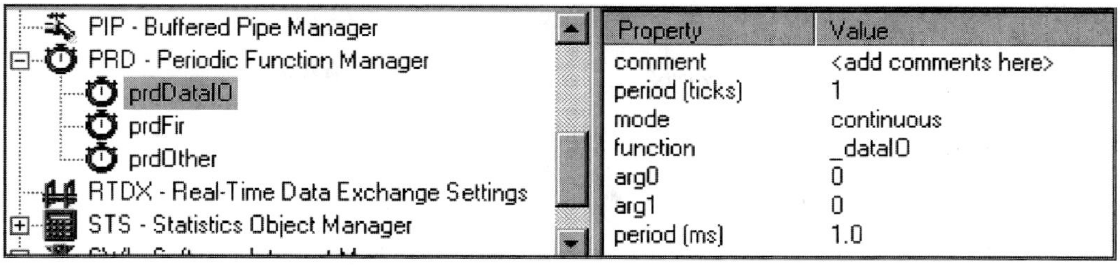

L7.3 Multithread Scheduling 237

(a)

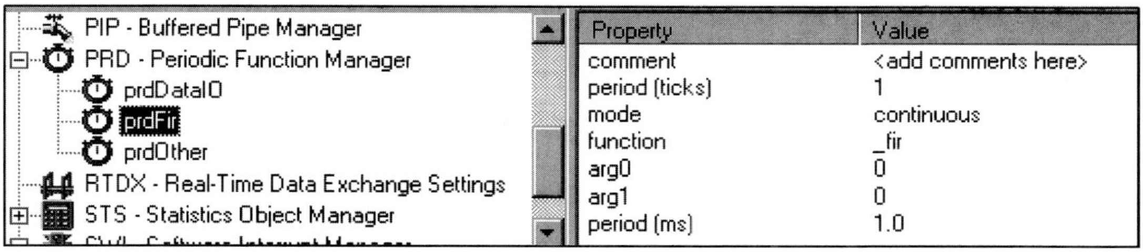

(b)

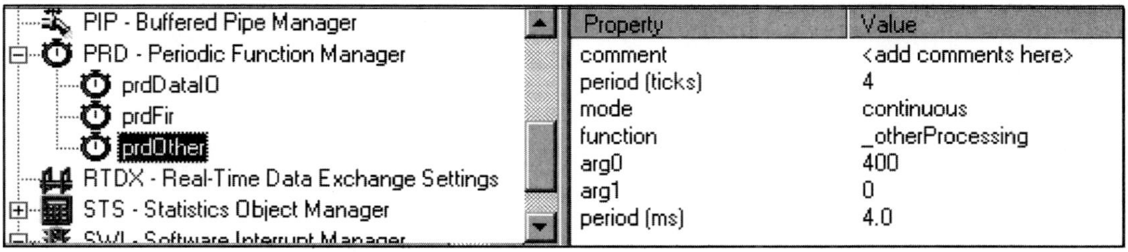

(c)

FIGURE L7-6 PRD objects for real-time analysis.

After building the project, to see whether the threads meet their real-time deadlines, choose Tools−>DSP/BIOS−>RTA Control Panel and place check marks in the boxes as indicated in Figure L7-7. Also, enable the global tracing option. Then, invoke the Execution Graph by choosing Tools−>DSP/BIOS−>Execution Graph. Right-click on the RTA Control Panel and choose Property Page from the pop-up menu. Run the program. The Execution Graph should look like the one shown in Figure L7-8.

Any missed deadline error appears in the Assertion row of the Execution Graph. From Figure L7-8, it can be seen that there are such errors in the preceding multithread program. Another way to see the same information is via the message log window Execution Graph Details, by choosing Tools−>DSP/BIOS−>Message Log. Inside the window,

238 Chapter 12   Lab 7:   DSP/BIOS

FIGURE L7-7   RTA Control Panel.

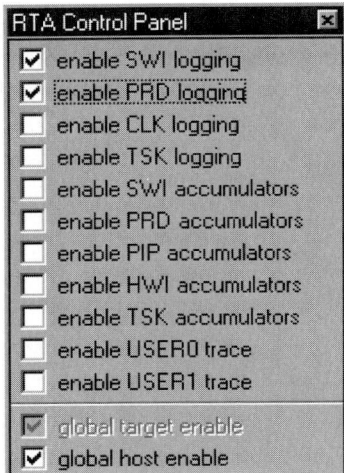

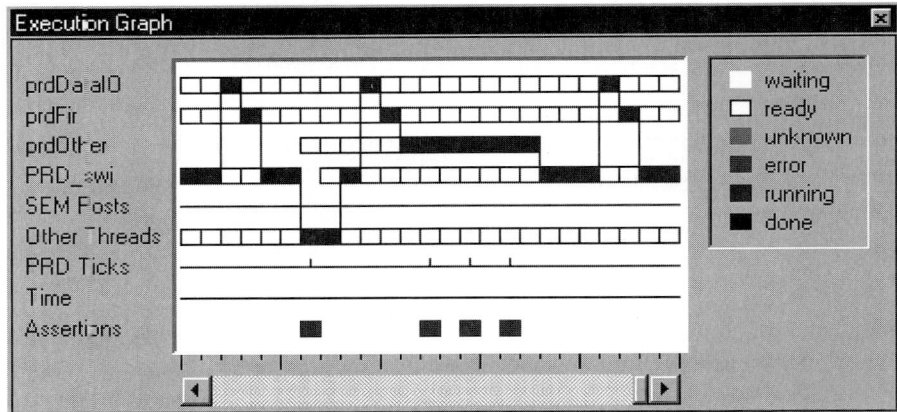

FIGURE L7-8   Execution Graph.

right-click and select Property Pages. In the Name field, choose Execution Graph Details and click OK. An Execution Graph Details window should appear as shown in Figure L7-9. The information in this window indicates that *prdFir()* is missing its real-time deadlines. Figure L7-10 shows the CPU load when the program is running. To invoke this window, choose the menu item Tools->DSP/BIOS->CPU Load Graph.

To overcome real-time errors, the scheduling of threads needs to be changed by assigning different priorities to them. As shown in Figure L7-9, *prdFir()* is missing its real-time deadlines. This is due to the fact that periodic functions execute at the same priority level, since they run as part of the same software interrupt PRD_swi. This scheduling problem is overcome by allowing each periodic object to post a software interrupt (SWI) object, which then calls the appropriate thread or function.

L7.3 Multithread Scheduling

```
Execution Graph Details
243  PRD: begin prdFir (0x80000164)
244  PRD: begin prdOther (0x80000184)
245  PRD: tick count = 50797 (0x0000c66d), ERROR: prdFir missed real-time
246  PRD: tick count = 50798 (0x0000c66e), ERROR: prdFir missed real-time
247  PRD: tick count = 50799 (0x0000c66f), ERROR: prdFir missed real-time
248  PRD: end
249  PRD: begin prdDataIO (0x80000144)
250  PRD: begin prdFir (0x80000164)
251  PRD: end
252  PRD: tick count = 50800 (0x0000c670), ERROR: prdFir missed real-time
253  PRD: begin prdDataIO (0x80000144)
254  PRD: begin prdFir (0x80000164)
255  PRD: begin prdOther (0x80000184)
```

FIGURE L7-9 Execution Graph Details.

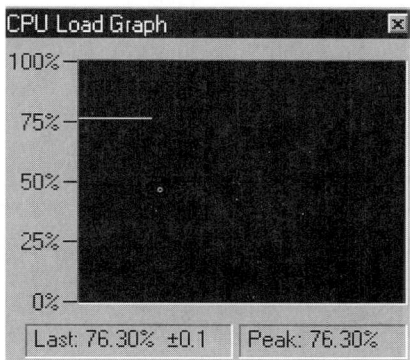

FIGURE L7-10 CPU Load Graph.

An SWI object has five properties: `priority`, `function`, `arg0`, `arg1`, and `mailbox`. The property `function` causes a specified function to be called when the SWI object is posted. The arguments `arg0` and `arg1` are passed to the function. The property `priority` stores the priority level assigned to the SWI object. The mailbox property will be covered in Lab 8. In this lab, three SWI objects are created: `swiIO`, `swiFir`, and `swiOther`. Instead of the PRD objects, the SWI objects are used to run the threads *dataIO()*, *fir()*, and *otherProcessing()*. The `swiIO` object runs *dataIO()*, the `swiFir` object *fir()*, and the `swiOther` object *otherProcessing()*. Assuming that the real-time constraint of *otherProcessing()* is not as demanding as *dataIO()*, the priority of `swiIO` is set to 3 and the priority of `swiOther` to 1. The property settings of the SWI objects are shown in Figure L7-11.

Now that the SWI objects are ready to call the threads or functions, three PRD objects need to be set up to post the software interrupts. This is achieved by changing the properties of the original PRD objects, as shown in Figure L7-12. The `PRD_swi` thread runs the PRD functions associated with `prdDataIO` and `prdFir` every 1 msec and those associated with `prdOther` every 4 msec. In other words, the PRD functions post the software interrupts associated with the SWI objects. For instance, the `prdDataIO` object runs the

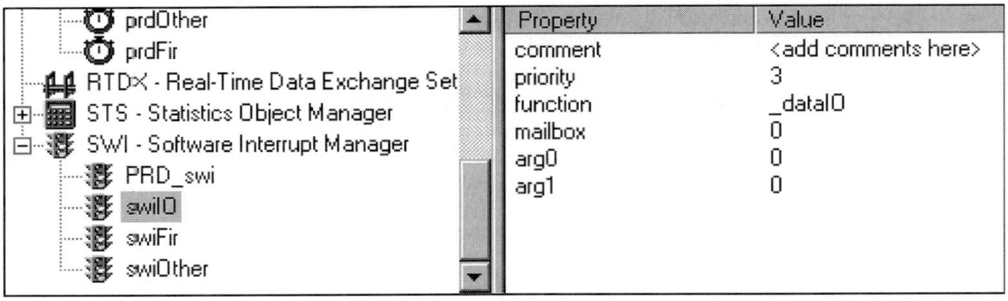

(a)

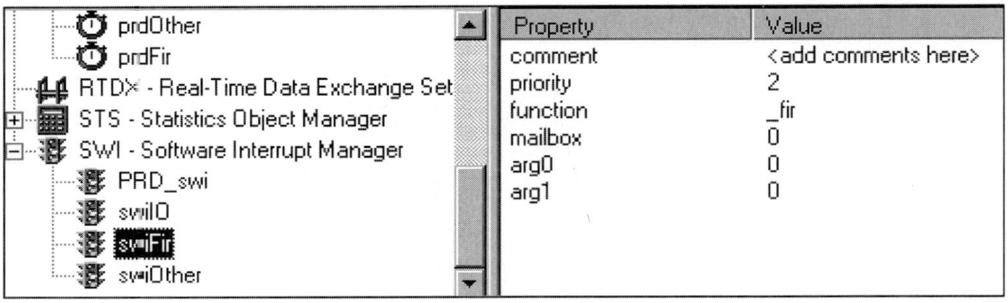

Property	Value
comment	<add comments here>
priority	3
function	_dataIO
mailbox	0
arg0	0
arg1	0

(b)

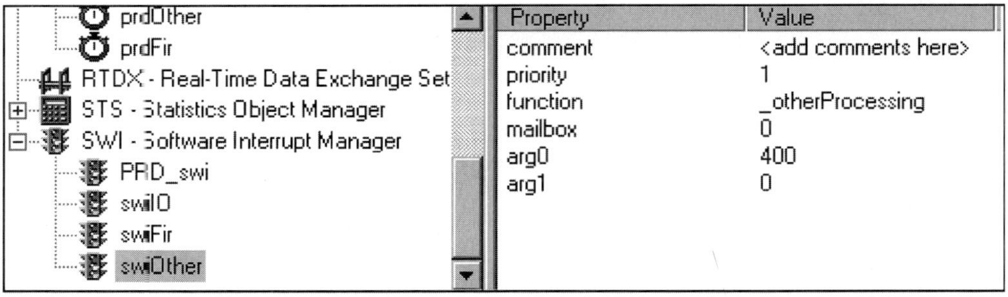

Property	Value
comment	<add comments here>
priority	2
function	_fir
mailbox	0
arg0	0
arg1	0

(c)

Property	Value
comment	<add comments here>
priority	1
function	_otherProcessing
mailbox	0
arg0	400
arg1	0

FIGURE L7-11 Properties of SWI objects.

function *SWI_post*(swiIO), which posts the software interrupt that in turn runs the function *dataIO()*. Although the software interrupts for both swiIO and swiFir are posted every 1 msec, the function *dataIO()* runs first because the associated swiIO has a higher priority than swiFir. After the *dataIO()* function is completed, the *fir()* function runs. The software interrupt for swiOther is posted every 4 msec, causing its associated func-

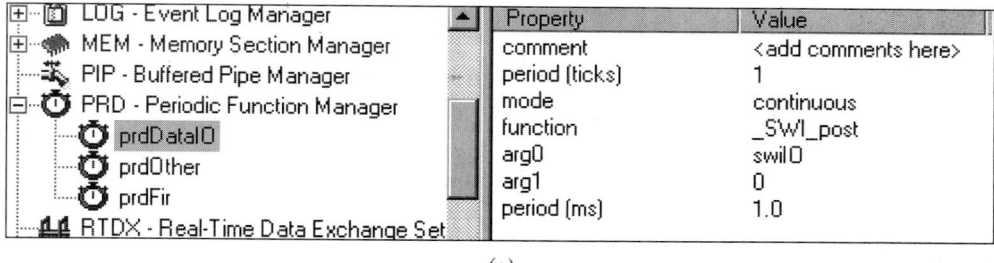

(a)

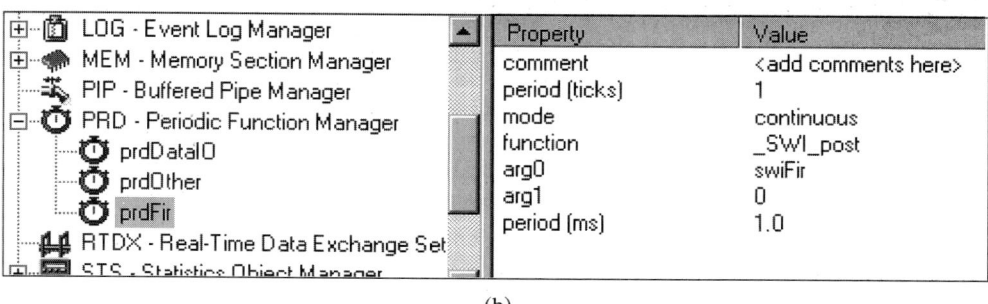

(b)

(c)

FIGURE L7-12 Properties of PRD objects for real-time scheduling.

tion, *otherProcession()*, to run. However, when a higher priority thread becomes ready, *otherProcessing()* is preempted. Figure L7-13 shows the scheduling of the periodic and software threads.

As illustrated in Figure L7-13, no real-time error is observed in the Assertion row because the threads are scheduled in such a way that they all meet their real-time deadlines. In general, critical and frequent events such as sampling should be assigned a higher priority. Next, let us go back to the ISR in the FIR filtering program. Based on the aforementioned real-time scheduling, the *otherProcessing()* thread can be moved into function when the mailbox value becomes zero. Initially, the mailbox is set to have a nonzero value of 3 (or 11 in binary). In order to run the FFT function, all the bits in the mailbox should be reset to zero. One possibility is to reset bit 0 to zero when the output frame is empty and reset bit 1 to zero when the input frame is full. The swiFFT object is therefore configured as shown in Figure L8-2. In order to create this object, right-click on SWI-Software Interrupt Manager inside the Configuration Tool window and select Insert SWI from the pop-up menu. A SWI0 object will be generated. Rename it swiFFT. Change the properties by right-clicking on the swiFFT object and selecting Properties. Inside the dialog box, change the properties as shown in Figure L8-2. The *fft()* function is assigned to the property function so that it is executed when the mailbox value becomes zero.

The property settings in Figure L8-2 show that the function *fft()* takes two arguments: pipReceive and pipTransmit. These are PIP objects, which are used as the input frame and the output frame for the *fft()* function. The PIP module manages these frames. The interrupt service routine *codec_isr()* copies data from the DRR to the input frame via the pipReceive object and from the output frame to the DXR via the pipTransmit object. The PIP objects need to be created and configured so that they can reset appropriate bits in the swiFFT's mailbox to zero, causing the *fft()* function to run with a full input frame and an empty output frame. To create the pipReceive object, right-click on PIP-Buffered Pipe Manager inside the Configuration Tool window, and choose Insert PIP from the pop-up menu. A PIP0 object will be created. Rename it pipReceive by right-clicking on it and selecting Rename. The pipTransmit object is created in the same manner. The properties are then modified to meet the synchronization between the swiFFT and PIP objects. Figure L8-3 shows the properties of the pipReceive object, which is

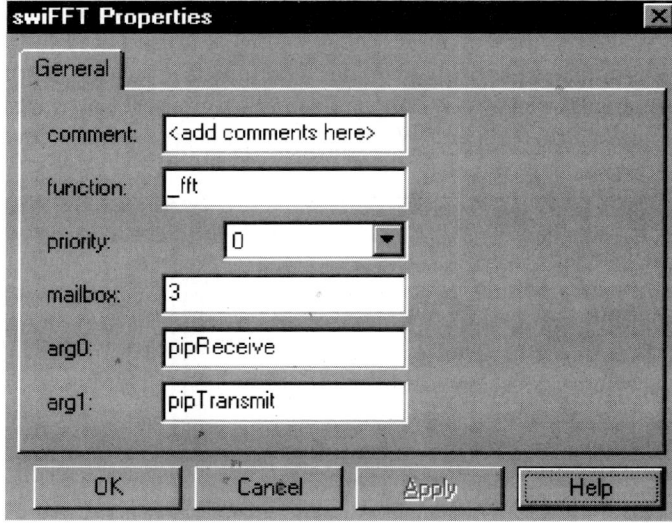

FIGURE L8-2   Properties of the swiFFT object.

Introduction

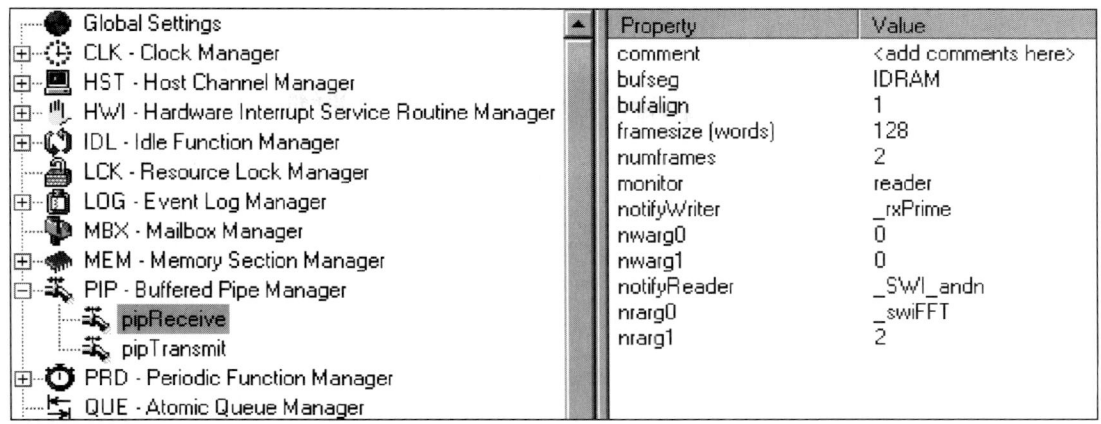

FIGURE L8-3 Properties of the `pipReceive` object.

configured to clear bit 1 in the `swiFFT`'s mailbox when the input frame is full and ready to be processed. To change properties of the `pipReceive` object, right-click on it and select `Properties` from the pop-up menu. A dialog box will appear in which new property values can be entered. The properties of the `pipTransmit` object are changed in a similar manner.

As shown in Figure Figure L8-3, the function `_SWI_andn` is specified in the property field `notifyReader`. This property assigns the function to run when the input frame buffer is full and ready to be processed. As a result, whenever the input frame is full, the `_SWI_andn` function is executed. The `_SWI_andn` function clears the bits in the mailbox and posts a software interrupt. Its first argument, `nrarg0`, specifies the SWI object to be applied, and its second argument, `nrarg1`, denotes the mask. The mailbox value is reset by the bitwise logical AND NOT operator: `mailbox = mailbox AND (NOT mask)`. Because the `pipReceive` object has the mask value of 2 (or 10 in binary), it resets bit 1 in the mailbox to zero. Consequently, since `swiFFT` is the first argument of `_SWI_andn`, the `pipReceive` object resets bit 1 of the `swiFFT`'s mailbox whenever the input frame is full and ready to be processed.

Now that bit 1 of the `swiFFT`'s mailbox is reset to zero by the `pipReceive` object, the only condition to run the FFT function is to reset bit 0 considering that 3 was the initial value in the mailbox. The `pipTransmit` object completes the synchronization process. Figure L8-4 shows the properties of the `pipTransmit` object, which is configured to clear bit 0 of the `swiFFT`'s mailbox when an empty frame is available. Note that the property field `notifyWriter` is set to `_SWI_andn`. This property specifies the function to run when an empty frame is available. Under such a condition, the `_SWI_andn` function runs with the mask value of 1 (or 01 in binary) and resets bit 0 of the `swiFFT`'s mailbox to zero. Hence, when the input frame is full and the output frame is empty, `pipReceive` and `pipTransmit` reset the `swiFFT`'s mailbox to zero, causing the FFT function to run.

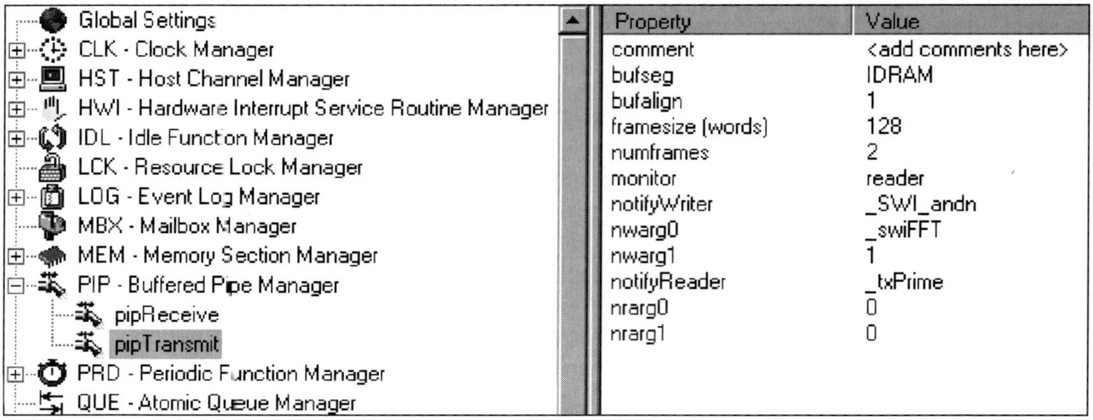

Property	Value
comment	<add comments here>
bufseg	IDRAM
bufalign	1
framesize (words)	128
numframes	2
monitor	reader
notifyWriter	_SWI_andn
nwarg0	_swiFFT
nwarg1	1
notifyReader	_txPrime
nrarg0	0
nrarg1	0

FIGURE L8-4 Properties of the `pipTransmit` object.

Next let us see how the FFT function makes use of data frames in the PIP objects. The following piece of code shows the sequence of events:

```
Void fft(PIP_Obj *in, PIP_Obj *out)
{
  int *src, *dst;
  ... ...
  PIP_get(in);
  PIP_alloc(out);
  src = PIP_getReaderAddr(in);
  dst = PIP_getWriterAddr(out);
  size = PIP_getReaderSize(in);
  PIP_setWriterSize(out,size);
  /* FFT */
  for (n=0; n<NUMPOINTS; n++)
  {
        x[n].imag = src[2*n + 1]>>16;
        x[n].real = src[2*n]>>16;
        ... ...
  }
  ... ...
  for (; size > 0; size--)    //------Copy input data into output
      *dst++ = *src++;
  PIP_put(out);
  PIP_free(in);
}
```

The first argument of the FFT function, in, is the pipReceive object and the second argument, out, is the pipTransmit object, as indicated in Figure L8-2. In order to use the frames in the PIP objects, first the *PIP_get()* and *PIP_alloc()* API functions should be called. The *PIP_get()* API gets a full frame from the `pipReceive` object and the *PIP_alloc()* API allocates an empty frame from the `pipTransmit` object. Normally, the *PIP_get()* API is followed by the *PIP_getReaderAddr()* API, which returns the address for the reading process. Similarly, the *PIP_alloc()* API is followed by the *PIP_getWriterAddr()* API, which returns the address for the writing process. A pointer `src` is therefore used to read from the input frame and a pointer `dst` to write to the output frame. In this program, the FFT function processes the data stored in `src`. Before leaving the *fft()* function, the *PIP_put()* API is called to put the full frame into the `pipTransmit` object. Normally, this API is used together with the *PIP_alloc()* API because *PIP_put()* puts a frame allocated by *PIP_alloc()* into a PIP object after the frame is full. Similarly, the *PIP_free()* API is used together with the *PIP_get()* API because *PIP_free()* releases the frame for *PIP_get()* after it is read. The released frame is recycled so that it can be reused by the *PIP_alloc()* API.

Note that the value of the property `notifyWriter` in the `pipReceive` object is set to `_rxPrime`, as shown in Figure Figure L8-3. Therefore, the function *rxPrime()* is called when a frame of free space is available in the `pipReceive` object. The following piece of code shows the relevant part in the *rxPrime()* function:

```
void rxPrime(void)
{
  PIP_Obj     *rxPipe = &pipReceive;
  ... ...
  if (rxCount == 0 && PIP_getWriterNumFrames(rxPipe) > 0) {
      PIP_alloc(rxPipe);
      rxPtr = PIP_getWriterAddr(rxPipe);
      rxCount = PIP_getWriterSize(rxPipe);
  }
  ... ...
}
```

The global variable `rxCount` keeps track of the remaining number of words for filling up the current `rxPipe` (or `pipReceive`) frame. In the *codec_isr()* function, `rxCount` is decreased by one whenever a sample is read from the DRR and copied into the `rxPipe` frame. When this frame becomes full and ready to be put into `rxPipe`, `rxCount` becomes zero. Then this function allocates the next empty frame from `rxPipe` by calling *PIP_alloc(rxPipe)*. The address of the frame is set to the global variable `rxPtr` so that *codec_isr()* can copy the content of the DRR into `rxPtr` by calling *PIP_getWriterAddr(rxPipe)*. In the *codec_isr()* function, `rxPtr` is increased by one to point to the next location whenever the DRR content is copied.

As shown in Figure L8-4, `_txPrime` is written in the property field `notifyReader`. Therefore, the function *txPrime()* runs when a frame is full and ready to be used. The following piece of code shows the relevant part in the *txPrime()* function:

```
void txPrime(void)
{
  PIP_Obj     *txPipe = &pipTransmit;
  ... ...
  if (txCount == 0 && PIP_getReaderNumFrames(txPipe) > 0) {
      PIP_get(txPipe);
      txPtr = PIP_getReaderAddr(txPipe);
      txCount = PIP_getReaderSize(txPipe);
  }
  ... ...
}
```

The global variable txCount keeps track of the remaining number of words for transmitting the current txPipe (or pipTransmit) frame. In the function *codec_isr()*, txCount is decreased by one whenever a sample is copied from the txPipe frame and written to the DXR. When all the samples in this frame are written, txCount becomes zero. Then, this function gets the next full frame from txPipe by calling *PIP_get(txPipe)*. The address of the frame is set to the global variable txPtr by calling *PIP_getReaderAddr(txPipe)* so that *codec_isr()* can copy the content of txPtr into the DRR. The pointer txPtr is increased by one to point to the next location whenever a sample is written to the DXR.

After properly configuring the PIP and SWI objects, the program is built. The entire DSP/BIOS version of the FFT files are provided on the attached CD. In order to verify the operation of the DSP/BIOS-based FFT program, connect a function generator to the EVM board and run the program. Figure L8-5 shows a snapshot of the CCS animation feature. This is done by setting a breakpoint at the end of the FFT function and by opening a graphical display window via the menu item View->Graph->time/Frequency. Place the global variable mag in the field Start Address to display the FFT magnitude values. Then select the menu item Debug->Animate to start the animation. As the input frequency from the function generator is changed, the peaks in the graphical display window should move accordingly.

FIGURE L8-5 FFT magnitude.

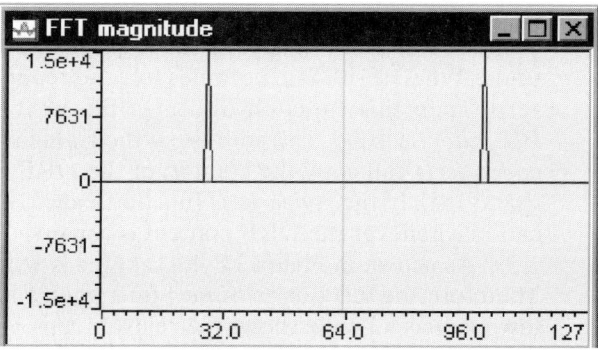

The FFT magnitude can be sent to the host PC by using the DMA. The following function *WriteFIFO_DMA()* can be used to perform this operation:

```
void WriteFIFO_DMA()
{
        dma_reset();

        dma_init(2,                                 //Channel
                    0x0A000010u,                    //Primary Control Register
                    0x0000000Au,                    //Secondary Control Register
                    (unsigned int) mag,             //Source Address
                    0x01710000u,                    //Destination Address
                    0x00010080u);                   //Transfer Counter Register

        DMA_START(2);
}
```

The DMA uses the global variable mag as the source address and the memory location 0x0171000, which is dedicated for FIFO access in PCI bus transfers, as the destination address. The program *host.exe* provided on the attached CD is written to display the FFT magnitude on the PC screen. This program makes use of the EVM API *evm6x_read()*. This API transfers data from the DSP to the host. In this lab, a PRD object prdDMA is created and configured to run the DMA transfer function *WriteFIFO_DMA()* every 8 msec, as shown in Figure L8-6. After adding the *WriteFIFO_DMA()* function to the FFT program, build the program and run it. To observe the FFT magnitude in real-time, also run *host.exe*. The CPU load Graph plug-in can be used to verify that the DMA transfers the contents of mag independently of the CPU. As shown in Figure L8-7, the CPU load remains almost the same while the DMA transfer is running. To invoke the CPU Load Graph plug-in, choose the menu item Tools->DSP/BIOS->CPU load Graph.

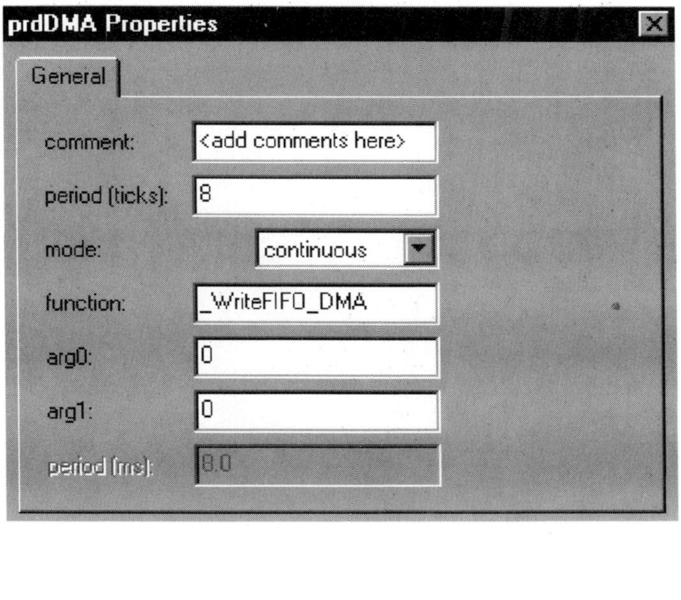

FIGURE L8-6 Properties of prdDMA.

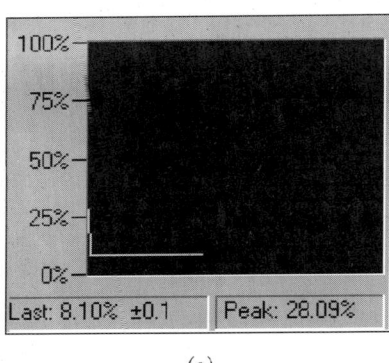 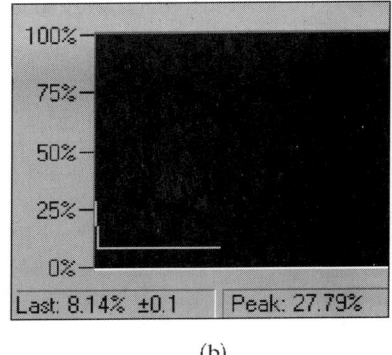

(a)   (b)

FIGURE L8-7 CPU Load Graph: (a) CPU load before DMA transfer; (b) CPU load while DMA transfer is running.

L8.1 PRIORITIZATION OF THREADS

Instead of using the function generator to generate input samples, a CD player can be connected to the EVM audio-in jack. Of course, a pair of amplified speakers needs to be connected to the EVM audio-out jack to hear the sound. Now let us examine the effect on the sound quality by changing the CPU load. To change the CPU load, a PRD object prdLoad is created and configured to run the function *changeload()* every 8 msec, as illustrated in Figure L8-8. The function *changeload()* calls the *otherProcessing()* function. The CPU load is determined by the global variable loadVal, which is passed to the *otherProcessing()* function. The function *changeload()* is shown next:

```
Void changeload(Int prd_ms)
{
  ... ...
  if (loadVal)
       otherProcessing(loadVal);
}
```

The global variable loadVal can be set by choosing Edit->Edit Variable to invoke the Edit Variable dialog box. In this dialog box, write loadVal in the field variable and the desired number in the field value. The prdLoad object will run the *changeload()* function every 8 msec.

FIGURE L8-8 CPU load with loadVal = 900.

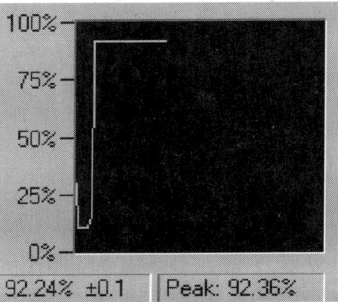

```
118  PRD: tick count = 32002 (0x00007d02)
119  SWI: post  swiFFT (0x80000184)
120  SWI: begin swiFFT (0x80000184)
121  PRD: tick count = 32003 (0x00007d03)
122  SWI: end   swiFFT (0x80000184) state = done
123  PRD: tick count = 32004 (0x00007d04)
124  SWI: post  PRD_swi (0x800001b0)
125  SWI: begin PRD_swi (0x800001b0)
126  PRD: end
127  SWI: end   PRD_swi (0x800001b0) state = done
```

(a)

```
118  SWI: end   PRD_swi (0x800001b0) state = still ready
119  SWI: begin swiFFT (0x80000184)
120  SWI: post  swiFFT (0x80000184)
121  SWI: end   swiFFT (0x80000184) state = still ready
122  SWI: begin PRD_swi (0x800001b0)
123  PRD: begin prdDMA (0x80000164)
124  PRD: end
125  SWI: end   PRD_swi (0x800001b0) state = done
126  SWI: begin swiFFT (0x80000184)
127  SWI: end   swiFFT (0x80000184) state = done
```

(b)

FIGURE L8-9 Execution Graph Details: (a) for `loadVal = 0`; (b) for `loadVal = 900`.

To observe the impact of the CPU load, let us build the program and run it while playing a CD. When the `loadVal` is changed to 900, the CPU load increases to about 92%, as shown in Figure L8-8, and the sound quality is degraded. The reason for this degradation can be seen by activating the Execution Graph Details window. As shown in Figure L8-9 (a), when `loadVal` is zero, the `swiFFT` and `PRD_swi` threads complete their tasks without any problem. However, when `loadVal` is 900, these threads cannot complete their tasks and frequently go into the ready state, as shown in Figure L8-9 (b). The Execution Graph in Figure L8-10 provides a graphical display of this situation. Since the `swiFFT` thread is competing with the `PRD_swi` thread, for large load values the

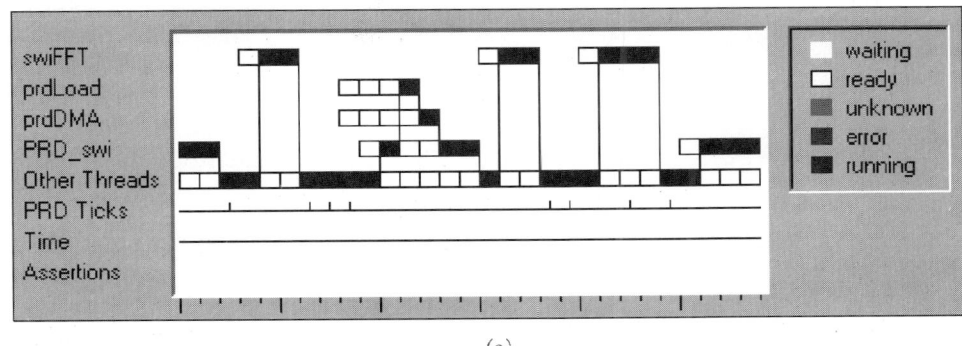

(a)

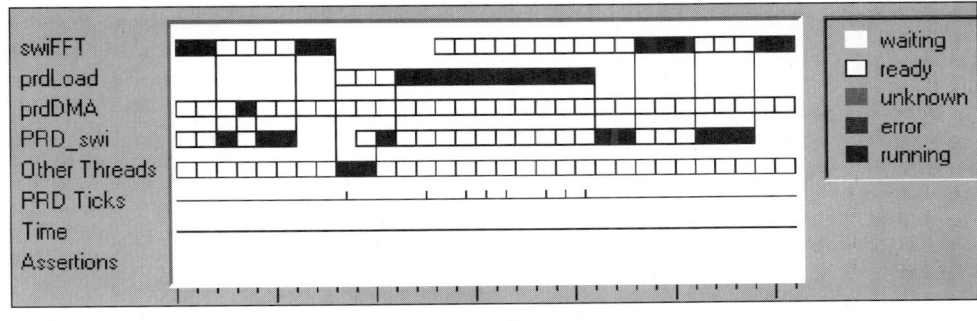

(b)

FIGURE L8-10 Execution Graph: (a) for `loadVal = 0`; (b) for `loadVal = 900`.

252   Chapter 12   Lab 8:   Data Synchronization and Communication

fft() function sits waiting to be executed. The PRD_swi thread executes all the PRD objects, so it eventually runs the CPU load function *otherProcessing()*. Consequently, since the audio from the CD player is copied into the output frame by the *fft()* function as part of the swiFFT thread, the sound quality suffers.

To solve this problem, the threads need to be properly prioritized. Let us assign a higher priority to the swiFFT thread. One simple way to do this is via the drag-and-drop method. Click on SWI_Software Interrupt Manager in the Configuration Tool. On the right side of the window, click and hold the left mouse button on the PRD_swi icon, drag it to Priority 1, and release the button to drop it. This way the priority of the PRD_swi thread becomes 1. Similarly, move the swiFFT icon to Priority 2 so that it gets a higher priority than PRD_swi. Now build the program and run it. The sound quality remains unaffected even though loadVal = 900. The DSP/BIOS plug-ins in Figure L8-11 illustrate that the swiFFT thread no longer waits to be executed. Notice that the CPU load is now about 94%.

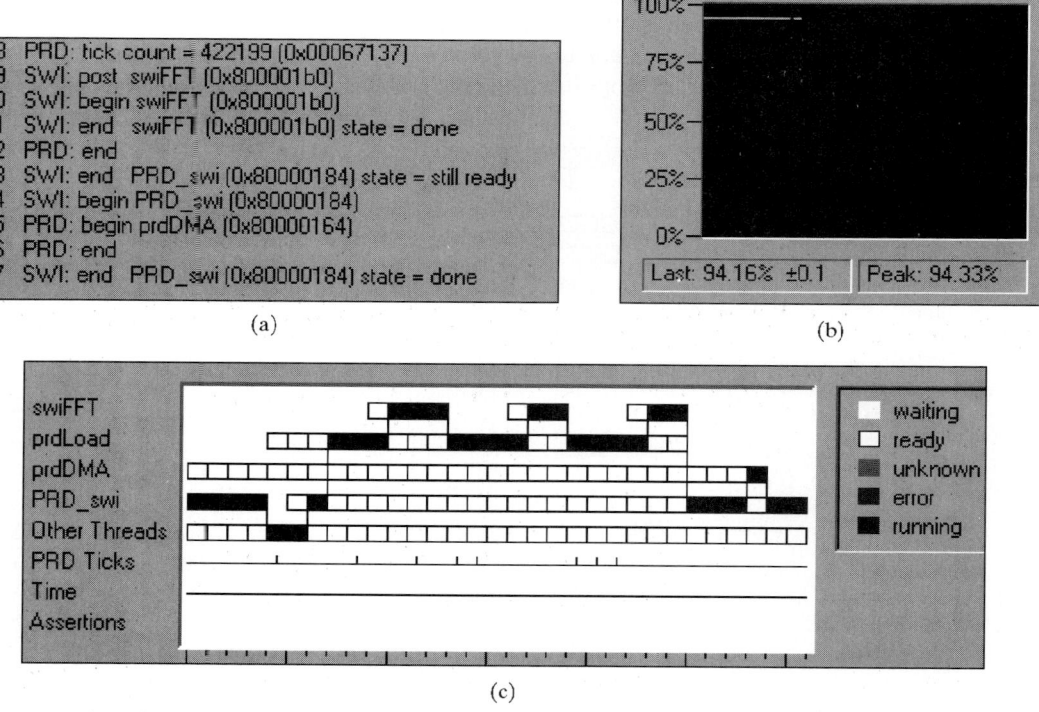

FIGURE L8-11   DSP/BIOS plug-ins with loadVal = 900 after the prioritization: (a) Execution Graph Details, (b) CPU Load, and (c) Execution Graph.

L8.2 RTDX

The CPU load can be changed by using the RTDX module. This module allows `loadVal` to be transferred from the PC host to the DSP while the program is running. The following piece of code shows the parts to be added to the original program to access the RTDX module (the entire program is provided on the attached CD):

```
#include <rtdx.h>
RTDX_CreateInputChannel(writeload);
RTDX_CreateOutputChannel(readload);
main()
{
    ... ...
    RTDX_enableInput(&writeload);
    RTDX_enableOutput(&readload);
    return;
}
Void changeload(Int prd_ms)
{
    ... ...
    if (!RTDX_channelBusy(&writeload)) {      // Reads new loadVal sent from
                                              // the PC host.
        RTDX_readNB(&writeload, &loadVal, sizeof(loadVal));
        if (oldLoad != loadVal ) {
                    oldLoad = loadVal;
                    RTDX_write(&readload, &loadVal, sizeof(loadVal));
                    LOG_printf(&logTrace,
                      "CPU load: new load = %d000 instructions every
                       %d ms", loadVal, prd_ms);
        }
    }
    if (loadVal)
    otherProcessing(loadVal);
}
```

In order to use the indicated RTDX APIs, the program should include the header file *rtdx.h*. The RTDX input channel structure is declared and initialized by the macro `RTDX_CreateInputChannel`. Since it is declared as a global variable, it can be accessed anywhere in the program. Similarly the `RTDX_CreateOutputChannel` macro defines and initializes the RTDX output channel structure. Because these channels are disabled during the initialization, they need to be enabled in *main()* by using the *RTDX_enableInput()* and *RTDX_enableOutput()* APIs. The input channel is examined by the *RTDX_channelBusy()* API to see whether it is busy or not. If it is not busy, data is read from the input channel by using the *RTDX_readNB()* API. This API posts a request to the RTDX host library indicating that the DSP application program is ready to receive data. The DSP program keeps running at this point. When the *RTDX_read()* API is used, the DSP program stops until it receives data from the input channel. The *RTDX_write()* API is used to write a new loadVal to the output channel in order to notify the PC host that such a value is received and used.

On the PC host side, an OLE application is written in Visual Basic to receive and send data. Build the program and run it. Then run the OLE application embedded in *rtdx.doc* (provided on the attached CD). It can be observed that the CPU load changes as a new loadVal is sent from the host OLE program to the DSP.

Appendix A: Quick Reference Guide

A-1: LIST OF C6X INSTRUCTIONS

.L Unit

Instruction	Description
ABS	Integer absolute value with saturation
ADD(U)	Signed(Unsigned) addition without saturation
AND	Bitwise AND
CMPEQ	Integer compare for equality
CMPGT	Signed integer compare for greater than
CMPGTU	Unsigned integer compare for greater than
CMPLT	Signed integer compare for less than
CMPLTU	Unsigned integer compare for less than
LMBD	Leftmost bit detection
MV	Move from register to register
NEG	Negate
NORM	Normalize integer
NOT	Bitwise NOT
OR	Bitwise OR
SADD	Integer addition with saturation to result size
SAT	Saturate a 40-bit integer to a 32-bit integer
SSUB	Integer subtraction with saturation to result size
SUBU	Unsigned integer subtraction without saturation
SUBC	Conditional integer subtract and shift–used for division
XOR	Exclusive OR
ZERO	Zero a register

.M Unit

Instruction	Description
MPY	Signed integer multiply 16lsb×16 lsb
MPYU	Unsigned integer multiply 16lsb×16lsb
MPYUS	Integer multiply (unsigned) 16lsb×(signed) 16lsb
MPYSU	Integer multiply (signed) 16lsb×(unsigned) 16lsb
MPYH	Signed integer multiply 16msb×16msb
MPYHU	Unsigned integer multiply 16msb×16msb
MPYHUS	Integer multiply (unsigned) 16msb×(signed) 16msb
MPYHSU	Integer multiply (signed) 16msb×(unsigned) 16msb
MPYHL	Signed multiply high low 16msb×16lsb
MPYHLU	Unsigned multiply high low 16msb×16lsb
MPYHULS	Multiply high unsigned low signed (unsigned) 16msb×(signed) 16lsb
MPYHSLU	Multiply high signed low unsigned (signed) 16msb×(unsigned) 16lsb
SMPY	Integer multiply with left shift and saturation
SMPYHL	Integer multiply high low with left shift and saturation
SMPYLH	Integer multiply low high with left shift and saturation
SMPYH	Integer multiply high with left shift and saturation

.S Unit

ADD(U)	Signed (unsigned) addition without saturation
ADDK	Integer addition using signed 16-bit constant
ADD2	Two 16-bit integer adds on upper and lower register halves
AND	Bitwise AND
B disp	Branch using a displacement
B IRP	Branch using an interrupt return pointer
B NRP	Branch using a NMI return pointer
B reg	Branch using a register
CLR	Clear a bit field
EXT(U)	Extract and sign-extend(zero-extend) a bit field
MV	Move from register to register
MVC	Move between the control file and register file
MVK	Move a 16-bit signed constant into a register and sign extend
MVKH	Move 16-bit constant into the upper bits of a register
MVKLH	Move 16-bit constant into the upper bits of a register
NEG	Negate
NOT	Bitwise NOT
OR	Bitwise OR
SET	Set a bit field
SHL	Arithmetic shift left
SHR	Arithmetic shift right
SHRU	Logical shift right
SHRL	Logical shift left
SUB(U)	Signed (unsigned) Integer subtraction without saturation
SUB2	Two 16-bit Integer subtracts on upper and lower register halves
XOR	Exclusive OR
ZERO	Zero a register

.D Unit

ADD(U)	Signed (unsigned) integer addition without saturation
ADDAB/ADDAH/ADDAW	Integer addition using addressing mode
DB(U)/LDH(U)/LDW	Load from memory with a 5-bit unsigned constant offset or register offset
LDB(U)/LDH(U)/LDW (15-bit offset)	Load from memory with a 15-bit constant offset
MV	Move from register to register
STB/STH/STW	Store to memory with a register offset or 5-bit unsigned constant offset
STB/STH/STW (15-bit offset)	Store to memory with a 15-bit offset
SUB	Signed integer subtraction without saturation
SUBAB/SUBAH/SUBAW	Integer subtraction using addressing mode
ZERO	Zero a register

A-2: LIST OF C67X FLOATING-POINT INSTRUCTIONS

.L Unit

Instruction	Description
ADDDP	Double-precision floating-point addition
ADDSP	Single-precision floating-point absolute value
DPINT	Convert double-precision floating-point value to integer
DPSP	Convert double-precision floating-point value to single-precision floating-point value
INTDP	Convert integer to double-precision floating-point value
INTDPU	Convert integer to double-precision floating-point value (unsigned)
INTSP	Convert integer to single-precision floating-point value
INTSPU	Convert integer to single-precision floating-point value (unsigned)
SPINT	Convert single-precision floating-point value to integer
SPTRUNC	Convert single-precision floating-point value to integer with truncation
SUBSP	Single-precision floating-point subtract
SUBDP	Double-precision floating-point subtract

.M Unit

MPYSP	Single-precision floating-point multiply
MPYDP	Double-precision floating-point multiply
MPYI	32-bit integer multiply–result in lower 32 bits
MPYID	32-bit integer multiply–64-bit result

.S Unit

ABSSP	Single-precision floating-point absolute value
ABSDP	Double-precision floating-point absolute value
CMPGTSP	Single-precision floating-point compare for greater than
CMPEQSP	Single-precision floating-point compare for equality
CMPLTSP	Single-precision floating-point compare for less than
CMPGTDP	Double-precision floating-point compare for greater than
CMPEQDP	Double-precision floating-point compare for equality
CMPLTDP	Double-precision floating-point compare for less than
RCPSP	Single-precision floating-point reciprocal approximation
RCPDP	Double-precision floating-point reciprocal approximation
RSQRSP	Single-precision floating-point square-root reciprocal approximation
RSQRDP	Double-precision floating-point square-root reciprocal approximation
SPDP	Convert single-precision floating-point value to double-precision floating-point value

.D Unit

ADDAD	Integer addition using doubleword addressing mode
LDDW	Load doubleword from memory with an offset

A-3: REGISTERS AND MEMORY-MAPPED REGISTERS†

Addressing Mode Register (AMR)

31	26	25	21	20	16
Reserved		BK1		BK0	
R, +0		R, W, +0		R, W, +0	

15	14	13	12	11	10	9	8	7 6	5 4	3 2	1	0
B7 mode		B6 mode		B5 mode		B4 mode		A7 mode	A6 mode	A5 mode	A4 mode	

R, W, +0

Control Status Register (CSR)

31	24	23	16
CPU ID		Revision ID	
R		R, W, +0	

15	10	9 8	7	5 4	2	1	0
PWRD		SAT	EN	PCC	DCC	PGIE	GIE
R, W, +0		R,C,+0	R,+x		R, W, +0		

Interrupt Flag Register (IFR)

15	14	13	12	11	10	9 8	7	6	5	4	3	2	1	0
IF15	IF14	IF13	IF12	IF11	IF10	IF9	IF8	IF7	IF6	IF5	IF4	rsv	rsv	NMIF	0

R,+0

Interrupt Set Register (ISR)

15	14	13	12	11	10	9 8	7	6	5	4	3	2	1	0
IS15	IS14	IS13	IS12	IS11	IS10	IS9	IS8	IS7	IS6	IS5	IS4	rsv	rsv	rsv	rsv

W

Interrupt Clear Register (ICR)

15	14	13	12	11	10	9 8	7	6	5	4	3	2	1 0
IC15	IC14	IC13	IC12	IC11	IC10	IC9	IC8	IC7	IC6	IC5	IC4	rsv	rsv	rsv	rsv

W

Interrupt Enable Register (IER)

15	14	13	12	11	10	9 8	7	6	5	4	3	2	1	0
IE15	IE14	IE13	IE12	IE11	IE10	IE9	IE8	IE7	IE6	IE5	IE4	rsv	rsv	NMIE	1

R, W, +0   R,1

Interrupt Service Table Pointer (ISTP)

31	10 9	5 4 3	2 1	0
ISTB		HPEINT	0 0 0 0 0
R, W, +0		R, +0	

NMI Return Pointer (NRP)

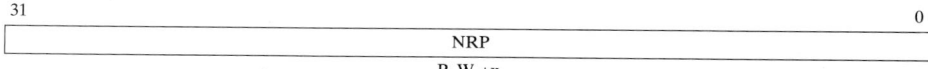

Interrupt Return Pointer (IRP)

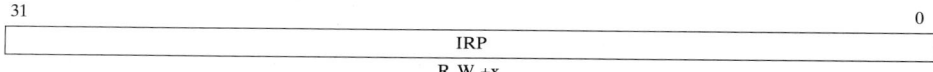

Note: Bits not shown are **reserved**.

A-3: Registers and Memory-Mapped Registers

EMIF

CE0 Space Control	1800008
CE1 Space Control	1800004
CE2 Space Control	1800010
CE3 Space Control	1800014
Global Control	180000C
SDRAM Control	1800018
SDRAM Refresh Period	180001C

HPI

Control Register	1880000

Interrupts

Multiplexer High	19C0000
Multiplexer Low	19C0004
External Interrupt Polarity	19C0008

DMA	Ch. 0	Ch. 1	Ch. 2	Ch. 3
Primary Control	1840000	1840040	1840004	1840044
Secondary Control	1840008	1840048	184000C	184004C
Source Address	1840010	1840050	1840014	1840054
Destination Address	1840018	1840058	184001C	184005C
Transfer Counter	1840020	1840060	1840024	1840064
Global Reload A	1840028			
Global Reload B	184002C			
Global Index A	1840030			
Global Index B	184003C			
Global Index C	1840068			
Global Index D	184006C			
Auxiliary Control	1840070			

McBSP	0	1
DRR	18C0000	1900000
DXR	18C0004	1900004
Control Register	18C0008	1900008
Receive Control Register	18C000C	190000C
Transmit Control Register	18C0010	1900010
Sample Rate Generator Register	18C0014	1900014
Multichannel Register	18C0018	1900018
Receive Channel Enable Register	18C001C	190001C
Transmit Channel Enable Register	18C0020	1900020
Pin Control Register	18C0024	1900024

Timers	0	1
Control	1940000	1980000
Period	1940004	1980004
Counter	1940008	1980008

C6x_MMR†

Memory-mapped Registers

You must include this file in each .asm that references
one of these mmr register names. It can be included by adding the following
line to the top of your .asm file:

.include c6x_mmr.asm

Using the names that follow simplifies access to peripheral mmr registers.
Here is an example for writing all F's into the CE1 and CE2 EMIF
space control registers:

MVK	.S1	0FFFFh, A0
MVKLH	.S1	0FFFFh, A0
MVK	.S1	EMIF, A1
MVKH	.S1	EMIF, A1
STW	.D1	A0, *+A1[CE1]
STW	.D1	A0, *+A1[CE2]

;Peripheral	Addr/Offset		Register
EMIF	.equ	01800000h	; EMIF global control
CE1	.equ	1	; EMIF CE1 space control
CE0	.equ	2	; EMIF CE0 space control
CE2	.equ	4	; EMIF CE2 space control
CE3	.equ	5	; EMIF CE3 space control
SDRAM	.equ	6	; EMIF SDRAM control
REFRESH	.equ	7	; EMIF SDRAM refresh period
DMA	.equ	01840000h	; Top of DMA registers
DMA0pc	.equ	0	; DMA0 primary control
DMA2pc	.equ	1	; DMA2 primary control
DMA0sc	.equ	2	; DMA0 secondary control
DMA2sc	.equ	3	; DMA2 secondary control
DMA0src	.equ	4	; DMA0 source address
DMA2src	.equ	5	; DMA2 source address
DMA0dst	.equ	6	; DMA0 destination address
DMA2dst	.equ	7	; DMA2 destination address
DMA0tc	.equ	8	; DMA0 transfer counter
DMA2tc	.equ	9	; DMA2 transfer counter
DMAcountA	.equ	10	; DMA global count reload register A
DMAcountB	.equ	11	; DMA global count reload register B
DMAindexA	.equ	12	; DMA global index register A
DMAindexB	.equ	13	; DMA global index register B
DMAaddrA	.equ	14	; DMA global address register A
DMAaddrB	.equ	15	; DMA global address register B
DMA1pc	.equ	16	; DMA1 primary control
DMA3pc	.equ	17	; DMA3 primary control
DMA1sc	.equ	18	; DMA1 secondary control

(Continues)

A-3: Registers and Memory-Mapped Registers

;Peripheral		Addr/Offset	Register
DMA3sc	.equ	19	; DMA3 secondary control
DMA1src	.equ	20	; DMA1 source address
DMA3src	.equ	21	; DMA3 source address
DMA1dst	.equ	22	; DMA1 destiantion address
DMA3dst	.equ	23	; DMA3 destination address
DMA1tc	.equ	24	; DMA1 transfer counter
DMA3tc	.equ	25	; DMA3 transfer counter
DMAaddrC	.equ	26	; DMA global address register C
DMAaddrD	.equ	27	; DMA global address register D
DMAaux	.equ	28	; DMA auxiliary control register
HPI	.equ	01880000h	; HPI control register
McBSP0	.equ	018C0000h	; McBSP0 DRR
McBSP1	.equ	01900000h	; McBSP1 DRR
DRR	.equ	0	; McBSP DRR
DXR	.equ	1	; McBSP DXR
SPCR	.equ	2	; McBSP control register
RCR	.equ	3	; McBSP receive control register
XCR	.equ	4	; McBSP transmit control register
SRGR	.equ	5	; McBSP sample rate generator register
MCR	.equ	6	; McBSP multichannel register
RCER	.equ	7	; McBSP receive channel enable register
CER	.equ	8	; McBSP transmit channel enable register
PCR	.equ	9	; McBSP pin control register
Timer0	.equ	01940000h	; Timer 0
Timer 1	.equ	01980000h	; Timer 1
TimCR	.equ	0	: Timer control register
TimTP	.equ	1	; Timer period
TimTC	.equ	2	; Timer counter
Interrupts	.equ	019C0000h	; Interrupts
IMH	.equ	0	; Interrupt Multiplexer high
IML	.equ	1	; Interrupt Multiplexer low
IP	.equ	2	; External interrupt polarity

A-4: COMPILER INTRINSICS†

C Compiler Intrinsic	Assembly Instruction	Description	Device
`int _abs (int src2);` `int _labs (long src2);`	**ABS**	Returns the saturated absolute value of *src2*.	
`int _add2 (int src1, int src2);`	**ADD2**	Adds the upper and lower halves of *src1* to the upper and lower halves of *src2* and returns the result.	
`uint _clr (uint src2, uint csta, uint cstb);`	**CLR**	Clears the specified field in *src2*. The beginning and ending bits of the field to be cleared are specified by *csta* and *cstb*, respectively.	
`unsigned _clrr (uint src1, int src2);`	**CLR**	Clears the specified field in src2. The beginning and ending bits of the field to be cleared are specified by the lower 10 bits of the source register.	
`int _dpint (double);`	**DPINT**	Converts 64-bit double to 32-bit signed integer, using the rounding mode set by the CSR register.	'C67x
`int _ext (uint src2, uint csta, int cstb);`	**EXT**	Extracts the specified field in src2, sign extended to 32 bits. The extract is performed by a signed shift right; csta and cstb are the shift left and shift right amounts, respectively.	
`int _extr (int src2, int src1);`	**EXT**	Exctracts the specified field in *src2*, sign extended to 32 bits.	
`uint _extu (uint src2, uint csta, uint cstb);`	**EXTU**	Extracts the specified field in *src2*, zero extended to 32 bits.	
`uint _extur (uint src2, int src1);`	**EXTU**	Extracts the specified field in *src2*, zero extended to 32 bits.	
`uint _ftoi (float);`		Reinterprets the bits in the float as an unsigned integer.	'C67x
`uint _hi (double);`		Returns the high 32 bits of a double as an integer.	'C67x
`double _itod (uint, uint);`		Creates a new double-register pair from two unsigned integers.	'C67x
`float _itof (uint);`		Reinterprets the bits in the unsigned integer as a float.	'C67x
`uint _lmbd (uint src1, uint src2);`	**LMBD**	Searches for a leftmost 1 or 0 of *src2* determined by the LSB of *src1*. Returns the number of bits up to the bit change.	
`uint _lo (double);`		Returns the low (even) register of a double register pair as an integer.	'C67x

(Continues)

A-4: Compiler Intrinsics

C Compiler Intrinsic	Assembly Instruction	Description	Device
int _mpy (int src1, int src2); int _mpyus (uint src1, int src2); int _mpysu (int src1, uint src2); uint _mpyu (uint src1, uint src2);	MPY MPYUS MPYSU MPYU	Multiplies the 16 LSBs of *src1* by the 16 MSBs of *src2* and returns the result. Values can be signed or unsigned.	
int _mpyhl (int src1, int src2); int _mpyhuls (uint src1, int src2); int _mpyhslu (int src1, uint src2); uint _mpyhlu (uint src1, uint src2);	MPYHL MPYHULS MPYHSLU MPYHLU	Multiplies the 16 MSBs of *src1* by the 16 LSBs of *src2* and returns the result. Values can be signed or unsigned.	
int _mpylh (int src1, int src2); int _mpyluhs (uint src1, int src2); int _mpylshu (int src1, uint src2); int _mpylhu (int src1, uint src2);	MPYLH MPYLUHS MPYLSHU MPYLHU	Multiplies the 16 LSBs of *src1* by the 16 MSBs of *src2* and returns the result. Values can be signed or unsigned.	
void _nassert (int);		Generates no code. Tells the optimizer that the expression declared with the assert function is true; this gives a hint to the optimizer as to what optimizations might be valid.	
uint _norm (int src2); uint _lnorm (long src2);	NORM	Returns the number of bits up to the first nonredundant sign bit of src2.	
double _rcpdp (double);	RCPDP	Computes the approximate 64-bit 'C67x double reciprocal.	
float _rcpsp (float);	RCPSP	Computes the approximate 32-bit 'C67x float reciprocal.	
double _rsqrdp (float);	RSQRDP	Computes the approximate 64-bit 'C67x double reciprocal square root.	
float _rsqrsp (float);	RSQRSP	Computes the approximate 32-bit 'C67x float reciprocal square root.	
int _sadd (int src1, int src2); long _lsadd (int src1, long src2);	SADD	Adds *src1* to *src2* and saturates the results.	
int _sat (long src2);	SAT	Converts a 40-bit value to an 32-bit value and saturates if necessary.	
uint _set (uint src2, uint csta, uint cstb);	SET	Sets the specified field in *src2* to all 1s and returns the *src2* value. The beginning and ending bits of the field to be set are specified by *csta* and *cstb*, respectively.	
unsigned _setr (unsigned, int);	SET	Sets the specified field in *src2* to all 1s and returns the *src2* value. The beginning and ending bits of the field to be set are specified by the lower 10 bits of the source register.	
int _smpy (int src1, int src2); int _smpyh (int src1, int src2); int _smpyhl (int src1, int src2); int _smpylh (int src1, int src2);	SMPY SMPYH SMPYHL SMPYLH	Multiplies src1 by src2, left shifts the result by one, and returns the result. If the result is 0×8000 0000, saturates the result to 0×7FFF FFFF.	

(Continues)

C Compiler Intrinsic	Assembly Instruction	Description	Device
`int _spint (float);`	**SPINT**	Converts 32-bit float to 32-bit signed 'C67x integer, using the rounding mode set by the CSR register.	
`uint _sshl (uint src2, uint src1);`	**SSHL**	Shifts *src2* left by the contents of *src1*, saturates the result to 32 bits, and returns the result.	
`int _ssub (int src1, uint src2);` `long _lssub (int src1, long src2);`	**SSUB**	Subtracts src2 from src1, saturates the result size, and returns the result.	
`uint _subc (uint src1, uint src2);`	**SUBC**	Conditional subtract divide step.	
`int _sub2 (int src1, int src2);`	**SUB2**	Subtracts the upper and lower halves of *src2* from the upper and lower halves of *src1*, and returns the result. Any borrowing from the lower half subtract does not affect the upper half subtract.	

Note: instructions not specified with a device apply to all 'C6x devices.

A-5: OPTIMIZATION CHECKLIST†

Phase	Description
1	Compile and profile native C code. • Validates original C code. • Determines which loops are most important in terms of MIPS requirements.
2	Add const declarations and loop count information. • Reduces potential pointer aliasing problems. • Allows loops with indeterminate iteration counts to execute epilogs.
3	Optimize C code using intrinsics and other methods. • Facilitates use of certain C6x instructions to be used. • Optimizes data flow bandwidth.
4	Write linear assembly. • Allows control in determining exact C6x instruction to be used. • Provides flexibility of hand-coded assembly without pipelining, parallelism, or register allocation.
5	Add partitioning information to the linear assembly. • Can improve partitioning of loops when necessary. • Can avoid bottlenecks of certain hardware resources.

Bibliography

[1] S. Haykin, *Adaptive Filter Theory,* Upper Saddle River, NJ: Prentice Hall, 1996.

[2] The Mathworks, *Matlab Reference Guide*, 1999.

[3] S. Mitra, *Digital Signal Processing: A Computer-Based Approach*, New York: McGraw-Hill, 1998.

[4] S. Norsworthy, R. Schreier, and G. Temes, *Delta-Sigma Data Converters, Theory, Design, and Simulation*, Washington, DC: IEEE Press, 1997.

[5] J. Proakis and D. Manolakis, *Digital Signal Processing: Principles, Algorithms, and Applications*, Upper Saddle River, NJ: Prentice Hall, 1996.

[6] B. Razavi, *Principles of Data Conversion System Design*, Washington, DC: IEEE Press, 1995.

[7] Texas Instruments, *Application Report SLAA 034*, 1998.

[8] Texas Instruments, *Application Report SPRA 291*, 1997.

[9] Texas Instruments, *Code Composer Studio User's Guide*, Literature ID# SPRU 328, 1999.

[10] Texas Instruments, *Technical Training Notes on TMS320C6x*, TI DSP Fest, Houston, 1998.

[11] Texas Instruments, *TMS320C6000 Assembly Language Tools User's Guide*, Literature ID# SPRU 186E, 1998.

[12] Texas Instruments, *TMS320C6x C Source Debugger User's Guide*, Literature ID# SPRU 188D, 1998.

[13] Texas Instruments, *TMS320C6000 Code Composer Studio Tutorial*, Literature ID# SPRU 301A, 1999.

[14] Texas Instruments, *TMS320C62x/67x CPU and Instruction Set Reference Guide*, Literature ID# SPRU 189C, 1998.

[15] Texas Instruments, *TMS320C6000 DSP/BIOS User's Guide*, Literature ID# SPRU 303, 1999.

[16] Texas Instruments, *TMS320C6x Evaluation Module Reference Guide*, Literature ID# SPRU 269, 1998.

[17] Texas Instruments, *TMS320C6000 Optimizing C Compiler User's Guide*, Literature ID# SPRU 187E, 1998.

[18] Texas Instruments, *TMS320C6201/6701 Peripherals Reference Guide*, Literature ID# SPRU 190B, 1998.

[19] Texas Instruments, *TMS320C62x/C67x Programmer's Guide*, Literature ID# SPRU 198B, 1998.

[20] R. Van de Plassche, *Integrated Analog-to-Digital and Digital-to-Analog Converters*, AA Dordrecht, Netherlands: Kluwer Academic Publishers, 1994.

Index

A

A/D. *See* Analog-to-digital (A/D)
Adaptive filter
  C and assembly, 201–205
Adaptive filtering
  lab, 196–205
Adaptive FIR
  implemented in C, 198–200
Adaptive FIR filter, 200–205
Address cross-path, 158
Addressing Mode Register (AMR), 195, 203–204, 258
ADSL, 4
Aliasing, 11
  of frequency, 9
Ambiguity, 11
Amplification
  signal conditioning, 35
Amplitude statistics, 13–16
AMR (Address Mode Register), 195, 258
  adaptive FIR filter, 203–204
  setup example, 195
Analog and digital domains
  Fourier transform pairs, 8–9
Analog and digital frequencies
  difference, 8–33
Analog signal
  sampling
    program, 133–135
Analog signals, 1
  periodic
    Fourier series, 12
Analog sinewave
  reconstruction, 30
Analog-to-digital (A/D)
  architectures, 60–70
  dynamic metrics, 49
  signal conversion, 8–33
  static metrics, 43–49
Analog-to-digital (A/D) converter, 1
  characteristic, 18–19
  LSB, 19
  quantization, 20
  SNR, 23
Antialiasing input filter
  specification, 68
API (application programmer interface), 127
Application programmer interface, 127
Arctangent function
  floating C6x DSP, 150
Arithmetic operations
  C6x, 146–150
Assembler optimizer
  code linear assembly, 191
Assembler optimizer software-pipelined assembly, 190–193
Assembler options, 114
Assembly file, 96–99
Assembly initialization, 105
Assembly loop codes, 161
Asymmetric version of DSL (ADSL), 4
Audio codec CS4231A, 127
Audio signal
  manipulate, 136–137
Audio signal sampling
  lab, 127–137

B

Big endian, 109
Binary fractional multiplication, 140
BIOS sections, 220
Boot Loader, 80
Branches, 159
Build Error, 117
Build options, 115–117

C

C and assembly
  mixing, 161
CCS. *See* Code Composer Studio (CCS)
CDF, 43
Cellular phone wireless
  communication DSP system, 4
C62 EVM, 127
Charge redistribution
  D/A converter, 72
Circular buffering, 194–195
CLK manager
  PRD module, 228
CLK module
  property, 234
Codec, 127
  adjust parameters, 132
  initialization, 128–133
  initializing
    code, 132–133
Codec library, 127
Code Composer Studio (CCS), 7, 94, 110
  complete lab 1 program, 123
  files indicated, 221
  IDE, 219
  lab, 110–124
  simulator
    lab programs, 157
Code development flow, 192
Code efficiency
  vs. coding effort, 94
Code initialization, 104–109
Code optimization, 158–174
Coding schemes
  efficient codes, 162
COFF, 98
Command file, 102
Comments, 190
Common Object File Format (COFF), 98
Communication
  lab, 243–254
Comparators limits
  flash A/D converters, 61
Compile options, 103
Compiler intrinsics, 262–264

267

268 Index

Compiler options, 114, 115–116
Compiler sections, 99
Compiler utility, 102–104
Composite signal
  frequency components, 176
  two cycles, 175
Configuration Tool
  files generated, 221
Configuration tool
  DSP/BIOS, 230
Constant alignment example, 108
Continuous-time signal
  Fourier transform, 10
Control Status Register (CSR), 125, 258
Cosine
  floating-point C67x DSP, 149
CPU load, 250–252
  changed, 253–254
CPU Load Graph, 239
CPU operation, 85–87
Creating projects
  CCS code, 110–115
Cross paths, 158
CS code development process, 110
CSR, 125, 258
Cumulative density function (CDF), 43
Current steering
  D/A converter, 72
C64x
  special-purpose instructions, 92
C6x datatypes, 144
C64x DSP
  optimize dot-product example, 170–174
  TMS320C6000 architecture, 90–92
C67x floating-point instructions
  list, 257
C6x instructions
  list, 255–256
C6x internal buses, 84
C6x memory map, 100
C6x MMR, 260–261
C64x packed-data
  instructions
    processing capability, 91
C64x pipelined code, 172
C6x processor
  peripherals, 80
  pipelined CPU, 88
C6x software tools, 95

D

D/A. *See* Digital-to-analog (D/A)
Data alignment, 105–109
Data communication
  DSP/BIOS, 243–254
  PIP module, 228
Data converters
  architectures of, 60–79
  DSP systems, 76–79
  specifications, 34–59
Data cross-path, 158
Data synchronization
  lab, 243–254
  PIP module, 228
Debugging tools, 114, 115–124
Decimation filter
  implementation, 70
Delayed branch, 159
Delays, 158
Dependency graph
  handwritten software-pipelined code, 187–189
  multicycle loop, 171
  software pipelining, 169
Dependency graph terminology, 166
Devices
  DSP systems, 6
DFT, 12–15
DFT implementation, 212–215
Differential nonlinearity
  description, 45
Differential nonlinearity (DNL), 43
  D/A converter, 55
Digital decimation filter
  sigma-delta A/D converters, 69
Digital Signal Processing (DSP), 1–2, 2
  host communication, 208
  processors
    characteristics, 2
    programming, 94
  single function VLSI
    implementation
    differences, 2
  system
    components, 1
    data converters, 76–79
    examples, 3–7
Digital Subscriber Line modem, 2
Digital-to-analog (D/A)
  architectures, 70–75
  data converters, 1

dynamic metrics
  D/A converter, 57
  static metrics, 54–59
Digitization process, 1
  FFT, 21
Digitizing, 1
Directives
  assembly code sections, 98
Direct Memory Access, 206–207
Discrete Fourier transform (DFT), 12–15
Discrete signals
  periodic
    DFT, 12
Discrete time signal, 8
Discrete time version
  Fourier transform, 10
Distorted sinewave, 18
Distortion
  of frequency, 9
Division
  floating-point C67x DSP, 146–148
Division types
  different databases, 147
DMA, 80
  global variable, 249
DMA API function, 211
DMA channels control registers, 207
DMA transfer exchange, 208
DNL, 43
  D/A converter, 55
Dot product C code, 104
Dot product code
  hand-coded software pipelined, 166
  linear assembly
    software pipelining, 169
  word-wide optimized version, 161
Dot product dependency graph, 166
Dot product example, 85–87
  linear assembly code, 163
  optimized, 159
Dot product loop
  word wide
    hand-coded pipelined code, 170
Dot product scheduling table, 166
Double precision (DP)
  floating-point processors, 142–143
Double-word packed datatype code, 174
DP
  floating-point processors, 142–143

Index

DR
  performance metric, 49
DRAM
  DSP processor, 99
DSK (DSP Starter Kit), 7–8
  board, 98
  C6x instructions, 137
  map, 100
DSL (Digital Subscriber Line)
  modem, 2
DSP. *See* Digital Signal Processing (DSP)
DSP/BIOS, 219–222
  analysis and instrumentation, 232–234
  API modules, 220
  CCS complements, 219
  data synchronization and communication, 243–254
  lab, 230–242
  program example, 231–232
  scheduling, 223
DSP Starter Kit. *See* DSK
Dynamic memory (DRAM)
  DSP processor, 99
Dynamic range (DR)
  performance metric, 49

E

Editing
  variable, 137
Effective number of bits (ENOB)
  performance metric, 50
EMIF, 80, 130
EMIF (External Memory Interface), 80, 130
Enabling cache feature, 159
ENOB
  performance metric, 50
Epilogue instruction
  elimination, 168
Error
  Build, 117
  gain
    A/D converter, 49
  real-time, 238
Ethernet DSP system, 5
Evaluation model. *See* EVM
EVM (evaluation model), 7–8
  board
    functional diagram, 96–97
    setting up, 127

codec
  initialization, 128–133
  map, 100
  stereo audio interface, 127
  target C6x board, 96
Execution Graph, 238
Execution Graph Details, 239
Extended sign bit, 139
External Memory Interface, 80, 130
External memory interface (EMIF), 80, 130
External memory ranges, 99

F

Fast Fourier transform (FFT), 12–15
  algorithm
    real-time, 216–217
  implementation, 215–216
  lab, 209–218
  magnitude, 248
  program, 243
  real-time, 216
Fetching
  phases, 88
Fetch packet
  C6x, 90
FFT. *See* Fast Fourier transform (FFT)
FIFO (first in first out) registers, 208
Filter. *See also* FIR filter; IIR filter
  adaptive, 196–205
  antialiasing input
    specification, 68
  decimation
    implementation, 70
  digital decimation
    sigma-delta A/D converters, 69
  interpolation
    D/A architecture, 73–74
  linear phase
    DSP, 1
  lowpass
    designed, 176
  steep-cutoff notch, 1
  switched-capacitor comb, 75
Filter coefficients
  scaling, 156
Fine-resolution converter, 63
Finite word length
  effects on fixed-point DSPs, 141–142

FIR code
  linear assembly, 191
FIR dependency graph, 188
FIR filter
  design, 175–178
  implementation, 178–193
FIR filter assembly code, 204
FIR filter coefficients, 177
FIR filters, 141
  scale coefficients, 145
FIR scheduling table, 189
First in first out registers, 208
Fixed point
  vs. floating point, 138–150
Fixed-point DSPs
  finite word length, 141–142
Flash architecture
  A/D architectures, 61
Floating-point
  vs. fixed-point, 138–150
  implementation, 193
  number representation, 142–143
Folding
  A/D converter, 64
Fourier transform pairs
  for analog signals, 8–9
Fractional division
  floating point C67x DSP, 147–148
Fractional representation
  fixed-point DSPS, 138–141
Frame processing, 206–208
Frequency representation
  filtering operation, 179
FS voltage, 19
FTT twiddle factors
  scaling, 156
Full-scale (FS) voltage, 19

G

Gain
  D/A converter, 57
Gain error
  A/D converter, 49
Generic C6x architecture, 83
GIE, 126
Gigabit Ethernet DSP system, 5
Glitch area
  D/A converter, 58
Global Index Register, 207
Global Interrupt Enable bit (GIE), 126

Graphical Display window, 118
Graph Property Dialog box, 118

H

Hand-coded software pipelined code
  steps, 165–166
Hand-coded software pipelining,
  164–166
Handwritten software-pipelined
  assembly, 187–189
Hard disk drive DSP system, 6
Hard real-time
  defined, 222
Harmonics
  distorted sinewave, 16–18
Host Port Interface, 80
HPI (Host Port Interface), 80
HWI object property setting, 243

I

ICR, 258
IDE, 94, 110
IER, 126, 258
IFR, 126, 258
IIR filter, 141
  coefficients, 197
  design of, 196–198
  implementation, 198–200
  response, 198
  scale coefficients, 145
IMD
  output code, 53
Implementation
  DFT, 212–215
  DSP and a single function VLSI
    differences, 2
  FFT, 215–216
  FIR filter, 193
  floating-point, 193
  IIR filter, 198–200
  LMS algorithm, 201
Infrastructure
  telecommunication, 2
Initialization
  assembly, 105
  code, 104–109
    for dot-product example, 106
    reset, 104
  of EVM and codec, 128–133
INL, 43, 46, 56
Input impedance
  A/D converters, 41–42
Input/output characteristic
  analog output, 54

Input signal
  time and frequency domains, 213
Integer arithmetic
  lab, 151–157
Integer division
  floating point C67x DSP, 147–148
Integral nonlinearity (INL), 43
  D/A converter, 56
  description, 46
Integrated development
  environment (IDE), 94, 110
Integrated Services Digital Network,
  4
Interleaved
  A/D converter products, 66
Interleaved architecture
  A/D converter products, 66
Intermodulation Distortion (IMD)
  output code, 53
Interpolating
  A/D converter, 64
Interpolation filter
  D/A architecture, 73–74
Interrupt
  data processing, 125–126
  defined, 125
  priorities, 125
  service routine, 133–137
Interrupt Clear Register (ICR), 258
Interrupt Enable Register (IER),
  126, 258
Interrupt Flag Register (IFR), 126,
  258
Interrupt mapping, 125
Interrupt Multiplex registers, 126
Interrupt Return Pointer (IRP), 258
Interrupt service routine (ISR), 7–8,
  125, 126
Interrupt Service Table Base (ISTB)
  bits, 126
Interrupt Service Table Pointer
  (ISTP), 258–259
  register, 126
  initialize, 133
Interrupt Set Register (ISR), 258
Intrinsics, 103
IRP, 258
ISDN (Integrated Services Digital
  Network), 4
ISR, 7–8, 125, 126, 258
ISTB bits, 126
ISTP, 258–259
  register, 126
  initialize, 133

L

Label, 190
Latency
  A/D converter, 49
  D/A converter, 58
Least mean square (LMS)
  algorithm, 196
Least significant bit (LSB), 19
Limiting bandwidth
  converter specifications, 36
Linear assembly
  efficient codes, 162
Linear assembly code line
  fields, 190
  general syntax, 190
Linear assembly directives, 164
Linear assembly dot product code
  software pipelining, 169
Linearization
  signal conditioning, 35
Linear phase filter
  DSP, 1
Linker options, 114
Linking
  memory management, 101
Little endian, 109
LMS algorithm, 196
  adaptive FIR filter, 201
Lookup table
  floating C6x DSP, 150
Loop code
  multicycle loop, 171
Loop example, 165
Lowpass filter
  designed, 176
LSB, 19

M

Magnitude response
  DFT, 215
Mailbox property, 243
Matlab function
  A/D converter, 19–21
  electronic noise, 43
  filter, 197
  interpolation of shifted since
    functions, 29
  jitter, 27
  sampling, 12
  SFDR *versus* frequency, 52
  T&H circuit, 39
  time jitter, 27–29

Index   271

Matlab toolbox
    A/D toolbox browser, 33
    data conversion, 33
McBSP (Multichannel Buffered
        Serial Port), 80–81, 127
    activated, 130
    API functions, 130
Memory management
    DSP processor, 99
Memory-mapped registers, 258–261
Memory Window
    Option
        dialog box, 118, 152
        showing array values, 153
MFLOPS, 80
Million floating-point operations per
        second (MFLOPS), 80
Million instructions per second, 80
MIPS (million instructions per
        second), 80
Mnemonic
    linear assembly code, 190
Modulator
    sigma-delta A/D converters, 69
Monotonicity, 43
    D/A converter, 57
    description, 48
Motor control
    DSP, 5, 6
Moving-window effect, 194–195
Multichannel Buffered Serial Port,
        80–81, 127, 130
Multicycle loops
    code optimization, 168–170
Multithread programming, 225
Multithread scheduling, 235–242

N

Naming conventions
    distinguishes functions, types, and
        constants, 222
NMI, 125
NMI Return Pointer (NRP), 258
No delay slots, 158, 159, 190
Nonmaskable interrupt (NMI), 125
Nonmonotonic static transfer function
    A/D converter, 48
Non-pipelined code, 165
NOPs, 158, 159, 190
Normalized frequency, 177
NRP, 258
Nyquist criterion
    analog signal bandwidth, 36

Nyquist rate
    description, 9
    sigma-delta converters, 67–70
Nyquist rate condition, 16
Nyquist theorem
    signal reconstruction, 29

O

Offset
    D/A converter, 56
    description, 48
Offset voltages
    flash A/D converters, 61
Operands, 190
Optimization checklist, 264
Optimization levels
    selection, 182
Optimization methods cycles, 170
Optimized dot product example, 159
Output buffer
    A/D converter, 37
Output characteristic
    analog output, 54
Overflow
    floating-point, 143–145
    handling, 151–152
    signal distorted, 153
Oversampling, 12
    architectures of data converters,
        68
    D/A architecture, 73–74
    sigma-delta converters, 67–70

P

Packed datatype code, 173
Parallel instructions, 158
Passing arguments convention, 163
PCC, 159
PCM voiceband DSP system, 5
Pdf, 13–16
Performance metrics
    A/D converters, 42–53
    D/A converters, 53–59
Periodic analog signals
    Fourier series, 12
Periodic discrete signals
    DFT, 12
Periodicity conditions
    sampling, 13
Pipelined
    A/D architectures, 63–64
Pipelined code, 165
Pipelined CPU, 87–89

PipReceive object
    properties, 245
PipTransmit object
    properties, 246
Power Down unit, 80–85
PrdDMA
    properties, 249
PRD module, 228
PRD objects
    properties, 241
    property settings, 236–237
PrdPrintf object
    property, 233
Pressure sensor
    signal conditioning circuit, 35
Primary Control Register, 207, 212
Printf()
    vs. LOGprintf(), 224
Probability density function (pdf),
        13–16
Profile Statistics window, 122
Program Cache Control (PCC), 159
Programmability
    digital processing, 1
Programming approach, 103

Q

Q-format precision loss example, 141
Q-format representation
    fixed-point DSPS, 138–141
Quantization
    defined, 18
    noise, 18
Quantizing, 1

R

Randomness
    flash A/D converters, 61
Real-time, 2
Real-time analysis, 219–229
    PRD objects, 237
Real-time constraints
    types, 222
Real-time data exchange, 229
Real-time errors, 238
Real-time FFT, 216
Real-time FFT magnitude response,
        218
Real-time fitting
    lab, 175–193
Real-time scheduling, 223–228
    PRD, 241
Registers, 258–261

Index

Reset initialization code, 104
Resistor ladder
   D/A architectures, 72
RTDX (Real-time data exchange), 229–230
   target/host dataflow, 229

S

Sample-and-hold (S&H)
   A/D converter, 37, 38–45
Sampled sinusoidal signal
   Fourier transform, 11
Sampling, 1, 8–18
   analog signal
      program, 133–135
   time jitter, 25–29
SAR
   A/D converter, 65
Scaling
   approach, 152–156
   example, 156
   floating-point, 143–145
   floating-point processors, 142
Scheduling, 219–229
   software pipelining, 166
   table
      multicycle loop, 171
      software pipelining, 169
SDK
   target C6x board, 96
SDRAM (Synchronous DRAM), 99
Sensors, 1
   DSP systems, 6
Setting time
   D/A converter, 57
Setup code
   dot-product routine, 106
SFDR
   performance metric, 50–51
S&H
   A/D converter, 37, 38–45
Shifting data, 194–195
Sigma-delta
   A/D architectures, 67–70
   D/A architecture, 73–75
   D/A converter
      for digital audio, 75
Signal conditioning
   A/D converters, 34–37
   DSP systems
      illustrated, 34

Signal conversion
   analog-to-digital, 8–33
Signal frequency content
   plotted, 176
Signal reconstruction, 29–32
Signals
   processing of, 1
Signal-to-noise-and-distortion ratio (SNDR), 38–39
   D/A converter, 58–59
   performance metric, 49
Signal-to-noise ratio (SNR), 22–25
   performance metric, 49
Simulator, 157
Simulator installation, 157
Sinc function
   approximation of, 30
Sine
   floating-point C67x DSP, 149
Single
   floating-point processors, 142–143
SNDR, 38–39, 49, 58–59
SNR, 22–25, 49
Soft real-time, 222
Software interrupt (SWI)
   object, 238
      properties, 239, 240
Software-pipelined version, 187–189
Software pipelining, 161–170
Software tools
   DSP, 94–124
Source file
   creating, 112
Spurious-free-dynamic-range (SFDR)
   performance metric, 50–51
Square root
   floating-point C67x DSP, 149
SRAM, 99
Static memory (SRAM), 99
Static nonlinearity
   A/D converter, 47
Static performance metrics
   A/D converter, 48
Statistics view, 234
Steep-cutoff notch filter, 1
Step size
   floating-point processor, 201
StsPrintf object properties, 233
Subranging
   A/D architectures, 61–63

Successive-approximation-register (SAR)
   A/D converter, 65
SWI
   object, 238–240
SwiFFT object
   properties, 244
SwiFFT's mailbox, 245
Switched-capacitor comb filter, 75
Synchronous DRAM, 99
System identification
   adaptive FIR filter, 196

T

Telecom applications
   TI data converters, 79
Telecommunication infrastructure, 2
T&H, 37–45, 69
Third generation wireless, 2
Thread priorities, 225
Threads
   prioritization, 250–252
   scheduled, 223–228
Thread scheduling rules, 226
TI chip set
   ADSL wired communication DSP system, 4
TI high-speed data converters, 76–79
Time domain representation
   filtering operation, 179
Time jitter
   sampling, 25–29
Timer, 80
Timer Control Register, 226–228
Timer Period Register, 226–228
Timing cycles
   different memory options, 205
   for optimizations, 187
Timing cycles for different builds, 183
TI successive-approximation
   A/D converter products, 65
TI TLC320AD75, 69
TI website
   C64x speedups, 91
TMS320C6000
   architecture, 80–93
   executable files, 7
TMS320C62x
   product specifications, 81